Introductory Concepts for Abstract Mathematics

Introductory Concepts for Abstract Mathematics

Kenneth E. Hummel

CRC Press
Taylor & Francis Group
Boca Raton London New York

CRC Press is an imprint of the
Taylor & Francis Group, an **informa** business

A TAYLOR & FRANCIS BOOK

CRC Press
Taylor & Francis Group
6000 Broken Sound Parkway NW, Suite 300
Boca Raton, FL 33487-2742

First issued in paperback 2019

ISBN-13: 978-1-58488-134-6 (hbk)
ISBN-13: 978-0-367-39883-5 (pbk)

Library of Congress Cataloging-in-Publication Data

Hummel, Kenneth E.
    Introductory concepts for abstract mathematics / by Kenneth E. Hummel.
        p. cm.
    Includes index.
    ISBN 1-58488-134-8 (alk. paper)
    1. Mathematics. I. Title.
QA39.2 .H845 2000
510—dc21
                                                                00-021301
                                                                CIP

Library of Congress Card Number 00-021301

Publisher's Note
The publisher has gone to great lengths to ensure the quality
of this reprint but points out that some imperfections in the original may be apparent.

Visit the Taylor & Francis Web site at
http://www.taylorandfrancis.com

and the CRC Press Web site at
http://www.crcpress.com

CONTENTS

PREFACE

This text is designed to help students move into more abstract junior/senior mathematics classes with more confidence, knowledge, and skills and help them learn how to read and write mathematics with more clarity and ease. Most students taking courses such as abstract algebra and real analysis experience difficulties with abstract concepts and writing mathematics in a clear and organized fashion. Partly, this difficulty arises because of the tendency in recent years to make mathematics courses (calculus, for example) more "user friendly." In some treatments of calculus the focus is directed away from dealing with the more abstract or tedious parts and even from doing a few algebraic manipulations. As a consequence, many students are not acclimated to studying a subject for its innate beauty, while others may think they do not have an interest in a subject that appears to have no applications. This text addresses these issues.

Nevertheless, a calculus course should be a springboard for success in this text. Calculus does provide experience in working with concepts that are expressed symbolically, so such a background helps ensure a certain level of comfort and expertise with mathematical ideas. In addition, some ideas from calculus are used in examples and exercises at various points.

There are different ways one might design a course to get students from "calculus mode" to "beyond." However, the subject matter chosen and the manner in which it is treated in this text are primarily based on the following ideas and the author's observations of students' problems in more advanced undergraduate mathematics courses.

- Before entering real analysis or algebra, students benefit from a treatment of various concepts such as functions in the abstract or partitions. They are also well served by learning about equivalence relations, number systems, etc. Many of these fundamental concepts are useful in advanced courses. Thus a course with the content of this text provides foundations for several junior/senior classes. In addition, facility with abstraction, understanding definitions, and theorem-proving capability are very beneficial skills for students later in their undergraduate studies. Today these abilities are inadequately attained in calculus and even in some linear algebra courses. Thus it is important to develop and hone the skills of dealing with abstract objects in order to be prepared for more advanced courses. We believe this text is well suited to accomplish these goals.

- One should write mathematics in such a way that the intent of the writer is clear. The objective is to enlighten, not cloud. We certainly hope this work achieves, in some measure, this lofty goal, and that it encourages students to strive for clarity of exposition.

- We think, and certainly hope, this text gives students a feeling for the beauty of mathematics and its internal harmony. We also hope students come away from this course with the understanding that a subject can be appreciated not only for what one can do with it, but also for the wonderful way that the ideas can be interwoven into a unified fabric of knowledge. Beyond that, our hope is that students come out of this course with eagerness and increased enthusiasm to continue the study of mathematics.

- The subject of geometry, which is thousands of years old, and calculus, which is hundreds of years old, may engender the belief that mathematics is unchanging. A student may not realize that mathematics is being created today. At several points in this text, the subject matter is appropriate for mentioning some 20th and late 19th century developments in set theory.

- Some of the topics found in this text are not often treated elsewhere in a typical undergraduate curriculum. For example, an undergraduate who has not taken a course based on subject matter treated herein may not, at graduation, have any organized ideas about infinite sets and almost certainly not about transfinite arithmetic. This material broadens a student's body of mathematical knowledge.

Typically, when teaching from this text, we are able to complete most of the first 26 chapters, although in few rare instances we have been able to do some of the subsequent chapters on partially ordered sets and the axiom of choice. The reason for the inclusion of Section VI in this book is to provide a challenge for those exceptional classes or students.

Throughout this text, insertions, parenthetical and otherwise, are found in the midst of certain arguments. Some of these are identified with "Why?'s"–appended to encourage the reader to find the actual supporting reason earlier in the book. Other parenthetical comments are included in order to indicate connections to related ideas elsewhere in the text or simply to explain the rationale in more depth, but they are not necessarily part of the argument. The hope is that these questions and comments will get students to appreciate how the various parts of this subject dovetail together in a harmonious manner to achieve a total body of material.

As mentioned above, some of the main objectives of this text are to get students to deal with abstraction and become familiar with a theorem and proof format that they will encounter in subsequent courses. In Section I, the various logical forms that mathematical assertions take are discussed, and strategies for tackling them are treated. For example, students learn how to structure proofs of different forms including, but certainly not limited to: "If (P or S) then Q," and, "If P then (Q iff R)," and, "If P then (Q or S)." Strategies for proving these, as well as other logical forms, are considered. The means for crafting a proof either directly or indirectly are also included. In addition, strategies for dealing with quantified statements are pursued to some depth. In particular, statements with mixed quantifiers are carefully treated because these situations arise at important points in real analysis and abstract algebra. Numerous examples are given fairly early, along with techniques for their proofs. These specific examples of logical forms are drawn from calculus and elementary number theory. Finally, Section I concludes with a chapter on mathematical induction.

Section II develops naive set theory. Although this development is not strictly axiomatic, it has an axiomatic flavor. It is a very orderly treatment that derives a complete array of properties of unions, intersections, and complements. In this chapter, frequent mention is made of the parallel evolution of material that appeared in the first section on logical structures. For example, since the logical connective "and" distributes over "or," one may parlay this knowledge into "∩" distributes over "∪." Many such parallels are indicated in subsequent sections to further engender the sense of unity these notions possess and to identify the common themes that keep running through the subject. Also included in this section is a study of generalized unions and intersections of families of sets. These topics further aid the student in coming to grips with abstractions and generalizations of the more familiar union and intersection of two sets.

Section III centers on relations, functions, equivalence relations, and partitions. These concepts assiduously employ the set theoretic material developed in Section II and the strategies discussed in Section I. Functions are treated as relations, and the notions of composition, inverses, etc. are carefully developed. Also, the intimate connection between partitions and equivalence relations is explored. This section establishes a strong framework for succeeding chapters.

Section IV focuses on algebraic and order properties of number systems, beginning with the natural and whole numbers. The whole numbers are developed as cardinal numbers of sets because of the connection with topics considered in the chapter on sets. Again, an emphasis is made on

how new ideas are integrated and interrelated with earlier ones to develop
new concepts. The system of reals is crafted via axioms immediately after
the rationals are formed from the integers. The important least upper
bound property is explored and students learn how the system of real
numbers differs from the rational numbers in this critical way.

Section V is an excursion through infinite sets, denumerable and non-
denumerable sets, and transfinite cardinal numbers. For almost all students this is a totally new mathematical experience. Here is where they
really see that finite–based intuition is not always a reliable predictor of
what might be true for infinite sets. As in other chapters, this one also
stresses how careful development can tame some of the ideas that may
seem hard to believe and even harder to grasp.

Throughout, there are exercises that present students with bogus arguments similar to some written by former students. The objective is to
have students find the things that are wrong. Then, perhaps, they will
see errors they are making as they solve the problems. There are a few
exercises asking students to look at a series of examples, make a conjecture, then prove it. Some exercises ask the student to prove or disprove
a certain statement.

Most chapters may be covered in one 50–minute class. Some exceptions include Chapter 9 on properties of sets. Chapter 9 is divided into
2 parts with 2 exercise groups and might be better done in 2 or more
classes. Perhaps Chapters 12 and 13 on equivalence relations and partitions are more easily done in 3 lectures. Chapter 19, on the derivation of
the integers, might be better spread out over at least $1\frac{1}{2}$ days. And almost certainly, Chapter 26, on transfinite cardinal numbers, will require
2 days.

Foreword to the students:

In this text you will experience a vastly different course from your
previous mathematics courses. It will certainly be different from calculus.
The foundations upon which mathematics are built are explored using
the topics of sets and functions. Although some of the topics may be
familiar, they are treated with some care because of their use later in
this text and in subsequent courses. Mostly you will learn many new
things. For example, you will see the strange and remarkable concepts of
infinite sets and transfinite cardinal numbers developed by Georg Cantor
and you will learn exactly why there are infinite sets of different infinite
sizes. Most importantly, you will learn to work effectively with rather
abstract ideas in an organized and careful manner.

When reading mathematics in any context, **read it with pencil in**

hand and paper in front of you. Work through what the author is doing; don't just read it. This helps you with learning to write mathematics as well. A main objective of this text is for you to learn how to read and write mathematics and to organize mathematical ideas in a meaningful way. This means you must write so the reader can follow what you are doing. Hence, writing in a manner that is difficult to misunderstand is of paramount importance. The objective is to write so that anyone else with the same level of training can read what you have written and understand what you have done. The reader should not have to guess what you meant because too many details were left out or misleading comments were made. In addition to making the intent clear to the reader, careful writing also helps the writer synthesize his or her own thoughts and it ensures that the arguments hold under the scrutiny of logic.

In working through this material, notice the questions which occasionally appear in the middle of a paragraph. Most of these parentheticals are placed there to encourage you to find the exact justification by way of a theorem, property, or exercise. The more capable you are of finding the reasons, the easier it will be to know how to proceed when you are staring at the blank space beneath the theorem you are expected to prove.

In addition to the "Why?'s", there are other parenthetical comments interspersed in the text. They are placed in an argument using parentheses because they are not part of that argument. Instead, the inserted remark is there to indicate a connection to another property or similar idea elsewhere in the text. The purpose of these comments is to help you see the interrelationships between various ideas. Thus it may be fruitful to refer to the reference to see that connection.

The text is divided into six major sections. Each section deals with one general category of material. For example, Section V centers on finite and infinite sets. The text is also divided by chapters, which are more specifically focused on more limited topics. The 30 chapters are numbered consecutively from the beginning to the end of the whole text. Theorems and examples, and exercises are identified by chapter number and theorem, or example or problem number. It is easy to find an item if you know its identification. For example, if Exercise 12.16b is cited, you will know it is the b part of Exercise 16 in Chapter 12.

Many of the problems and exercises have "answers" in the back. Actually, what you will find there is usually quite condensed. Do not assume that a simple number or expression, such as cited in the Solutions section, is actually a satisfactory response. The brief answer is there to help you determine if you are on the right track. In most cases, there will be work to be done to arrive at the answer, and your answer should show that work. It is a part of trying to be "impossible to <u>mis</u>understand." In most cases you should explain your solution; don't just write a simple "yes" or "no" or a number or expression.

In addition to a Solutions section in the back of the book, you will find a Reading List of a bibliographic nature. Many of the references in the list are mentioned in the main text. You may want to consult these to get new ideas, alternate approaches, or elaborations on concepts mentioned in this text in order to expand your knowledge on the subjects treated in this book.

Finally, we hope that you develop an excitement for this material. It truly embodies what the rest of mathematics is based upon. Open your mind to new ideas and new ways to do things. Get involved. Thrill in constructing a proof on your own. Develop a sense of making certain that all cases have been considered in an argument and take pride in the polished final product.

Acknowledgments:

This text grew out of course notes for a foundations course I have taught at Trinity University for years and students' receptions of the ideas and the manner of presenting them have helped guide the work to its present state. I wish to acknowledge that support and thank them for their feedback. I would also like to thank Mr. Robert Stern and Ms. Michele Berman of Chapman & Hall/CRC Press for helping make this work a reality. Also, I want to thank Andrew Arana, Manuel Berriozábal, Allen Holder, Darwin Peek, and students in this course, for making very useful suggestions. A very big thanks should go to Donald Knuth, who developed TeX, thereby making this difficult process easier than it might otherwise be. But most importantly, I would like to thank my family, particularly my wife, for putting up with me for the many months of preparing the manuscript.

Kenneth E. Hummel

SECTION I

LOGIC AND PROOF

CHAPTER 1

LOGIC AND PROPOSITIONAL CALCULUS

Introduction:

Mathematics is a discipline which, like most disciplines, deals with ideas which are expressed in sentences. In the English language, sentences can be imprecise or even ambiguous. In fact, it is not unusual to be able to interpret a sentence in more than one way. This is perfectly acceptable or even desirable for casual conversation, but it is totally unacceptable in mathematics or in any serious discussion.

In mathematical discussions there must be precision. Ambiguity must not be present. Since English is the underlying language used to convey mathematical ideas in this country, and since there is ambiguity and imprecision in everyday English usage, some care must be taken to recognize and deal with possible ambiguous expressions when they arise. Thus our mathematical ideas, expressed in English sentences, must not be ambiguous; they must have very clear, precise meanings. We can be sure this happens by limiting the discussions to those sentences which are suitable for mathematical purposes. Those sentences are called statements.

DEFINITION: A declarative sentence which is unambiguously either true or false (but not both) is called a statement; otherwise it is just a sentence. If a statement is true, we say it has a truth value of true and conversely, if a statement is false, we say its truth value is false.

It may be possible to determine that a given sentence is a statement without determining that it is true (or maybe false). For example, the sentence, "Mary Ellen Williams was a calculus student at XYZ University in the spring of 1999," is a statement because it is known that an examination of the mathematics class rolls at that time would reveal whether she was (or was not) enrolled in a mathematics class. The class rolls do not have to be physically checked to make this determination.

A feature that results from the careful use of statements is that any two reasonable and knowledgeable persons would agree on the truth value of a given statement. Certainly, there would be a fundamental problem if two persons reading the same material came up with conflicting interpretations. It is the use of statements in mathematical literature that achieves the necessary precision and reliability that is not characteristic of everyday language.

3

To illustrate the terminology introduced above, consider the following: "Two plus three equals five," and "All continuous functions on $[a, b]$ have derivatives on (a, b)." The first statement is true while the second is false. "The 19–story ABC Building in downtown San Antonio is a beautiful building," while declarative, is NOT a statement. Its truth value would be impossible to determine. So it is clear that in mathematical writing, it is desirable to limit discussion to statements. Certain exceptions to this practice will be considered later in more detail; however, to illustrate one way to make some sentences into statements, consider the declarative sentence, "*Unfailing Spirit* is a fast horse." The problem with its truth value centers on the word *fast*. It is imprecise. It is possible to eliminate the confusion by defining *fast* in some manner. Perhaps one could define a horse to be *fast* if it has run a 0.5 km race at 14 m/s. Then either *Unfailing Spirit* has done this or not. In this way a disagreement could be settled.

Consider the sentences: "Two is less than three," "Six is greater than seven," "How old are you?" and "BMW is an expensive car." The first two sentences are statements since they have truth values of true and false, respectively. "How old are you?" is not a declarative sentence, so it is not a statement. "BMW is an expensive car" is a declarative sentence, but cannot be classified as true or false since *expensive* is a vague term for automobiles. If *expensive* could be suitably defined, and other limitations imposed, then the truth value could be determined.

Logical Connectives:

The ideas that are expressed in one or more statements may be modified or combined by the use of logical connectives. The connectives considered in the following section are: *and, or, negation, conditional,* and *biconditional.* The sentence that results from the application in a proper manner, of connectives to one or more statements, is called a compound statement.

For example, the compound statement, "Two is less than three and six is greater than seven" is the result of joining simpler statements by the connective *and.* This compound statement is false, as you will see. If, on the other hand, the simpler statements were joined by the logical connective *or,* then a new compound statement would be obtained. It would be, "Two is less than three or six is greater than seven." By the conventional meaning of "or" the truth value of this statement is true.

Before getting to the exact effect that connectives like *and* and *or* have on statements, a bit of symbolism is considered to shorten such compound statements. Just as letters of the alphabet are used in algebra and calculus to represent various mathematical entities from numbers to functions, letters may be used to represent statements. For example, P and Q could be used to denote two of the previously mentioned statements. This may be done by employing what might be called a *dictionary*, because it describes what the symbols stand for.

P : Two plus three equals five.

Q : All continuous functions on $[a, b]$ have derivatives on (a, b).

As is customary in mathematical symbolism, the use of P and Q does not necessarily mean that the statements they represent are different, nor is the use of letters other than P and Q precluded. However, in a given discussion only one statement may be represented by a given symbol, otherwise ambiguity would result. Letters such as P and Q are sometimes called statement variables.

Truth Tables:

Now that statements have been symbolized by letters, we continue by symbolizing connectives by characters. The characters used for the connectives *and* and *or* and the interrelationships with the member statements are given below.

Let P and Q be statements. The conjunction of P and Q is the compound statement, "P and Q" and is denoted symbolically by

$$P \wedge Q.$$

The disjunction of P and Q is the compound statement, "P or Q" and it is denoted symbolically by

$$P \vee Q.$$

The effect of the individual truth values of the individual statements on the compound statement may be displayed by using a *truth table*. All possible cases of truth values of the individual components must be considered. Since P is either true or false and likewise Q is either true or false, then all possibilities are displayed in *Table 1.1*, with T representing *true* and F representing *false*.

Using *Table 1.1* as a starting point, the truth tables for $P \wedge Q$ and $P \vee Q$ are given in *Tables 1.2* and *1.3*. In mathematics, any conjunction or disjunction has truth values as given in the tables, according to the truth values of the simpler component statements.

Table 1.1

P	Q
T	T
T	F
F	T
F	F

Table 1.2

P	Q	$P \wedge Q$
T	T	T
T	F	F
F	T	F
F	F	F

Table 1.3

P	Q	$P \vee Q$
T	T	T
T	F	T
F	T	T
F	F	F

Let P be the statement, "The sun is shining now" and Q the statement, "The grass is green now." The only way for the compound statement $P \wedge Q$ to have a truth value of TRUE is for each of the components P and Q to have a truth value of TRUE. If either of the component statements is false, the compound statement $P \wedge Q$ is false. This is the way the word *and* is interpreted in mathematics as well as everyday English, and it should agree with your thinking concerning that connective.

Again, using the particular statements above for P and Q, or any statements, it is clear from *Table 1.3* that the disjunction $P \vee Q$ has FALSE truth value only when both P and Q are false. It is in this *inclusive* way that *or* should be interpreted in mathematics. It should be emphasized that $P \vee Q$ is true when BOTH P and Q are true. Actually, this is really how the word *or* is normally interpreted in English. For example, both: "The sun is shining," and "The grass is green," may be true and as a result the compound statement, "The grass is green or the sun is shining" is true. Sometimes from the context of the individual statements P and Q, there is an exclusion, (say P excludes Q), but that is not due to the word *or*. The word *or* in mathematics is ALWAYS an INclusive *or*.

The other principal logical connectives, negation, conditional, and biconditional are now considered.

If P is a statement, then the <u>negation</u> of P, denoted by $\sim P$ is a statement which has the opposite truth value to P. That is, $\sim P$ is true when P is false and $\sim P$ is false when P is true. One reads $\sim P$ as: "It

Table 1.4

P	$\sim P$
T	F
F	T

Table 1.5

P	Q	$P \longrightarrow Q$
T	T	T
T	F	F
F	T	T
F	F	T

Table 1.6

P	Q	$P \longleftrightarrow Q$
T	T	T
T	F	F
F	T	F
F	F	T

is not true that *P*" or more simply *"not P."* As an example, suppose *P* denotes: "Now the sun is shining." Then $\sim P$ is the statement: "It is not true that now the sun is shining," or equivalently, "Now the sun is not shining." The truth table for negation is given in *Table 1.4*.

One should be cautious, however, about the location of the word *Not* in negating a statement. For example, let *P* be the statement: "All students are clever." The statement: "Not all students are clever," is the negation of *P*, while, "All students are not clever," is NOT the negation of *P*. More about negating complex sentences will be discussed later in Chapter 3 on quantifiers. The point is, be careful about the placement of the word *not* when negating a statement.

Let *P* and *Q* be statements. The compound statement of the form: "If *P* then *Q*," is called a <u>conditional</u> statement and is denoted by: $P \longrightarrow Q$ and may also be read as: "*P* <u>implies</u> *Q*," "*Q* <u>if</u> *P*," "*Q* <u>whenever</u> *P*," "*Q* <u>is necessary for</u> *P*," or "*P* <u>is sufficient for</u> *Q*." The statement: "*P* <u>if and only if</u> *Q*" is called a <u>biconditional</u> and is abbreviated by: "*P* <u>iff</u> *Q*" and denoted by $P \longleftrightarrow Q$. The truth tables for the conditional and biconditional are given in *Tables 1.5* and *1.6*. Those tables interrelate the truth values of the conditional and biconditional with *P* and *Q*.

When writing a conditional, be careful NOT to write "If $P \longrightarrow Q$" or "*P* then *Q*" for "$P \longrightarrow Q$." Keep in mind that $\longrightarrow$ is read as "implies" in an English sentence, and is NOT read as "then," so the sentence: "If $P \longrightarrow Q$" would be read as: "If *P* implies *Q*," which is not the same as: "*P* implies *Q*."

A point that is frequently troublesome concerning the truth table for conditionals is that $P \longrightarrow Q$ is true whenever *P* is false (see the third and fourth rows of *Table 1.5*). An example may serve to illustrate this. Suppose that John Doe is having a building built by a contractor and a clause in their agreement reads, "If it does not rain, then the job will be finished in 180 days." To secure the agreement, the contractor agrees to pay John Doe a sum of money if that clause is not upheld. Letting P be "It does not rain," and Q be "The job will be finished in 180 days," one sees that there are four ways the truth values interplay, as described by the values under P and Q in *Table 1.5*. Clearly the clause is upheld and the agreed sum need not be paid if P and Q are both true. On the contrary, if P is true and Q is false, i.e., it does not rain and the job is NOT finished in 180 days, then the contractor has not lived up to the agreement. In this case, the agreement was violated so the contractor must pay. However, if it rains the contractor is "off the hook," because whether he finishes on time or not, the clause is not violated. The clause makes no assertion concerning what will happen if it rains. Thus $P \longrightarrow Q$ is true in each case except when P is true and Q is false.

At the risk of belaboring the conditional, note that the truth table assigns TRUE in the two cases where P is false. In these cases the implication is said to be vacuously true. This is an important thing to remember when we start proving theorems.

To illustrate the biconditional, suppose P denotes: "Triangle ABC is equilateral" and Q denotes: "Triangle ABC is equiangular." Consider the four possible arrangements for T and F. One sees that the table must be as described in *Table 1.6*. You will recall from geometry, that the statements: "Triangle ABC is equilateral," and "Triangle ABC is equiangular," for any given triangle ABC are either both true or both false. The two cases in which one of P and Q is true and the other is false makes $P \longleftrightarrow Q$ a false statement. The latter condition is illustrated by the middle two rows of *Table 1.6*.

Definitions frequently appear as *if and only if* statements. Actually they can always be expressed in this way except some writers choose to express definitions in a variety of other forms, including the use of just the word *if*. Just remember, a definition is interpreted "both ways" i.e., as *if and only if* (which is abbreviated as *iff*).

DEFINITION: A statement P is atomic iff there are NO connectives of the types: $\wedge$, $\vee$, $\longrightarrow$, or $\longleftrightarrow$.

Heretofore, P and Q appeared to be atomic statements, but that was not required. Consider P to be the conditional: $R \longrightarrow S$ and Q to be $R \wedge W$ where R, S, and W are statements. Then $P \vee Q$ becomes the

Logic and Propositional Calculus

9

Table 1.7

R	S	W	$R \longrightarrow S$	$R \wedge W$	$(R \longrightarrow S) \vee (R \wedge W)$
T	T	T	T	T	T
T	T	F	T	F	T
T	F	T	F	T	T
T	F	F	F	F	F
F	T	T	T	F	T
F	T	F	T	F	T
F	F	T	T	F	T
F	F	F	T	F	T

compound statement $(R \longrightarrow S) \vee (R \wedge W)$. How does one determine the truth values of such a statement? One way is to use a truth table. This time there are eight possible arrangements of T and F for the three statements: R, S, and W. After completing the eight rows under R, S, and W describing all possible arrangements of truth values, one moves columnwise to the right, assessing in turn, the truth values under the column headings: $R \longrightarrow S$, $R \wedge W$, $(R \longrightarrow S) \vee (R \wedge W)$ by using the basic truth tables for $\longrightarrow$, $\wedge$, and $\vee$. For example, in the third row of *Table 1.7* (the row in which R, S, and W are True, False, and True, respectively) the truth values are as follows. For $R \longrightarrow S$ note that R is True and S is False, so from *Table 1.5* for $\longrightarrow$, the truth value F is assigned. Likewise since R and W are each true in this row, the truth value of $R \wedge W$ is True. The last column has True in its third row since $(R \longrightarrow S) \vee (R \wedge W)$ is a disjunction of a False statement and a True statement. See *Table 1.3* to verify this choice.

Now that statements have been put together using logical connectives, let us call the resulting compound statement a <u>proposition</u> if the compound statement is composed in an unambiguous way that makes sense.

EXAMPLE 1.1: Which of the following sentences are propositions?

 a. $\sim (P \longrightarrow Q) \longleftrightarrow (P \wedge (\sim Q))$ is a proposition. The atomic statements and the connectives are used in an unambiguous way.

 b. $\sim (PQ)$ is not a proposition. This expression makes no sense because P and Q are NOT separated by a connective. You cannot even say $\sim (PQ)$ for particular statements P and Q and make sense of it. Finally, we have no way to assess its truth value.

 c. $P \sim Q$ is not a proposition. A statement followed immediately by $\sim$ is nonsense. To reinforce the "nonsense" of it, try saying a sentence of this form in English and interpreting it.

d. $\sim P \wedge Q$ is not a proposition because it is ambiguous. Does it mean $\sim (P \wedge Q)$ or $(\sim P) \wedge Q$? This difficulty and more will be addressed soon.

e. $P \wedge Q \vee S$ is not a proposition since it is ambiguous. Does it mean $(P \wedge Q) \vee S$ or $P \wedge (Q \vee S)$? A quick tabulation of the truth tables for these two expressions will show they are different.

f. $P \wedge Q \wedge S$ is not a proposition since it might mean $P \wedge (Q \wedge S)$ or it might mean $(P \wedge Q) \wedge S$. However, a truth table may be constructed that shows the two alternatives have the same truth values, so after suitable conventions are adopted, this expression will be acceptable.

g. $P \longrightarrow Q \longrightarrow S$ is not a proposition. Does it mean $P \longrightarrow (Q \longrightarrow S)$ or does it mean $(P \longrightarrow Q) \longrightarrow S$ or something else? Actually it will be taken to be something else. See Convention 4 below.

Just as in algebra, or as in Example 1.1, parentheses can be used to dictate the order of operations with logical connectives. However, this may yield a cumbersome array of parentheses, so let us agree on some conventions which tend to reduce the number of parentheses and simplify the expressions.

Conventions for Logical Connectives:

1. Negation applies only to the statement immediately following the negation symbol unless parentheses indicate otherwise. Thus $\sim P \longrightarrow Q$ means $(\sim P) \longrightarrow Q$. It <u>doesn't</u> mean $\sim (P \longrightarrow Q)$. It works very much like the negation signs in the algebra of the reals. This resolves the difficulty mentioned in Example 1.1d.

2. The connectives $\wedge$ and $\vee$ are executed before the conditional and biconditional. So $P \wedge Q \longrightarrow P \vee Q$ means $(P \wedge Q) \longrightarrow (P \vee Q)$.

3. If the negation sign immediately follows another connective, such as $\longrightarrow$, then execute $\sim$ first. Thus $P \longrightarrow \sim Q$ means $P \longrightarrow (\sim Q)$ and $P \wedge \sim Q$ means $P \wedge (\sim Q)$.

4. $P \longrightarrow Q \longrightarrow S$ may be taken to mean $(P \longrightarrow Q) \wedge (Q \longrightarrow S)$.

A proposition with n atomic statements has 2^n rows in its full truth table. If there are 4 atomic statements then the truth table will have $2^4 = 16$ rows. Thus, since there are three atomic statements in the

Table 1.8

P	∧	Q	→	P	∨	Q
T		T		T		T
T		F		T		F
F		T		F		T
F		F		F		F

Table 1.9

P	∧	Q	→	P	∨	Q
T	T	T	T	T	T	T
T	F	F	T	T	T	F
F	F	T	T	F	T	T
F	F	F	T	F	F	F
1	3	2	5	1	4	2

expression: $(R \longrightarrow S) \vee (R \wedge T)$, the truth table will have eight rows as seen in *Table 1.7*. It is clear that if there are more than three or four atomic statements, then constructing a truth table becomes tedious.

One way to reduce this tedium is to use *abbreviated* truth tables. Consider the proposition $(P \wedge Q) \longrightarrow (P \vee Q)$. The ordinary truth table would be formed with $2^2 = 4$ rows, and separate columns headed by P, Q, $P \wedge Q$, $P \vee Q$, and $P \wedge Q \longrightarrow P \vee Q$. The abbreviated truth table for this proposition is formed by simply putting $P \wedge Q \longrightarrow P \vee Q$ at the top. Then under each occurrence of each atomic statement, with the necessary number of rows, put in all combinations of T and F. The result at this stage is *Table 1.8*. Notice that at each occurrence of P and of Q the truth values are the same.

Next, fill in truth values below the connectives in the same order you would use in an ordinary truth table, i.e., as dictated by the hierarchy of operations. *Table 1.9* is the completed table. You will notice that the columns are numbered at the bottom to indicate the order followed when filling out the table. The final column to be completed, in this case the column numbered 5, contains the truth values for the given proposition.

There is another shortcut to use. After completing the column numbered 3, note that only the first entry is true. Rows 2, 3, and 4 of column 3 are all false, so those rows need not be completed in column 4 because "false implies either true or false." Thus, after row 1 is completed, all other entries in column 5 must be true. Remember that if the "R" part of the conditional $R \longrightarrow S$ is False, then the conditional is always true.

DEFINITION: If P and Q are statements then the <u>converse</u> of $P \longrightarrow Q$ is $Q \longrightarrow P$, and the <u>contrapositive</u> of $P \longrightarrow Q$ is $\sim Q \longrightarrow \sim P$. If $P \longrightarrow Q$, P is called the <u>hypothesis</u> or <u>antecedent</u> and Q is called the <u>conclusion</u> or the <u>consequent</u>. And we say P is <u>sufficient for</u> Q and Q is <u>necessary</u> <u>for</u> P. If $P \longleftrightarrow Q$ then P is <u>necessary</u> <u>and</u> <u>sufficient</u> <u>for</u> Q.

Consider the implication: "If ABCD is a square then ABCD is a rectangle," which happens to be true. The converse, which is false, is: "If ABCD is a rectangle then ABCD is a square." The contrapositive is: "If ABCD is not a rectangle then ABCD is not a square." The contrapositive is true exactly when the original statement is true.

Exercises:

  1.1 Decide whether the following are statements.

   a. That (pointing to a certain car) is a fast car.

   b. $12 < 5$.

   c. Good morning, how are you?

   d. $\sim(a < 7) \longrightarrow (a > 7)$.

   e. $(6 < 7) \longrightarrow \sim(7 \leq 6)$.

   f. This sentence is false.

   g. The number 17,239,546,187 is a prime number.

   h. If John is 12 years old then Jane must be 15 years old.

  1.2 Whenever possible, determine which parts of Exercise 1 are true and which are false.

  1.3 Construct truth tables for the propositions below. Assume P, Q, and S are statements.

   a. $P \vee (\sim P)$.

   b. $(P \wedge Q) \longrightarrow P$.

   c. $(P \wedge (P \longrightarrow Q)) \longrightarrow Q$.

   d. $(P \vee Q) \longrightarrow (P \wedge Q)$.

   e. $(P \longrightarrow Q) \longrightarrow \sim(\sim Q \vee S)$.

  1.4 Assume P, Q, R, and S are statements. Determine whether or not the following are propositions. You may use the conventions of this section.

   a. $(P \wedge (Q \vee S)) \longrightarrow ((P \wedge Q) \vee (P \wedge S))$.

    b. $P \wedge Q \sim R$.

    c. $P \wedge \vee Q$.

    d. $\sim (P \longrightarrow Q) \longrightarrow P \vee \sim Q$.

    e. $P \wedge Q \vee R$.

1.5   Remove any unnecessary parentheses in accordance with the conventions mentioned in this section.

    a. $(\sim P) \wedge Q$.

    b. $\big(\sim (P \wedge Q) \wedge R\big) \longrightarrow (Q \vee R)$.

    c. $\big(\sim (\sim P) \wedge R\big) \vee Q$.

    d. $(P \wedge R) \longrightarrow (\sim Q)$.

1.6   Construct an abbreviated truth table for each proposition in Exercise 1.3.

1.7   Write the converse and the contrapositive of each of the following without using $\sim$.

    a. If F is a field then F is an integral domain.

    b. If T has an element then T is not empty.

    c. G is a group $\longrightarrow$ G is not empty.

    d. S is bounded $\longrightarrow$ S has an upper bound.

    e. If n is an even integer greater than 2 then n is not prime.

    f. If $0 < |x - a| < \delta$ then $|f(x) - L| < \epsilon$.

1.8   Write a correct negation of each of the following sentences. Try to do so in a nontrivial way.

    a. $2 < 3$.

    b. All rational functions are continuous.

    c. Kilroy was here.

    d. All mammals have hair.

    e. Some sequences converge.

1.9   Let P and Q be the statements indicated below. Then state whether P is a necessary condition, a sufficient condition, or a necessary and sufficient condition for Q based on what you know about the concepts.

    a. P is "a is an even integer" and Q is "a^2 is an even integer."

    b. P is "a is an odd integer" and Q is "a^2 is an odd integer."

    c. P is "f is a continuous function" and Q is "f is a differentiable function."

    d. P is "$-1 < x < 1$" and Q is "$0 < x^2 < 1$."

    e. P is "$-1 < x < 1$" and Q is "$0 \leq x^2 < 1$."

1.10 Use parentheses to indicate the order in which the statements are to be interpreted.

    a. "a is rational $\land\ 0 < a < 1$."

    b. "a is an integer and $a \neq 0 \longrightarrow a^2 > 0$."

CHAPTER 2

TAUTOLOGIES AND VALIDITY

In this chapter the focus is on propositions whose truth values are all true. These propositions will be singled out and given a name, but first it is convenient to have notations for them.

NOTATION: If P, Q, R, etc. are statements then any proposition involving connectives and the given P, Q, R, etc. may be denoted by $\mathcal{F}(P, Q, R, \ldots)$ or simply by $\mathcal{F}$.

For example $\mathcal{F}$ may be $P \longrightarrow Q$, which may or may not always be true. However, it is easy to see the compound statement $P \wedge Q \longrightarrow P$ is always true because whenever $P \wedge Q$ is true then P must be true. The statement $P \wedge Q \longrightarrow P$ is called *simplification*.

EXAMPLE 2.1: $P \wedge (P \longrightarrow Q) \longrightarrow Q$ is a frequently used compound statement, called the *law of detachment*. This proposition is also always true as you can see from column 5 in *Table 2.1*.

Since some compound statements are always true, as in Example 2.1, and some are not, as in *Table 1.7*, let us associate a name with those compound statements that are always true.

DEFINITION: A compound statement $\mathcal{F}(P, Q, R, \ldots)$ that is true for all possible combinations of True and False in the individual component statements P, Q, R, etc. is called a <u>tautology</u>.

The example of the law of detachment, illustrated in *Table 2.1*, is a tautology. The statement $P \wedge Q \longrightarrow P \vee Q$ in *Table 1.9* and the statement $P \wedge Q \longrightarrow P$ are also tautologies.

DEFINITION: A proposition or compound statement $\mathcal{F}(P, Q, R, \ldots)$ that is false for all possible combinations of True and False for the component statements P, Q, R, etc. is called a <u>contradiction</u>.

EXAMPLE 2.2: Consider the statement $P \wedge (\sim P)$. Its abbreviated truth table is given in *Table 2.2*. Since the final column to be done, column 3, is all false, this proposition is a contradiction. In fact, it is the classic form for a contradiction.

Table 2.1

P	$\wedge$	$(P$	$\longrightarrow$	$Q)$	$\longrightarrow$	Q
T	T	T	T	T	T	T
T	F	T	F	F	T	F
F	F	F	T	T	T	T
F	F	F	T	F	T	F
1	4	1	3	2	5	2

Table 2.2

P	$\wedge$	$(\sim$	$P)$
T	F	F	T
F	F	T	F
1	3	2	1

Some compound statements may be determined to be tautologies because they have been derived from other tautologies by replacing some of the individual or atomic component statements by other propositions. This is illustrated in the following example.

EXAMPLE 2.3:   What happens to the tautology $P \longrightarrow P \vee Q$ if the component statement P is replaced by $R \longrightarrow S$ and Q by M? A glance at the two parts of *Table 2.3* reveals that $(R \longrightarrow S) \longrightarrow ((R \longrightarrow S) \vee M)$ is also a tautology.

Sometimes two compound statements, $\mathcal{F}$ and $\mathcal{G}$ in the statement variables P, Q, etc., have the same truth values for each of the different combinations of T and F.

EXAMPLE 2.4:   An example comparing the truth values of the statements $P \longrightarrow Q$ and $\sim P \vee Q$ is given in *Table 2.4*.

Since columns 4 and 5 represent truth values for the compound statements $P \longrightarrow Q$ and $(\sim Q) \vee P$ respectively, and since those two compound statements have the same truth values, then $P \longrightarrow Q$ and $\sim Q \vee P$ are somehow equivalent. Here is a definition.

DEFINITION: The two compound statements: $\mathcal{F}$ and $\mathcal{G}$ are <u>logically equivalent</u>, denoted $\mathcal{F} \Longleftrightarrow \mathcal{G}$, if and only if the biconditional $\mathcal{F} \longleftrightarrow \mathcal{G}$ is a tautology. Another way to denote the logical equivalence of $\mathcal{F}$ and

<u>Table 2.3 a</u>

P	→	P	V	Q
T	T	T	T	T
T	T	T	T	F
F	T	F	T	T
F	T	F	F	F
1	4	1	3	2

<u>Table 2.3 b</u>

(R	→	S)	→	(R → S)	V	M
T	T	T	T	T	T	T
T	T	T	T	T	T	F
T	F	F	T	F	T	T
T	F	F	T	F	F	F
F	T	T	T	T	T	T
F	T	T	T	T	T	F
F	T	F	T	T	T	T
F	T	F	T	T	T	F
1	2	1	4	2	3	1

$\mathcal{G}$ is $\mathcal{F} \equiv \mathcal{G}$. Then $\mathcal{F} \equiv \mathcal{G}$ iff $\mathcal{F}$ and $\mathcal{G}$ have the same truth values for all of the possible combinations of T and F.

Thus, from Example 2.4, $(P \longrightarrow Q) \equiv (\sim P \lor Q)$.

DEFINITION: If $\mathcal{F}$ and $\mathcal{G}$ are statements and if $\mathcal{F} \longrightarrow \mathcal{G}$ is a tautology we say, $\mathcal{F}$ <u>logically</u> <u>implies</u> $\mathcal{G}$. The notation for, "$\mathcal{F}$ logically implies $\mathcal{G}$" is $\mathcal{F} \Longrightarrow \mathcal{G}$. We also say $\mathcal{G}$ is a <u>valid</u> <u>consequence</u> of $\mathcal{F}$.

EXAMPLE 2.5: If the statements P and Q are: "Triangle ABC is equilateral" and "Triangle ABC is equiangular," respectively, then $P \longleftrightarrow Q$ is a tautology, i.e., $P \equiv Q$. On the other hand, if R is the statement: "Quadrilateral ABCD is a square" and S is the statement: "Quadrilateral ABCD is a rectangle," then $R \longrightarrow S$ is tautology, but $R \longleftrightarrow S$ is not. Thus $R \Longrightarrow S$, but $R \not\equiv S$.

For example, $P \longrightarrow P$ is a tautology, so we can say P logically implies P, i.e., $P \Longrightarrow P$. Previously, we saw $P \land Q$ logically implies P, i.e., $P \land Q \Longrightarrow P$.

<u>Table 2.4</u>

(P	$\longrightarrow$	Q)	$\longleftrightarrow$	($\sim P$)	$\vee$	Q
T	T	T	T	F	T	T
T	F	F	T	F	F	F
F	T	T	T	T	T	T
F	T	F	T	T	T	F
1	4	2	6	3	5	2

Table 2.5 is a list of tautologies that are used hereafter in this work. References will be made to these tautologies by name or number from the table. Each tautology in *Table 2.5* may be established by the construction of a truth table. You may want to do some of them for practice.

Although the list in *Table 2.5* is long, there are other tautologies that could have been added to the list. For example, generalizations of the associative, commutative, and distributive laws for $\wedge$ and $\vee$ are true and convenient to have. In addition, there is another principle that is useful. If $P \equiv Q$ then any compound statement $\mathcal{F}$, having P as a component statement in it, can be rewritten with P replaced everywhere by Q. The resulting compound statement $\mathcal{G}$ is equivalent to $\mathcal{F}$.

Since $P \wedge (Q \wedge R) \equiv (P \wedge Q) \wedge R$ and analogously, $P \vee (Q \vee R) \equiv (P \vee Q) \vee R$, there is no ambiguity in writing $P \wedge Q \wedge R$ for $P \wedge (Q \wedge R)$. Using this, we incur no difficulty in interpreting the expression: $P \wedge Q \wedge R \wedge S$. Thus a generalization of the associative law for conjunction is the conjuction of any number of statements that need not be grouped by parentheses. The same is true for disjunction.

Using the generalized associative law it is easy to verify more general distributive laws.

$$P \wedge (Q \vee R \vee S) \equiv (P \wedge Q) \vee (P \wedge R) \vee (P \wedge S), \text{ and}$$
$$P \vee (Q \wedge R \wedge S) \equiv (P \vee Q) \wedge (P \vee R) \wedge (P \vee S).$$

The details are left as exercises (see Exercise 2.12).

A more general commutative law holds. In a conjunction of more than two statements, any proposition, obtained by reordering the statements, is equivalent to the original proposition. For example,

$$P \wedge Q \wedge R \equiv P \wedge R \wedge Q \equiv R \wedge P \wedge Q \equiv R \wedge Q \wedge P$$
$$\equiv Q \wedge R \wedge P \equiv Q \wedge P \wedge R.$$

Table 2.5

Here is a list of standard tautologies. Some assert logical equivalence while others are logical implications.

1.		$P \wedge (P \longrightarrow Q) \Longrightarrow Q$	Law of detachment
2.		$P \wedge Q \Longrightarrow P$	Law of simplification
3.		$P \Longrightarrow P \vee Q$	Law of addition
4.		$(P \vee Q) \wedge (\sim P) \Longrightarrow Q$	Disjunctive syllogism
5.		$(P \longrightarrow Q) \wedge (Q \longrightarrow P) \equiv (P \longleftrightarrow Q)$	
6.		$(P \wedge \sim P) \Longrightarrow Q$	
7.		$P \equiv \sim(\sim P)$	Double negation (D.N.)
8.	a.	$(P \longrightarrow Q) \equiv (\sim Q \longrightarrow \sim P)$	Law of contrapositive
	b.	$(P \longleftrightarrow Q) \equiv (\sim Q \longleftrightarrow \sim P)$	
9.	a.	$\sim(P \wedge Q) \equiv (\sim P \vee \sim Q)$	DeMorgan's law
	b.	$\sim(P \vee Q) \equiv (\sim P \wedge \sim Q)$	DeMorgan's law
10.	a.	$(P \wedge Q) \equiv (Q \wedge P)$	Commutative law
	b.	$(P \vee Q) \equiv (Q \vee P)$	Commutative law
11.	a.	$P \wedge (Q \wedge R) \equiv (P \wedge Q) \wedge R$	Associative law
	b.	$P \vee (Q \vee R) \equiv (P \vee Q) \vee R$	Associative law
12.	a.	$P \wedge (Q \vee R) \equiv (P \wedge Q) \vee (P \wedge R)$	Distributive law ($\wedge$ over $\vee$)
	b.	$P \vee (Q \wedge R) \equiv (P \vee Q) \wedge (P \vee R)$	Distributive law ($\vee$ over $\wedge$)
13.		$(P \longrightarrow Q) \equiv (\sim P \vee Q)$	
14.		$\sim(P \wedge (\sim P))$	Law of non–contradiction
15.		$P \vee (\sim P)$	Law of excluded middle
16.		$\big((P \wedge \sim Q) \longrightarrow (R \wedge \sim R)\big) \Longrightarrow (P \longrightarrow Q)$	Reductio ad absurdum
17.		$(P \longrightarrow Q) \wedge (Q \longrightarrow R) \Longrightarrow (P \longrightarrow R)$	Transitivity of $\longrightarrow$
18.	a.	$(P \longrightarrow Q) \wedge (R \longrightarrow Q) \Longrightarrow \big((P \vee R) \longrightarrow Q\big)$	
	b.	$(P \longrightarrow Q) \wedge (P \longrightarrow R) \Longrightarrow (P \longrightarrow Q \wedge R)$	
19.		$(P \longrightarrow Q) \wedge (P \longrightarrow \sim Q) \Longrightarrow (\sim P)$	Law of absurdity
20.	a.	If P is a tautology then $(P \wedge Q) \equiv Q$	
	b.	If Q is a contradiction then $(P \vee Q) \equiv P$	
21.	a.	$(P \longrightarrow (Q \longrightarrow R)) \Longrightarrow \big((P \wedge Q) \longrightarrow R\big)$	Law of importation
	b.	$(P \wedge Q \longrightarrow R) \Longrightarrow \big(P \longrightarrow (Q \longrightarrow R)\big)$	Law of exportation
22.		$(\sim Q \wedge (P \longrightarrow Q)) \Longrightarrow \sim P$	Modus tollendo tollens
23.		$(P \longrightarrow (Q \wedge \sim Q)) \Longrightarrow \sim P$	Law of absurdity
24.	a.	$(P \longrightarrow Q) \Longrightarrow (P \vee R \longrightarrow Q \vee R)$	
	b.	$(P \longrightarrow Q) \Longrightarrow (P \wedge R \longrightarrow Q \wedge R)$	
25.		$\big((P \longleftrightarrow Q) \wedge (Q \longleftrightarrow R)\big) \Longrightarrow (P \longleftrightarrow R)$	
26.		$(P \longleftrightarrow Q) \equiv (Q \longleftrightarrow P)$	
27.		$(P \vee Q) \equiv (P \vee (Q \wedge \sim P))$	

A similar law applies to disjunction.

Valid Arguments:

Some compound statements are of the type $P_1 \wedge P_2 \longrightarrow Q$ or more generally, $P_1 \wedge P_2 \wedge P_3 \longrightarrow Q$, and they are called <u>arguments</u>. In the latter argument the individual statements P_1, P_2, and P_3 are called <u>premises</u>, <u>hypotheses</u>, or <u>antecedents</u> and Q is called the <u>conclusion</u> or <u>consequent</u>. The argument $P_1 \wedge P_2 \wedge P_3 \longrightarrow Q$ is said to be <u>valid</u> if and only if, the proposition $P_1 \wedge P_2 \wedge P_3 \longrightarrow Q$ is a tautology, that is $P_1 \wedge P_2 \wedge P_3 \Longrightarrow Q$.

How can the validity or invalidity of such an argument be determined? In determining validity, care must be taken to avoid focusing on the truth or falseness of the individual component statements. Rather, the real test of validity must focus on the structure of the argument. For example, is the argument:

$$(P \vee Q) \wedge \sim P \longrightarrow Q$$

a valid argument? A quick check of the list of tautologies reveals that this is just Tautology 4 in *Table 2.5*, so it is a valid argument. In other words, the given conclusion Q is a valid consequence of the given premises: $P \vee Q$ and $\sim P$. Thus we may conclude that the argument: "If (either the sun always shines or turtles do not swim) and (the sun does not always shine) then (turtles do not swim)," is a valid argument, regardless of what you may think of the truth of the conclusion or the individual components. Another argument based on the general one just discussed is the argument: "Either 279 is divisible by 7 or 279 is divisible by 9, but 279 is not divisible by 7, so 279 is divisible by 9," is a valid argument. It is valid whether the conclusion is actually a true statement or not. Thus the validity of an argument is based on the structure of the argument, rather than the truth or falseness the individual parts may have. By replacing the statements: "279 is divisible by 7" by P, and "279 is divisible by 9" by Q, the argument, unadorned with possible misleading irrelevancies, is the statement: "$P \vee Q$ and $\sim P$ therefore Q," which is a valid argument by Tautology 4.

Consider the argument: "If $(P \longrightarrow Q) \wedge \sim P$ then $\sim Q$." A check of the list of tautologies reveals that there are none of this form. Did it just fail to be included in the list, is it an equivalent, but subtle variation of a tautology in the list, or is it simply an invalid argument? The student can check that this is NOT a tautology by making a truth table for it. Not all T's arise, so this is called a <u>fallacy</u>. It is invalid.

Some of the more common fallacies are listed below:

FALLACY OF DENYING THE ANTECEDENT: The preceeding argument, i.e., $((P \longrightarrow Q) \wedge \sim P) \longrightarrow \sim Q$, is invalid as indicated by a truth table. This can also be seen in *Table 1.5*. Since one of the premises, $\sim P$ of this argument, is "P is false," only the last two rows of that *Table 1.5* apply, because these are the only rows where P is false. But the truth value of Q may either be T or F in those rows, so it is NOT POSSIBLE to conclude that Q must be true when $P \longrightarrow Q$ is true and P is false. Any similarity to Tautology 1 or 22 is strictly superficial. To assume the argument above is valid, is committing the *fallacy of denying the antecedent*. The name arises from the negation of P which appears as a premise.

FALLACY OF AFFIRMING THE CONSEQUENT: $(P \longrightarrow Q) \wedge Q \longrightarrow P$ is an invalid argument. A review of *Table 1.5* reveals that even though Q is assumed true, and therefore you are on either row 1 or row 3 of that table, you CANNOT conclude P is true. This is because in row 1, P is true while in row 3 it is not. To assume Q is true and conclude P is true in $P \longrightarrow Q$ is committing the *fallacy of affirming the consequent*.

BEGGING THE QUESTION: This is one of the most common logical errors made in mathematics and one of the most vexing for students when the professor points it out on their papers. Simply, it is the mistake of assuming that which is to be concluded. For example, if you want to prove $P \longrightarrow Q$ and you assume Q is true, you are ignoring half the truth table (see *Table 1.5*). Imagine deleting rows 2 and 4. The resulting table, in effect, ignores the case that makes $P \longrightarrow Q$ false, namely row 2. So by assuming the conclusion Q to be true, you are assuming the argument to be valid, which it may not be. To do this is committing the fallacy of *begging the question*.

The Derivation Method:

We have seen some of the possible mistakes that may be made in testing validity, so how do you ensure that an assertion is valid? In other words, how do you do a valid proof? We could do it by truth tables, but we have seen that that can be tedious, particularly if there are three or more atomic statements. We need a more efficient method. Actually, you probably have already been exposed to such a technique. We call it the *derivation method*. Although the derivation method is less mechanical than filling out a truth table, it has the advantage of being more powerful and efficient than using truth tables. It also has the important feature of illustrating how some proofs of mathematical assertions may be organized. It may seem complicated at first, but it is not. In fact, it is quite sensible.

In practice a derivation is arranged with two columns. In the left column are logical statements. To their right, in the second column, are the reasons for the assertions in column 1. The derivation method is suitable for verifying the validity of an assertion of the form $P_1 \wedge P_2 \wedge \cdots \wedge P_n \Longrightarrow Q$.

THE DERIVATION METHOD: To verify that Q is a valid consequence of premises $P_1, P_2, \ldots, P_n$, a sequence $S_1, S_2, \ldots, S_m$ of propositions must be found such that S_m is Q and for each i such that $1 \leq i < m$, one of the following is true:

a. S_i is one of the premises P_1, P_2, etc. Rule P may be cited here to signify that it is a Premise.

b. S_i is a tautology, with the reason cited as Rule T, for Tautology.

c. S_i is a logical consequence of an S_j where $j < i$ (i.e., an earlier step in the derivation) or a logical consequence of several S_j's with $j < i$. Here you cite the step number(s) you are using from steps above and the tautology you are using.

EXAMPLE 2.6: Suppose A, B, and C represent statements. Verify that the argument $[(A \longrightarrow B) \wedge (B \longrightarrow C) \wedge A] \longrightarrow C$ is valid.

PROOF:
1.	$A \longrightarrow B$	Rule P
2.	$B \longrightarrow C$	Rule P
3.	$A \longrightarrow C$	Steps 1, 2 and Transitivity of $\longrightarrow$
4.	A	Rule P
5.	C	Steps 3, 4 and Detachment (Rule 1)

Note that each step in the derivation is a statement S_i subject to conditions of the derivation method and the last step is the conclusion of the proposition.

This type of argument is a formal proof as you might see in logic. We will investigate more of this type of proof later, but now let us explore an alternate procedure that is a little more algebraic in nature. We will employ a feature mentioned earlier. That is, in a compound statement, a component may be replaced by its logical equivalent. For example, if $P \longrightarrow Q$ is a part of a proposition then that part could be replaced by $(\sim P) \vee Q$, using Tautology 13. To use this technique BEGIN with the LEFT member of the $\equiv$ then work step by step toward the RIGHT member of the given $\equiv$ by replacing components by their logical equivalents.

The last step is the right side of the equivalence being established. This type of proof procedure uses what might be called the *algebra of statements* and it is appropriate for showing two statements are equivalent.

As an example let us prove $P \vee Q \equiv ((\sim P) \longrightarrow Q)$ using the algebra of statements.

THEOREM 2.1: $P \vee Q \equiv (\sim P) \longrightarrow Q$.

PROOF:
$$P \vee Q \equiv \sim(\sim P) \vee Q \qquad \text{Double negation}$$
$$\equiv (\sim P) \longrightarrow Q \qquad \text{Rule 13}$$

Sometimes in mathematical proof, when asked to prove a theorem of the form $P \longrightarrow Q$, it may be easier to prove the contrapositive of $P \longrightarrow Q$, i.e., prove the equivalent statement $\sim Q \longrightarrow \sim P$ instead. Next we show these are equivalent by the algebra of statements.

THEOREM 2.2: Law of contrapositive.

$$(P \longrightarrow Q) \equiv ((\sim Q) \longrightarrow (\sim P)).$$

PROOF:
$$(P \longrightarrow Q) \equiv \sim P \vee Q \qquad \text{Rule 13}$$
$$\equiv Q \vee \sim P \qquad \text{Commutativity of } \vee$$
$$\equiv \sim(\sim Q) \vee (\sim P) \qquad \text{Double negation}$$
$$\equiv (\sim Q) \longrightarrow (\sim P) \qquad \text{Rule 13}$$

This reverifies the equivalence of an implication and its contrapositive.

An example, mentioned earlier, of the use of the contrapositive is the equivalence of: "If ABCD is a square then ABCD is a rectangle," and "If ABCD is not a rectangle then ABCD is not a square." They are equivalent because the contrapositive of an implication is logically equivalent to that implication, as we have just verified.

When using the contrapositive, it is necessary to reverse the direction of the $\longrightarrow$ when negating the P and the Q. If one simply reverses the $\longrightarrow$ in $P \longrightarrow Q$ the result is called the converse of $P \longrightarrow Q$. That is, $Q \longrightarrow P$ is the converse of $P \longrightarrow Q$. The converse of an implication is NOT logically equivalent to the original implication. For example, the converse of: "If ABCD is a square then ABCD is a rectangle" is "If ABCD is a rectangle then ABCD is a square," which is certainly false.

Another logical equivalence relates a biconditional with two conditionals. This is Tautology 5 in the list.

THEOREM 2.3: $(P \longleftrightarrow Q) \equiv (P \longrightarrow Q) \wedge (Q \longrightarrow P)$.

PROOF: Do this as an exercise using truth tables.

When the conditionals $\longrightarrow$ and $\longleftarrow$ are tautologies, then $\longleftrightarrow$ is too, and conversely, thus Theorem 2.3 may be rewritten as $(P \Longleftrightarrow Q) \equiv (P \Longrightarrow Q) \wedge (Q \Longleftarrow P)$.

Theorem 2.3 will be used quite often. In fact it is usually used when showing that two statements are logically equivalent.

Although we have begun to see how *negation* interplays with other connectives, perhaps the most important relationships have yet to be discussed. Two of these are DeMorgan's laws.

THEOREM 2.4:   DeMorgan's laws   (See Rules 9a and 9b in *Table 2.5*.)

 a. $\sim (P \wedge Q) \equiv (\sim P \vee \sim Q)$

 b. $\sim (P \vee Q) \equiv (\sim P \wedge \sim Q)$

A way to say Part a of this theorem is: "The negation of the conjunction of two propositions is equivalent to the disjunction of the negations of those propositions." The following examples illustrate how reasonable this is. Also notice that Part b of Theorem 2.4 reverses the roles of $\wedge$ and $\vee$ in Part a. The two DeMorgan's laws may be proven by completing their truth tables (see Exercise 2.2).

EXAMPLE 2.7:   Suppose p is a prime number and P is the proposition "p divides a or p divides b." What is $\sim P$? It is: "$\sim (p$ divides a or p divides b)." According to DeMorgan's laws, "p does NOT divide a AND p does NOT divide b." Isn't this reasonable? If it is not true that "p divides a or p divides b" then both "p divides a" and "p divides b" must be false.

EXAMPLE 2.8:   Suppose P is the statement, "Either n is odd or n is divisible by 4." Then $\sim P$ is, "It is false that either n is odd or n is divisible by 4." According to DeMorgan's laws, "n is NOT odd AND n is NOT divisible by 4."

Here is another theorem that involves negations, disjunctions, conjunctions, and implications.

THEOREM 2.5:   Let P and Q be statements. Then

$$\sim (P \longrightarrow Q) \equiv (P \wedge (\sim Q)).$$

PROOF: Let P and Q be statements. Then

$$\sim(P \longrightarrow Q) \equiv \sim(\sim P \vee Q) \qquad \text{Rule 13}$$
$$\equiv \sim(\sim P) \wedge \sim Q \qquad \text{DeMorgan}$$
$$\equiv P \wedge (\sim Q) \qquad \text{Double negation}$$

EXAMPLE 2.10: If you have had a course in calculus, you will recall a theorem like the following.

A CALCULUS THEOREM: If f is a differentiable function then whenever $f'(x) > 0$ on $[a, b]$, f is increasing on $[a, b]$.

The form of this theorem is $P \longrightarrow (Q \longrightarrow R)$ where P is "f is a differentiable function on $[a, b]$," Q is "$f'(x) > 0$ on $[a, b]$," and R is "f is increasing on $[a, b]$." How would you interpret this in order to prove it? That is, what do you assume as true and what do you show as true? Recall that in order to prove a proposition of the form $S \longrightarrow T$, one supposes S is true and shows T has to be true. With this in mind, what is S for this theorem? S really has two parts; two assumptions: P and Q. That is because $P \longrightarrow (Q \longrightarrow R)$ is logically equivalent to $P \wedge Q \longrightarrow R$, by Tautology 21. Our hypothesis is $P \wedge Q$, and our conclusion is R. Thus to prove this theorem, we begin by assuming f is differentiable on $[a, b]$ and f is increasing on $[a, b]$. Then calculus methods are used to carry out the proof that f is increasing on $[a, b]$.

Exercises: Do problems 2.1–2.8 by truth tables or other suitable methods.

2.1 Verify the tautologies numbered 1, 8, 15, 22.

2.2 Verify the tautologies numbered 2, 9, 16, 23.

2.3 Verify the tautologies numbered 3, 10, 17, 24.

2.4 Verify the tautologies numbered 4, 11, 18, 25.

2.5 Verify the tautologies numbered 5, 12, 19.

2.6 Verify the tautologies numbered 6, 13, 20.

2.7 Verify the tautologies numbered 7, 14, 21.

2.8 Verify the following tautologies.

   a. $P \Longrightarrow P$

   b. $P \Longleftrightarrow P$

   c. $P \vee P \Longleftrightarrow P$

   d. $P \Longleftrightarrow P \wedge P$

2.9 Which of the following propositions are true and which are false based on the truth of the individual parts?

a. If Denver is the capital of Colorado, then $2 + 3 = 6$.

b. If Denver is the capital of Colorado, then $2 + 3 = 5$.

c. If Denver is the capital of Idaho, then $2 + 3 = 5$.

d. If Denver is the capital of Idaho, then $2 + 3 = 6$.

2.10 Let P be the statement "$2 > 3$" and Q the statement, "If $ABCD$ is a square then $ABCD$ is a rectangle." Determine the truth value of the following based on the truth or falseness of P and Q.

a. $P \wedge Q$

b. $P \vee Q$

c. $P \longrightarrow Q$

d. $P \longleftrightarrow Q$

e. $Q \longrightarrow P$

f. $\sim P \longleftrightarrow Q$

2.11 Find the valid arguments and indicate the tautologies which justify them. For the invalid arguments, indicate if possible, the type of fallacy committed.

a. If the sun shines then the temperature rises. The sun shines. Therefore the temperature rises.

b. If the sun shines then the temperature rises. The sun does not shine. Therefore the temperature does not rise.

c. If the sun shines then the temperature rises. The temperature does not rise. Therefore the sun does not shine.

d. Either the sun shines or the temperature does not rise. The temperature does not rise. Therefore the sun shines.

e. Either the sun shines or the temperature does not rise. The temperature does not rise. Therefore the sun does not shine.

f. Either the sun shines or the temperature does not rise. The temperature does not rise. Therefore the sun shines.

2.12 Verify the more general distributive law:

$$P \wedge (Q \vee R \vee S) \Longleftrightarrow (P \wedge Q) \vee (P \wedge R) \vee (P \wedge S).$$

2.13 Illustrate a more general kind of commutative law by verifying each of the equivalences

$$P \wedge Q \wedge R \Longleftrightarrow P \wedge R \wedge Q \Longleftrightarrow R \wedge P \wedge Q.$$

2.14 Let p be the sentence "R is a field," let q be "R is an integral domain," and let r be "R is finite." Then convert $q \land r \Longrightarrow p$ into a sentence, free of symbolic connectives and free of the symbols p, q, and r.

2.15 Convert the statement: "If T is a topological subspace of the reals and T is closed and T is bounded then T is compact" into symbolic form using logical connectives and letters P, Q, R, and S for the appropriate atomic statements. Include a dictionary in your solution.

2.16 Consider the statement: "If G is a group and N is a normal subgroup of G then G/N is a group."

    a. Convert the statement above to a statement using P, Q, and R for the atomic parts. Include a dictionary in your solution.

    b. Rewrite the negation of the converted statement in 2.16a, then rewrite the negation using the original phraseology.

    c. Write the contrapositive of the converted statement in 2.16a, then rewrite the contrapositive using the original terminology.

2.17 Consider the statement: "If a is an element of E_1 and $a \neq 0$ then a^{-1} is an element of E_1."

    a. Write the contrapositive of the given statement.

    b. Negate the statement given in 2.17a.

2.18 Complete the following by filling in the blanks. There may be more than one correct answer.

    a. $P \longrightarrow Q$ is false if P is _____ and Q is _____.

    b. $\sim P \longrightarrow Q$ is true if P is _____ and Q is _____.

    c. $P \lor \sim Q$ is false if P is _____ and Q is _____.

    d. $\sim(\sim P \lor Q)$ is true if P is _____ and Q is _____.

    e. $\sim(P \land \sim Q)$ is true if P is _____ and Q is _____.

2.19 Verify $\left[(J \Longrightarrow N) \land (\sim J \Longrightarrow D) \land (D \Longrightarrow \sim A)\right] \Longrightarrow (\sim A \lor N)$ by the derivation method.

2.20 Verify $\left[(D \lor P) \land (D \Longrightarrow \sim E) \land (P \Longrightarrow M)\right] \Longrightarrow (\sim E \lor M)$ by the derivation method.

2.21 Which tautology would aid you in devising a proof of a theorem of the form $P \Longrightarrow (Q \Longrightarrow R)$?

2.22 True or False? A statement must either be a tautology or a contradiction.

CHAPTER 3

QUANTIFIERS AND PREDICATES

Introduction:

Ordinary language is frequently imprecise and ambiguous. These traits, while adequate for relaxed and informal discussion, and even for poets and politicians, are totally inadequate for conveying mathematical ideas from person to person. In some instances it is perfectly appropriate to intentionally mislead, be vague, or to want many possible interpretations. For example, in the case of a best–selling murder mystery or a diplomatic response to a question involving sensitive international issues. But mathematicians, scientists, and others require precision and clarity. There is a need to be able to rely on mathematical results, not only in mathematics, but in many other disciplines. A principal means for attaining the objective of having reliable results is through the process of deductive or inferential reasoning. The Derivation method, developed in previous chapters, as well as the algebraic manipulation of statements, also discussed previously, provide efficient ways to tell whether certain arguments are valid. In addition, a list of valid arguments is given in *Table 2.5*. Those rules for deductive reasoning have long been used to assure that certain results in mathematics cannot get called into question. Those rules will be expanded in this chapter to allow us to deal with a greater variety of arguments.

Another necessity in achieving precision without ambiguity is the use of carefully defined terms. No doubt you have seen a heated debate which would not have occurred had there been common agreement on the meaning of certain words. This situation is avoided when definite meanings are attached to words through definitions. The careful and systematic defining of terms is a cornerstone in the development of any mathematical work. In fact, the presence of definitions is such a ubiquitous feature throughout mathematics, that it is unusual to find a section of a mathematics text that does not define at least one term. However as common as definitions are, students will still incur some difficulty in writing their own definitions, when they first begin. And when asked to define a term, the novice may be somewhat vague or imprecise or otherwise overlook important aspects which pin down the concept to be defined. This common difficulty is alleviated by studying similar definitions and observing the characteristics that definitions have of pinpointing meanings, then by writing definitions. With practice comes more confidence and skill.

Variables, Predicates, and Quantifiers:

The tools of logic, thus far developed, are still inadequate for our intended purposes. The propositional calculus of the previous sections is simply not sophisticated enough to deal with many everyday mathematical sentences such as: "x is a positive real number" or "For every integer x, x^2 is nonnegative." The first sentence is NOT even a statement, since the specific value of x is not known. The second sentence is a statement, but as yet our tools are too superficial to deal with it.

In this section, ways are found to deal with these problems. We will initiate that inquiry by discussing some new terminology.

No doubt you recall a mathematics teacher saying, "Define your terms," when you were working a stated problem. What your teacher meant was, "Let x be such and such and y so and so," before proceeding with the solution using these variables. Obviously there is a need for the reader to know what x and y represent, and this is accomplished by a definition. You wouldn't really expect the reader to just guess at what they meant. The x and y are also examples of the first notion to be considered: *term*.

In this chapter we treat three types of terms: variables, constants, and definite descriptions. A *variable* is frequently denoted by x or y, possibly subscripted, or another letter, usually from near the end of the alphabet. The letter is used to refer to one of a collection of objects in much the way that the pronoun, "she," might be used to refer to any individual in a classroom having all women. This feature is what we want a variable like x or y to have, in referring to (no particular) an individual member of a collection of objects such as the collection of real numbers. It is possible to say, for example, "For any real numbers x and y, $x + y = y + x$," and be using x and y as variables. The collection of objects from which the variables are selected is called <u>universe of discourse</u>. For example, the collection of women in the classroom mentioned above is a suitable universe of discourse for the first discussion. And the collection of real numbers is a suitable universe of discourse for the second discussion.

A second type of term is a *constant*. An example of this type of term is illustrated in the sentence

$$\text{Find } \int_a^b x^2 \, dx.$$

The terms a and b are used as constants, while x is used as a variable. A constant refers to a particular object that is left unspecified except for the symbol given to it, which implies that it doesn't vary, while the x does vary over the interval $[a, b]$. The usual choices for symbols for constants are letters from the beginning of the alphabet, while the symbols for variables are usually from near the end.

The final term is *definite description*. We may replace a phrase or sentence that describes an object by a letter, usually from the beginning of the alphabet. An example of such a term occurs in the following. Let

$$x^3 + 3x^2 + 5x - 15 = 0$$

be an equation with variable x. We may determine using calculus methods, that there is exactly one solution to the equation. Let d be that unique real number root to the given equation. In this case d is given by a definite description because we have described a condition it must meet. Another example follows. Let π denote the ratio of the circumference of the unit circle to its diameter. Again, π is given by a definite description.

The next notion to be considered is the notion of *predicate*. In English grammar a predicate is frequently thought of as what is said about the subject of the sentence, and it is not, by itself, a complete sentence. For our purposes, "____ is a mathematics student" is a predicate; there is no subject. Mathematically, this may be written as, "x is a mathematics student." And, in the customary notation for predicates, the expression, "x is a mathematics student at Trinity University," may be symbolized by, "$M(x)$." Here the x is an *object variable* and $M(x)$ is a *predicate variable*. A predicate variable is a symbol for the predicate, followed by the object variable in parentheses.

As mentioned at the beginning of this discussion, a predicate variable like $M(x)$, is NOT a statement since it has no truth value. There is however, a way to make a statement out of a predicate such as $M(x)$. Simply assign a constant value to the predicate variable. When this is done it is called a *substitution instance*. The substitution must be of a specific object appropriate for the predicate. For example, one may replace x in $M(x)$ above by Martha Ann Smith, to obtain, "Martha Ann Smith is a mathematics student at Trinity University." Clearly, this is a statement, since a check of the mathematics rolls would determine the truth or falseness of the sentence. We do not have to physically check the rolls for the presence or absence of the name of Martha Ann Smith; just recognize that such a check would determine the truth value of "M(Martha Ann Smith)." It is inappropriate to substitute, "Triangle ABC," for x in the $M(x)$ above, because "Triangle ABC is a mathematics student at Trinity

University" does not make good sense. To aid in limiting the choices for substitution into a predicate, one only considers specific objects in the universe of discourse. For example, a suitable universe of discourse for, "x is a mathematics student at Trinity University," might be the universe of all students at Trinity, or even all persons in the world.

One may join predicates by connectives to obtain *compound* predicates. For example, consider "x is a real number and $x^2 - x$ is positive." We could let $R(x)$ denote "x is a real number" and let $S(x)$ denote "$x^2 - x$ is positive." Then we would have $R(x) \wedge S(x)$ as a compound predicate. It is still not a statement, but it can be made into one, as before, by selecting a specific real number, say the constant a, and substituting a for x in $R(x) \wedge S(x)$ to get $R(a) \wedge S(a)$. THEN we would have a statement. Again we wouldn't know that it is true or false, only that we could determine truth value by whatever a represents.

Predicates can be made into statements by another means. To accomplish this, the predicate is prefaced by one of two phrases called *quantifiers*. The quantifiers are "for all x" and "there exists an x such that" which are denoted symbolically by $\forall x$ and $\exists x$, respectively. The predicate $P(x)$, with object variable x, when prefaced by $\forall x$ or $\exists x$, becomes

$$\forall x P(x) \quad \text{or} \quad \exists x P(x).$$

These are read as, "For all x, $P(x)$" and "There exists an x such that $P(x)$," respectively. In each case the new sentence is a statement, i.e., it has truth value.

To see that this use of quantifiers does indeed make a predicate into a statement, return to the example in which $P(x) \wedge S(x)$ denotes, "x is a real number and $x^2 - x > 0$." It is easy to see that $\forall x [P(x) \wedge S(x)]$ is false, because the substitution instance $P(\frac{1}{2}) \wedge S(\frac{1}{2})$ is false. (You can check it.) In other words, it is NOT true that $P(x) \wedge Q(x)$ is true for all x. On the other hand, the sentence $\exists x, P(x) \wedge S(x)$ is true, because $P(3) \wedge S(3)$ is true. All it takes to establish that $\exists x, [P(x) \wedge S(x)]$ is true is one substitution instance that makes $P(x) \wedge Q(x)$ true, which $x = 3$ does.

The prefacing phrases "for all" and "there exists," i.e., $\forall x$ and $\exists x$ are called the universal quantifier and existential quantifier, respectively. When it is important that the object variables be selected from the proper universe, a slight modification of this notation is used. For example, if the universe is U then $\forall x \in U$ or $\exists x \in U$ would limit the selection of x to an appropriate domain.

One may have several variables in a predicate. The sentence $x + y =$

$y + x$ is a predicate in two variables x and y. To use quantifiers to convert such a predicate into a statement, requires two quantifiers. This may be done with both universal, both existential, or mixed quantifiers. For example, if Z is the collection of all integers, one of these instances (using two universal quantifiers) is the familiar

$$(\forall x \in Z)(\forall y \in Z)[x + y = y + x]. \tag{1}$$

In such an instance $x+y = y+x$ may be denoted by $P(x,y)$. Equation (1) becomes the familiar commutative law: $(\forall x \in Z)(\forall y \in Z)[x+y = y+x]$, i.e., $(\forall x \in Z)(\forall y \in Z)[P(x,y)]$. It says every $x \in Z$ commutes with each $y \in Z$ under the operation of addition. It may be abbreviated as: $\forall x, y \in Z(x + y = y + x)$.

When using *mixed* quantifiers it is important to use them in the proper order. In the examples below you will see there is a <u>very</u> <u>distinct</u> <u>and</u> <u>important</u> <u>difference</u> <u>between</u> <u>the</u> <u>statements</u>:

$$\forall x \exists y \, P(x,y) \quad \text{and} \quad \exists y \forall x \, P(x,y).$$

EXAMPLE 3.1: Let $P(x,y)$ denote the predicate, "The person x has residence y." The statement $\forall x \exists y P(x,y)$ says, in a more casual manner, that "every person has a residence." On the other hand $\exists y \forall x P(x,y)$ says that "there is a residence such that every person has THAT residence." Clearly these two statements have vastly different meanings.

EXAMPLE 3.2: In the context of the system of integers, the statement $\forall x \exists y \, [x + y = 0]$ means, "for each integer x, an integer y can be found so that $x + y = 0$." A result of using this order is that y DEPENDS upon x and when x changes then y may change. Here y is really $-x$ so this dependence of y on x is clear. On the other hand, in the statement $\exists y \forall x [x + y = x]$ the y is fixed and NOT DEPENDENT upon x. In fact it says there is (at least) one y that makes $x + y = x$ for all x's. In this latter case, that y is 0 and clearly that one y works for all x and is not dependent on x.

EXAMPLE 3.3: In each of the following, determine whether you may put $\forall x \exists y$, $\forall x \forall y$, $\exists x \forall y$, or $\exists x \exists y$ in front of the given predicate $P(x,y)$ in the context of the given universe of discourse.

   a. $P(x,y)$: "$x + y = 2$." Suppose the universe of discourse is the collection Z of integers, i.e., $0, \pm 1, \pm 2, \ldots$

   b. $P(x,y)$: "$2xy = x$." Suppose the universe of discourse is the rational number system, i.e., numbers of the form: $\frac{p}{q}$ where p and q are integers and $q \neq 0$.

c. $P(x, y)$: "x is married to y." Suppose the universe of discourse is "all married persons."

Answer these questions on your own before reading the answers below. In Part a, the possible double quantifiers that may be placed in front of $x + y = 2$ are: $\forall x \, \exists y$ and $\exists x \, \exists y$ and not the other two. In Part b the quantifiers $\forall x \, \exists y$, $\exists x \, \forall y$ (what x?) and $\exists x \, \exists y$ may be legitimately placed in front of $2xy = x$. In Part c, $\forall x \exists y$ and $\exists x \exists y$ may be placed in front of the given predicate.

Now let us look at some hints for translating English sentences into quantifiers and predicates. As a general rule, if a statement says, "for some $\cdots$," "there is $\cdots$," or "there is at least one $\cdots$, " then the existential quantifier is used. If a sentence says, "all $\cdots$," "each," or "every," then expect a universal quantifier and (quite often) a conditional. The connectives $\wedge$ and $\vee$ arise, obviously from "and" and "or." In addition, the connective $\wedge$ is frequently used when the words, "but," "while," "however," etc. occur. Some examples are given below.

EXAMPLE 3.4:   In parts a – e let the universe of discourse be the collection of all persons.

a. "Some students are clever," is denoted by

$$\text{"}\exists x(\; x \text{ is a student and } x \text{ is clever}).\text{"}$$

b. "All students are clever," should be written as

$$\text{"}\forall x(\; x \text{ is a student } \longrightarrow x \text{ is clever})\text{"}$$

and not "$\forall x(\; x$ is a student and x is clever)." Why?

c. "All freshmen who like beer are science majors." Using $F(x)$, $B(x)$, and $S(x)$ for "x is a freshman,"   "x likes beer," etc. we get "$\forall x\big(F(x) \wedge B(x) \longrightarrow S(x)\big).$" Why is it $\longrightarrow$ and not $\wedge$?

d. "All senior men favor women or books." Using $S(x)$, $M(x)$, $W(x)$, and $B(x)$ this becomes "$\forall x\big[S(x) \wedge M(x) \longrightarrow W(x) \vee B(x)\big].$" Why is it $\longrightarrow$ and not $\wedge$?

e. "Some mathematicians are eccentric, but not all mathematicians are friendly." Using $M(x)$, $E(x)$, and $F(x)$ this becomes

$$\text{"}\exists x\big[M(x) \wedge E(x)\big] \wedge \; \sim \forall x\big[M(x) \longrightarrow F(x)\big].\text{"}$$

Why is there an implication here?

Some more mathematical examples are:

f. "For all real numbers x (i.e., $x \in R$) such that $0 \le x \le 1$, it is true that $x^2 \le x$," would be written as "$(\forall x \in R)\big[$ if $0 \le x$ and $x \le 1$ then $x^2 \le x\big]$," or more symbolically as

$$\text{"}(\forall x \in R)\big[0 \le x \wedge x \le 1 \longrightarrow x^2 \le x\big].\text{"}$$

Why is the connective "$\longrightarrow$" used here and not "$\wedge$"?

g. "Every differentiable function on the interval (a, b) is continuous on (a, b)." Let $\mathcal{F}$ denote all functions, and let $D(f)$ and $C(f)$ denote "f is differentiable on (a, b)" and "f is continuous on (a, b)," respectively. The given statement converted into quantifiers, predicates, and connectives is then "$(\forall f \in \mathcal{F})\big[D(f) \longrightarrow C(f)\big].$"

EXAMPLE 3.5: Determine the truth value of the following statements. Let N denote the natural numbers, Z the integers and R the reals. You may use results you know from calculus and the algebra of inequalities.

a. $(\forall x \in Z)[x^2 > 0]$. This is FALSE. It is not true that the square of each integer is positive, in particular: $0^2 \not> 0$. Remember, all it takes is one substitution instance that makes a predicate, $P(x)$ false for the statement $\forall x P(x)$ to be false. On the other hand $(\forall x \in Z)[x^2 \ge 0]$ is TRUE.

b. $(\exists x \in Z)[x^2 > 0]$. This is TRUE because $(-3)^2 > 0$. That is, we can say there is an x such that $x^2 > 0$ since -3 is just such a number. Remember, it takes only one substitution instance into $P(x)$ to make $\exists x[P(x)]$ true. If it is true more than once then $\exists x[P(x)]$ is still true.

c. $(\exists x \in R)[3x^3 - 5x^2 + 3x - 5 = 0]$. Here we need to determine if there is a particular $a \in R$ such that $3a^3 - 5a^2 + 3a - 5$ is 0. Let $f(x) = 3x^3 - 5x^2 + 3x - 5$ then we know that $f(0)$ is -5 and $f(2) = 3 \cdot 2^3 - 5 \cdot 2^2 + 3 \cdot 2 - 5$ is positive and we know that $f(x)$ is a continuous function. (Why?) By the *intermediate value theorem*, there is a real number a such that $f(a) = 0$, so the answer is true. Notice we did not identify the exact value of a, but we did determine that it existed.

EXAMPLE 3.6: Symbolize the assertion, "Every continuous function on the interval $(-1, 1)$ is differentiable on $(-1, 1)$," with quantifiers and predicates. From what you know from calculus, then determine whether it is true or false.

SOLUTION: Let U be the collection of all functions and let $C(f)$ denote f is continuous on $(-1,1)$ and $D(f)$ be f is differentiable on $(-1,1)$. The given assertion becomes $(\forall f \in U)[C(f) \longrightarrow D(f)]$. The function $f(x) = |x|$ is continuous on $(-1,1)$, but is not differentiable at 0, so it is not differentiable on $(-1,1)$. Thus $C(|x|)$ is true, but $D(|x|)$ is false. Since a *true* statement cannot imply a *false* statement (see *Table 1.4*), this means $C(f) \longrightarrow D(f)$ is NOT true for all f defined on $(-1,1)$. Thus the given statement, "Every continuous function on $(-1,1)$ is differentiable on $(-1,1)$," is FALSE.

We now consider two additional rules dealing with quantifiers. They have to do with negating a quantified statement. To get us into the right frame of mind, we reconsider an earlier example of the two statements:

P: "Not all students are clever."
Q: "There is at least one student who is not clever."

It should be clear from the earlier discussions that P and Q are logically equivalent statements, i.e., $P \equiv Q$. Let $S(x)$ denote "x is a student," and $C(x)$ denote "x is clever." What are the formulations of the statements above with quantifiers and the given predicates?

$$P \text{ is formulated by } \sim \forall x[S(x) \longrightarrow C(x)].$$
$$Q \text{ is formulated by } \exists\, x[S(x) \wedge \sim C(x)].$$
(2)

These are reasonable translations into quantified statements. So it would appear that

$$\sim \forall x[S(x) \longrightarrow C(x)] \equiv \exists\, x[S(x) \wedge \sim C(x)]. \tag{3}$$

Using Theorem 2.5, the right hand member of the $\equiv$ in Equation (3) may be further modified into $\exists\, x\big[\sim [S(x) \longrightarrow C(x)]\big]$. By rewriting Equation (3) we would get:

$$\sim \forall x[S(x) \longrightarrow C(x)] \equiv \exists\, x\Big[\sim [S(x) \longrightarrow C(x)]\Big]. \tag{4}$$

It appears that a general rule would be $\sim \forall x[P(x)] \equiv \exists x[\sim P(x)]$. This is exactly Part a of the next theorem.

THEOREM 3.1:   Let $P(x)$ be a predicate. Then

a. $\sim \forall x[P(x)] \equiv \exists x[\sim P(x)]$ and   b. $\sim \exists x[P(x)] \equiv \forall x[\sim P(x)]$.

PROOF: a. Suppose $\sim \forall x[P(x)]$. Then it is not true that $P(x)$ is true for all x. This means there is some substitution instance $P(a)$ which is false. That is $\sim P(a)$ is true. But this says $\exists x[\sim P(x)]$ is true. Since we began with $\sim \forall x[P(x)]$ and found that $\exists x[\sim P(x)]$ was necessarily true, we have shown that $\sim \forall x[P(x)] \Longrightarrow \exists x[\sim P(x)]$. Since we are to show these are equivalent, we must show the conditional also goes the other way. To do that, begin with the right side, i.e., suppose $\exists x[\sim P(x)]$. Then there is some substitution instance, a for x, for which $\sim P(a)$ is true. That means that $P(a)$ is false, for the specific a, but then $\forall x[P(x)]$ is not true. In other words, $\sim \forall x[P(x)]$ is true. Thus the implication does go the "other" way. Hence $\exists x[\sim P(x)] \Longrightarrow \sim \forall x[P(x)]$. Putting the two conditionals together, and using Theorem 2.3, we get that Part a is valid.

Part b is an exercise. Hint, use Part a and double negation.

Keep in mind that when negating a quantified statement, the quantifiers are exchanged and the predicate gets negated. Here are some further examples. Notice how the negating begins on the outside and works toward the innermost predicate in a step–by–step fashion.

EXAMPLE 3.7:   The following are examples of negating statements with various quantifiers.

a.
$$\sim \forall x\Big[\exists y, (xy \neq 0)\Big] \equiv \exists x\Big[\sim \exists y, (xy \neq 0)\Big]$$

$$\equiv \exists x\Big[\forall y(xy = 0)\Big].$$

b. The following three statements are logically equivalent.

$$\sim \Big[\forall x\big[(x \text{ is rational } \wedge\ 0 < x < 1) \longrightarrow x^2 < x)\big]\Big]$$

$$\exists x\Big[\sim \big[(x \text{ is rational } \wedge\ 0 < x < 1) \longrightarrow x^2 < x\big]\Big]$$

$$\exists x\Big[(x \text{ is rational } \wedge\ 0 < x < 1) \wedge x^2 \not< x\Big]$$

c. Statements such as $\forall \epsilon[\epsilon > 0 \Longrightarrow \exists x(0 < x < \epsilon)]$ are very common in mathematical literature. This proposition may also be interpreted and variously written as

$$(\forall \epsilon > 0)(\exists x)\big[(x > 0) \wedge (x < \epsilon)\big] \text{ or as } (\forall \epsilon > 0)(\exists x > 0)(x < \epsilon).$$

The denial of the latter expression is

$$\sim\!\big[(\forall\,\epsilon>0)(\exists\,x>0)(x<\epsilon)\big] \equiv (\exists\,\epsilon>0)\big[\sim(\exists\,x>0)(x<\epsilon)\big]$$
$$\equiv (\exists\,\epsilon>0)(\forall\,x>0)\big[\sim(x<\epsilon)\big]$$
$$\equiv (\exists\,\epsilon>0)(\forall\,x>0)(x\ge\epsilon).$$

d. Negate the expression $(\forall\,x\in R)\big[0\le x\wedge x\le 1 \longrightarrow x^2\le x\big]$ of Example 3.4f. It is $(\exists\,x\in R)\big[(0\le x\wedge x\le 1)\wedge x^2 > x\big]$.

e. The negation of $(\forall\,f\in \mathcal{F})\big[D(f)\longrightarrow C(f)\big]$ in Example 3.4g is
$$\sim(\forall f\in\mathcal{F})[D(f)\longrightarrow C(f)] \equiv (\exists\,f\in\mathcal{F})\big[D(f)\wedge\sim C(f)\big].$$

Scope, Free Variables, and Bound Variables:

Without trying to be precise, we shall agree that the scope of a quantifier is the quantifier and that part of the sentence to which the quantifier applies (usually indicated by some symbol of grouping).

EXAMPLE 3.8:

a. $\underline{\forall x[P(x)\longrightarrow Q(x)]}\wedge\big(R(x)\longrightarrow S(y)\big)$. The scope of the quantifier $\forall$ is underlined.

b. $\exists x\big[\underline{\forall y[P(x)\wedge Q(y)\longrightarrow R(y)]}\wedge S(x)\big]$. The scopes of $\exists$ and $\forall$ are underlined.

DEFINITION: An occurrence of a variable is bound if it is in the scope of a quantifier using that variable.

DEFINITION: An occurrence of a variable is said to be free if it is NOT in the scope of a quantifier using that variable.

EXAMPLE 3.9:

a. Referring to Example 3.8a observe that the occurrence of x in $Q(x)$ is bound, since that occurrence is in the scope of the universal quantifier. Also, the occurrence of x in $R(x)$ is free, since $R(x)$ is not in the scope of any quantifier. The variable y in that example is free.

b. Referring to Example 3.8b the occurrence of both x and y is bound since x and y are each in the scope of a quantifier.

EXAMPLE 3.10: Use standard symbols of algebra, calculus, and logic, along with a dictionary to define any predicates, to express the following symbolically.

a. If f is a continuous function on the interval $[a, b]$ then there exists c in (a, b) such that $\int_a^b f(x)\, dx = f(c)(b - a)$.

SOLUTION: Let $C(f)$ denote f is continuous on $[a, b]$. Then

$$C(f) \Longrightarrow \left(\exists\, c \in (a, b)\right) \left[\int_a^b f(x)\, dx = f(c)(b - a)\right].$$

b. Every even integer greater than 2 is the sum of two prime numbers.

SOLUTION: Let $x \in E$ denote x is an even integer and $a \in P$ denote a is a prime number. The given statement becomes

$$\forall c\left[(c \in E \wedge c > 2) \Longrightarrow (\exists\, a \in P)(\exists\, b \in P)(c = a + b)\right].$$

Exercises:

In Exercises 1–9, find a suitable universe of discourse and predicates and use quantifiers along with standard mathematical symbols of logic, algebra, and calculus, if needed, to symbolize the following (see Examples 3.4f,g and 3.10a,b). Use a dictionary in case you define any predicates. That is, if using a predicate $P(x)$ to denote conditions on x, such as $x > 0$, enumerate the meanings of these things in a list, i.e., as in a dictionary. Also let Z denote the collection of integers and R the reals.

3.1 For any x and any $y < x$ there is a z such that $y < z < x$.

3.2 There is no smallest integer.

3.3 For all non-negative integers x, $x^2 \geq x$.

3.4 Every triangle is a polygon.

3.5 There are some integers whose cubes are even.

3.6 No matter what the real number x is, $x^2 + 6x + 10$ is positive.

3.7 There is no real number x such that $x^2 + 1 = 0$.

3.8 If f is continuous on $[a, b]$ and if $f'(x)$ exists on (a, b) then there exists c in (a, b) such that

$$f'(c) = \frac{f(b) - f(a)}{b - a}.$$

3.9 If p is a prime and p divides $a \cdot b$, (denoted $p \mid (a \cdot b)$) then $p \mid a$ or $p \mid b$.

3.10 Answer the questions in Example 3.4b,c. For example, in 3.4b there is the statement, "$\forall x [x$ is a student $\longrightarrow x$ is clever $]$." Why is it incorrect to write, "$\forall x(x$ is a student and x is clever $)$"?

3.11 Prove $\sim \exists x[P(x)] \equiv \forall x[\sim P(x)]$. That is, prove Part b of Theorem 3.1. Hint, Use Part a and double negation.

3.12 What is the difference between the two doubly quantified sentences $\forall x \exists y P(x,y)$ and $\exists y \forall x P(x,y)$, when $P(x,y)$ is

a. The person x is a citizen of country y.

b. $x + y = 2$ where x and y are integers.

3.13 Underline the scope of the quantifiers and identify the free and bound variables by listing the free variables and the bound variables.

a. $\forall x[F(x) \wedge B(x) \longrightarrow S(x)] \vee C(x)$.

b. $\forall x[\exists y[S(x) \wedge R(y) \longrightarrow T(x)] \longrightarrow T(y)]$.

3.14 Determine the truth value of each of the following statements in the given universe of discourse. Explain your answers. Let Z denote the integers, R the real numbers, and N the natural numbers.

a. $(\forall x \in Z)[x^2 \leq 0]$.

b. $(\exists x \in Z)[x^2 \leq 0]$.

c. $(\forall x \in Z)[x < 3 \Longrightarrow (x^2 < 12) \vee (x = 0)]$.

d. $(\forall x \in R)[x > 1 \Longrightarrow x^2 > x]$.

e. $(\forall x \in R)[(0 < x) \wedge (x < 1) \Longrightarrow x^2 < x]$.

f. $(\exists x \in R)[(0 < x) \wedge (3x^3 - 5x^2 + 3x - 5 = 0)]$.

g. $(\exists n \in N)[(n + 1)^4 < 4^{n+1}]$.

3.15 Negate each part of Exercise 3.14.

CHAPTER 4

TECHNIQUES OF DERIVATION
AND
RULES OF INFERENCE

In Chapter 2, the derivation method was introduced and an example was given. Let us look a little more carefully at the derivation method and investigate other rules having to do with determining validity, first with statements, then with quantified predicates. For convenience, we restate the derivation method.

THE DERIVATION METHOD: To verify that Q is a valid consequence of premises P_1, P_2, ... , P_n, a sequence S_1, S_2, ... , S_m of propositions must be found such that S_m is Q and for each i such that $1 \leq i < m$ one of the following is true:

    a. S_i is one of the premises P_1, P_2, etc. Rule P may be cited here to signify that it is a premise.

    b. S_i is a tautology. With Rule T, for Tautology, cited as the reason.

    c. S_i is a logical consequence of an S_j where $j < i$ (i.e., an earlier step in the derivation) or a logical consequence of several S_j's with $j < i$. Here you cite the step number (or numbers) that you are using from steps above and the tautology used.

EXAMPLE 4.1: Here is an example of the derivation method.

$$\left[(\sim B \longrightarrow C) \wedge (C \longrightarrow \sim D) \wedge (D \vee O) \wedge (\sim O) \right] \implies B.$$

PROOF:
  1. $\sim B \longrightarrow C$ Rule P
  2. $C \longrightarrow \sim D$ Rule P
  3. $D \vee O$ Rule P
  4. $\sim O$ Rule P
  5. D Steps 3 and 4 and (Rules 4 and 10)
  6. $\sim B \longrightarrow \sim D$ Steps 1 and 2 and Transitivity
  7. $D \longrightarrow B$ Step 6, Contrapositive and D. N.
  8. B Steps 5 and 7, Detachment

Notice how each line of the derivation is a proposition S_i, and the last line is $S_8 = B$, which is the conclusion, as required by the derivation method.

In addition, each statement satisfies the criteria of the derivation method. It may be worth mentioning that if a truth table were constructed, since it has 4 atomic statements, there would be $2^4 = 16$ rows in that truth table. Which way would you rather do it?

Not all arguments have a consequent that is a simple atomic statement such as the argument just mentioned. Sometimes the consequent is a conditional. For example, refer to Example 2.10.

If f is a differentiable function then whenever $f'(x) > 0$ on $[a, b]$, f is increasing on $[a, b]$.

is an argument of the form $P \longrightarrow (Q \longrightarrow R)$. How do you use the derivation method on this or on a more general argument such as the following, where the conclusion is a conditional?

$$(P_1 \wedge P_2 \wedge P_3 \wedge \cdots \wedge P_k) \Longrightarrow (R \longrightarrow Q).$$

You may use the law of exportation, Rule 21, which says this statement is equivalent to

$$(P_1 \wedge P_2 \wedge P_3 \wedge \cdots \wedge P_k \wedge R) \longrightarrow Q.$$

<u>CONDITIONAL PROOF</u> (RULE C.P.): If Q is a valid consequence of the statement R and a set of premises then the implication $R \longrightarrow Q$ is a valid consequence of the premises alone.

So in a derivation, how do you implement this rule? It says you take the statement R as an additional premise. However, when citing the reason for that step, use "conditional premise" instead of Rule P. Then do the derivation as before and get Q as the "next to last" line in the argument. The last line of the argument is $R \longrightarrow Q$ and the reason may be cited as Rule C.P., for conditional proof. In the following example, the consequent Q is the conditional $A \longrightarrow \sim D$.

EXAMPLE 4.2: Verify the argument

$$\left[(A \longrightarrow B \vee C) \wedge (B \longrightarrow \sim A) \wedge (D \longrightarrow \sim C)\right] \Longrightarrow (A \longrightarrow \sim D).$$

PROOF: 1. $A \longrightarrow B \vee C$ Rule P
            2. $B \longrightarrow \sim A$ Rule P

3.	$D \longrightarrow \sim C$	Rule P
4.	A	Conditional premise
5.	$B \vee C$	Steps 1 and 4 and detachment
6.	$\sim B$	Steps 2 and 4 and Rule 22 and D.N.
7.	C	Steps 5 and 6, Rule 4
8.	$\sim D$	Steps 3 and 7, Rule 22 and D.N.
9.	$A \longrightarrow \sim D$	Steps 4 and 8, C.P.

A was treated as a conditional premise in Step 4, and as a result of that assumption we got $\sim D$. So by conditional proof we have $A \longrightarrow \sim D$.

Indirect Proofs:

The types of proofs considered so far may all be called *direct* proofs. This is because the premises are assumed and the proof proceeds toward the conclusion of the argument, i.e., in the direction of the arrow $\longrightarrow$. These types of arguments are the most commonly used; however, they sometimes do not develop as well as you might like. So it is nice to have an alternative method.

Two alternatives to direct proofs are now considered. The first is called *proof by contradiction* and the second, *proof by contrapositive*. These may be used when the direct approach of the derivation method is not fruitful.

The method of proof by contradiction depends upon the fact that, in a valid argument, a false statement cannot be concluded from true premises. Referring to Rule 16, *reductio ad absurdum*, we see that in order to prove $P \Longrightarrow Q$ we may show $(P \wedge \sim Q) \Longrightarrow (R \wedge \sim R)$. Since $R \wedge \sim R$ is a contradiction, $P \wedge \sim Q$ is false because it is the premise whose conclusion is a contradiction, but $(P \wedge \sim Q) \equiv \sim (P \Longrightarrow Q)$. Thus $\sim (P \Longrightarrow Q)$ is false. Therefore, $P \Longrightarrow Q$ is true.

An actual proof by contradiction is simpler than the above remarks may suggest. In an actual derivation of $P_1 \wedge P_2 \wedge P_3 \longrightarrow Q$, using proof by contradiction, suppose the premises, P_i are true, citing *Rule P* AND as a conditional premise, also assume $\sim Q$, citing *indirect proof* as a flag to say what is going on (actually we abbreviate it to I.P.). The task then is to seek a contradiction of the form $R \wedge \sim R$ for any statement R you can find. On the line following $R \wedge \sim R$, write $(\sim Q) \Longrightarrow R \wedge (\sim R)$ and cite Rule C.P. Up to this point the derivation proceeds as though it were a conditional proof. The next and final line is Q with the *law of absurdity* quoted as the reason. To aid in clarifying this procedure let us look at an example.

EXAMPLE 4.3: Reprove the argument of Example 4.2 by contradiction.

That is, use the derivation method to prove the following by contra-diction.

$$\left[(A \longrightarrow B \vee C) \wedge (B \longrightarrow \sim A) \wedge (D \longrightarrow \sim C)\right] \Longrightarrow (A \longrightarrow \sim D).$$

PROOF:
1.	$\sim(A \longrightarrow \sim D)$	I.P.
2.	$\sim(\sim A \vee \sim D)$	Rule __ ?
3.	$A \wedge D$	DeMorgan and D.N.
4.	$A \longrightarrow B \vee C$	P
5.	$B \longrightarrow \sim A$	P
6.	$D \longrightarrow \sim C$	P
7.	A	Step __? and Simplification
8.	$B \vee C$	Step__ ? and Detachment
9.	D	__ ?
10.	$\sim C$	__ ?
11.	B	Rule __ ?
12.	$\sim A$	Steps___ ? and Detachment
13.	$A \wedge \sim A$	Steps___ ? and Conjunction
14.	$\sim(A \longrightarrow \sim D) \longrightarrow (A \wedge \sim A)$	C.P.
15.	$A \longrightarrow \sim D$	Law of absurdity

The step numbers and some reasons were left out so the student can gain extra insight in working through this proof and supplying the missing items. Notice that some steps were consolidated as in step 3. Other consolidations may be done without serious breach of understanding.

A proof by the contrapositive of an argument of the form $R \Longrightarrow Q$ begins with $\sim Q$, citing I.P. as the reason. The derivation should then work toward $\sim R$. The step after $\sim R$ is $\sim Q \Longrightarrow \sim R$ and the reason is C.P. The final step in the derivation is $R \Longrightarrow Q$ and the reason is law of contrapositive.

EXAMPLE 4.4: A proof by contrapositive provides a simple way to verify the law of addition $P \Longrightarrow P \vee Q$ using the law of simplification.

1.	$\sim (P \vee Q)$	I.P.
2.	$\sim P \wedge \sim Q$	Step 1 and DeMorgan
3.	$\sim P$	Step 2 and simplification
4.	$\sim (P \vee Q) \Longrightarrow \sim P$	C.P.
5.	$P \Longrightarrow P \vee Q$	Step 4 Law of contrapositive

Unless instructed to use one of the indirect methods, try a direct proof first. If several attempts at a direct proof are made without success, perhaps an indirect proof may be more fruitful. However, a direct proof may be more desirable if an indirect proof is longer or has within it the rudiments of a direct proof.

If an argument is of the form $P_1 \wedge P_2 \implies Q$ having two or more premises, a proof by contrapositive will possibly involve the consideration of several cases. If an indirect proof is desired, working through it by using contradiction may be more profitable.

The Derivation Method with Quantifiers:

The derivation method discussed previously deals with statements or propositions and the means employed in taking premises and deriving conclusions. Although we have learned how to negate quantified statements, there are additional rules governing quantified predicates that we need to explore so that we may do derivations with quantified statements. These rules deal with the manner in which one may drop and add quantifiers in a derivation. Before proceeding you may want to review the notion of *free* variables from Chapter 3.

UNIVERSAL SPECIFICATION RULE (RULE U.S.): Suppose that $\forall x[P(x)]$ appears as a line in a derivation of some argument. A subsequent line may then be written as $P(y)$ where y is free in the universe of discourse. This rule drops the quantifier $\forall x$ and replaces $P(x)$ with $P(y)$. The $P(y)$ is considered as a proposition or statement in the derivation in the same way a proposition was treated earlier. When this rule is used in a derivation, the reason (cited as U.S.) is placed to the right side.

This rule makes a lot of sense. If $P(x)$ is true for all x, then certainly it is true for a specific value of x, say y, hence $P(y)$. For example, in the context of the integers, $\forall x[x^2 \geq 0]$ is certainly true. Since -17 is an integer, $(-17)^2 \geq 0$ follows. Similarly, in the context of the integers, $\forall x[x > 1 \implies x^2 > x]$ is true. So, if y is a specific integer, the implication $[y > 1 \implies y^2 > y]$ follows by Rule U.S.

The preceeding rule provides the procedure for removing the universal quantifier from the front of a predicate. The next rule spells out how to put the existential quantifier in front of a predicate.

EXISTENTIAL GENERALIZATION RULE (RULE E.G.): If $P_1, P_2, \ldots, P_n$ logically imply $S(y)$ where y is a free object in the universe of discourse [that is, $S(y)$ appears as a line in a derivation without the presence of $\forall y$ or $\exists y$], then a subsequent line of the derivation may be written as $(\exists x)[S(x)]$.

How can you know that you have satisfied the criteria for y being free? One way is to see if y appeared on a prior line of the proof in the "free" sense. If y appeared in some prior statement $P(y)$, without quantifiers applied to y, you would know y is free. Therefore, if you have $S(y)$ as a line in your derivation, you could write $\exists x P(x)$ as a subsequent line in the proof. Another question that may arise is, "How do you know when you will be using Rule E.G.?" The answer is if you are establishing an argument, whose conclusion is existentially quantified, you would likely use this rule.

As another example, suppose you know $2^3 - 7 \cdot 2^2 + 8 \cdot 2 + 4 = 0$. Then you may conclude that $(\exists x)[x^3 - 7x^2 + 8x + 4 = 0]$. You will have used Rule E.G.

Rule E.G. provides the requirements that must be met in order to conclude an existentially quantified statement in a derivation. The following rule provides the requirements for removing an existential quantifier. This rule is easily misused, so pay close attention to it.

EXISTENTIAL SPECIFICATION RULE (RULE E.S.): Suppose $\exists x[P(x)]$ is a line in a derivation and suppose y is not a free variable on any prior line of the derivation. Beneath $\exists x[P(x)]$ you may then write $P(y)$. The reason cited at this point is Rule E.S.

The reason that y must not be free prior to the application of this rule is that y must not already have a specific meaning. In other words, if y already meant something special, say $Q(y)$, we would not be able to say it also met the possibly different special condition $P(y)$ obtained from $\exists x P(x)$. For example, suppose y is the real solution to the equation $(x^3 + x^2 + x - 15 = 0)$ and suppose we have the statement $\exists x [x^3 + x^2 + 5x - 15 = 0]$ with a different equation. We don't want to use the same y to represent the x that we know exists from the second equation since y already represents a solution to $(x^3 + x^2 + x - 15 = 0)$. In other words, we would not say $y^3 + y^2 + 5y - 15 = 0$ since y was the solution to another equation which likely would NOT be the solution to the second equation. Following Rule E.S. we won't make this mistake.

The act of invoking Rule E.S. on the statement $\exists x[P(x)]$ to get $P(y)$ is said to free the variable y. To put it another way, y is said to be freed by the use of Rule E.S. That is, this rule drops the quantifier $\exists x$ and replaces the predicate $P(x)$ by $P(y)$ where $P(y)$ has truth value *True*. This y is now a particular object about which $P(x)$ is true. As in the case of Rule U.S., $P(y)$ is a statement and is used as such in the derivation method.

For example, suppose you know the statement, "There exists an x such that $x^2 < x$," is true. Then you could let y be the special value asserted to exist, as long as y didn't already represent a particular number. That is, y would be a free variable and $y^2 < y$ would be true. Now $y^2 < y$ would be treated as a true statement in the rest of the argument.

UNIVERSAL GENERALIZATION RULE (RULE U.G.): If $P_1, P_2, \ldots, P_n$ logically imply $Q(y)$ [that is, $Q(y)$ follows from the premises in a derivation] AND IF y is not free in any prior premise, NOR has it been freed by the use of Rule E.S. on any prior line of the derivation, then $\forall x Q(x)$ is a logical consequence of the premises. On the line having $\forall x Q(x)$ cite Rule U.G.

This rule is frequently used to establish theorems concerning sets as we shall see. Its use presumes y is NOT free in any premise and that y has not been freed by the use of Rule E.S. on a prior line in the derivation. One way we can know that we have satisfied these requirements is for y to not have been used in the proof prior to the point where it is to be used. Its absence is enough to guarantee we meet the requirements of Rule U.G. The reason y cannot have been free is that if y were free in a prior $P(y)$, that y would already have a special meaning. So $Q(y)$ may not be true for all y.

Rules E.S. and U.S. are the rules which remove the quantifiers $\exists x$ and $\forall x$, respectively, while Rules E.G. and U.G. are rules which put quantifiers on, or back on. Care must be taken to follow the rules completely, otherwise, false conclusions may be drawn.

EXAMPLE 4.5: Let us prove

$$[\exists x[S(x) \wedge W(x)] \wedge \forall x[T(x) \vee \sim W(x)]] \implies \exists x[T(x) \wedge S(x)].$$

PROOF:
1.	$\exists x[S(x) \wedge W(x)]$	Rule P
2.	$[S(y) \wedge W(y)]$	Rule E.S.
3.	$\forall x[T(x) \vee \sim W(x)]$	Rule P
4.	$[T(y) \vee \sim W(y)]$	Rule U.S.
5.	$W(y)$	Step 2, Simplification
6.	$T(y)$	Steps 4, 5, Rules 4 and 7
7.	$S(y)$	Step 2, Simplifiction
8.	$T(y) \wedge S(y)$	Steps 6, 7, Conjunction
9.	$\exists x[T(x) \wedge S(x)]$	Rule E.G.

It is important to use the rules for quantifiers correctly. In Example 4.5, Step 8 states $T(y) \wedge S(y)$ with y free. It had been freed by the use of Rule E.S. in Step 2. Had the two premises been used in the

<u>opposite</u> <u>order,</u> <u>the</u> <u>derivation</u> <u>could</u> <u>not</u> <u>have</u> <u>been</u> <u>concluded</u> as we see in the following

1.	$\forall x[T(x) \vee \sim W(x)]$	Rule P.
2.	$[T(y) \wedge \sim W(y)]$	Rule U.S.
3.	$\exists x[S(x) \wedge W(x)]$	Rule P.
4.	$[S(z) \vee W(z)]$	Rule E.S.

The reason that z had to be used as a variable in Step 4, <u>and not</u> y, is that the variable y is free in Step 2, and Rule E.S. requires, in order to use a variable, that variable be "not free" prior to invoking Rule E.S. After simplification of $S(z) \wedge W(z)$ to $W(z)$ and $T(y) \wedge \sim W(y)$ to $\sim W(y)$, the proof cannot proceed as it did before because $W(z)$ and $\sim W(y)$ are not contradictory. Since $W(z)$ and $W(y)$ may, or may not, have opposite truth values, $W(z) \wedge \sim W(y)$ may be true, or may be false. We cannot proceed to use Rule 4. The difficulty here is that we de-quantified $\forall x[T(z) \vee \sim W(x)]$ <u>first</u>, using the free variable y. Then, when we de-quantified $\exists x[S(x) \wedge W(x)]$, we had to use a different symbol z by the requirements of Rule E.S., because y was already free, i.e., already had a special meaning.

If an argument has two or more existentially quantified statements, a similar problem may arise. Rule E.S. will prevent the freeing of the same variable by two applications of that rule. There is no way to complete the derivation, if it is necessary to get the same variable.

The following example illustrates that some assertions having quantifiers <u>seem plausible, but are in fact false.</u> The example demonstrates the need to follow the rules.

EXAMPLE 4.6a: Let $A(x)$ and $B(x)$ be predicates.
$$\exists x[A(x) \wedge B(x)] \Longrightarrow (\exists x)[A(x)] \wedge (\exists x)[B(x)].$$

PROOF PART a:

1.	$\exists x[A(x) \wedge B(x)]$	Rule P
2.	$[A(y) \wedge B(y)]$	Rule E.S.
3.	$A(y)$	Simplification
4.	$B(y)$	Simplification
5.	$\exists x[A(x)]$	Rule E.G.
6.	$\exists x[B(x)]$	Rule E.G.
7.	$\exists x[B(x)] \wedge \exists x[B(x)]$	Steps 5,6, Conjunction

This establishes the assertion.

EXAMPLE 4.6b: Since we did so well on Part a and since it seems right, let's do the converse of the implication in Part a. That is, let us see if we can establish $\exists x[A(x)] \wedge \exists x[B(x)] \longrightarrow \exists x[A(x) \wedge B(x)]$.

"PROOF" PART b:

1.	$\exists x[A(x)] \wedge \exists x[B(x)]$	Rule P
2.	$\exists x[A(x)]$	Simplification
3.	$A(y)$	Rule E.S.
4.	$\exists x[B(x)]$	Simplification
5.	$B(z)$	Rule E.S.

Now why did we put z in at Step 5 instead of y? See Exercise 13. It is not possible to take $A(y)$ and $B(z)$ to get $\exists x[A(x) \wedge B(x)]$. Thus the converse to Example 4.6a is invalid, because A is true at y and B is true at z, but y and z are not known to be the same. We cannot get the indicated consequent when following the rules.

The procedure that has been followed in the derivations above can be summarized into a set of general steps that are usually used in writing a formal derivation.

- If the premises are not already in logical form with quantifiers, predicates, connectives, etc., symbolize the argument with these things.

- Drop the quantifiers, if any, by Rules E.S. or U.S., carefully following the rules to get statements with free variables.

- Apply the procedures of derivations to derive a preliminary conclusion without quantifiers.

- Quantify (or requantify) using Rules E.G. and U.G. as necessary, to obtain the final conclusion. Pay attention to the requirements of the Rules E.S., U.S., E.G., and U.G..

- Reinterpret the result in the original terminology if the original problem had to be symbolized.

Exercises:

4.1 Using the derivation method verify $\sim H$ is a valid consequence of

$$(M \Longrightarrow J) \wedge (J \Longrightarrow \sim H) \wedge (\sim H \Longrightarrow \sim J) \wedge (H \Longrightarrow M) \wedge J.$$

4.2 Use the law $Q \Longleftrightarrow Q \vee Q$ to verify the implication $P \Longrightarrow P \wedge P$ by the contrapositive.

4.3 Verify

$$[(F \longrightarrow G) \wedge (H \longrightarrow G) \wedge (K \longrightarrow F \vee H)] \Longrightarrow (K \longrightarrow G).$$

4.4 Derive the argument, "If $\forall x[A(x) \Longrightarrow B(x)]$ and $\forall x[B(x) \Longrightarrow C(x) \vee D(x)]$ and $\exists x[A(x) \wedge \sim C(x)]$ then $\exists xD(x)$."

4.5 Try the derivation method on the following argument. Is it valid? If not, why? "$\exists x[A(x) \wedge B(x)]$ and $\forall x[B(x) \Longrightarrow (C(x) \vee D(x))]$ and $\exists x[A(x) \wedge \sim C(x)]$ implies $\exists xD(x)$."

4.6 Convert the following sentence into a statement, using logical forms, then determine if it is a valid argument. "Either logic is difficult or not many students like it. If mathematics is easy then logic is not difficult. Therefore, if many students like logic then mathematics is not easy." Use D, L, and M for the statements.

4.7 Derive the proposition

$$\left[\forall x\big(A(x) \Longrightarrow B(x)\big) \wedge \forall x\big(B(x) \Longrightarrow C(x)\big)\right] \Longrightarrow \forall x[A(x) \Longrightarrow C(x)].$$

4.8 Derive the proposition

$$\left[\forall x\big(A(x) \Longrightarrow B(x)\big) \wedge \exists x\big(P(x) \wedge A(x)\big)\right] \Longrightarrow \exists x[P(x) \wedge B(x)].$$

4.9 Derive $\exists x[D(x)]$ from the premises $\forall x\big(B(x) \Longrightarrow C(x) \vee D(x)\big)$ and $\forall x\big(A(x) \Longrightarrow B(x)\big)$ and $\exists x\big(A(x) \wedge \sim C(x)\big)$.

4.10 Derive the argument, "If $\forall x\big(A(x) \Longrightarrow B(x)\big)$ and $\exists x[D(x) \vee A(x)]$ and $\forall x\big[B(x) \Longrightarrow (\sim C(x) \vee \sim A(x))\big]$ and $\forall x[\sim D(x) \Longrightarrow C(x)]$ then $\exists x[D(x)]$."

4.11 $\exists! x[P(x)]$ denotes the quantified statement, "There exists a <u>unique</u> x such that $P(x)$." Using calculus techniques, verify that

$$\exists! x[x^3 + x^2 + 5x - 15 = 0].$$

4.12 Negate each of the following statements using Theorem 3.1.

 a. $\forall x[x \text{ is real } \wedge x \neq 0 \Longrightarrow x^2 > 0]$.

 b. $\forall x[S(x) \wedge T(x)] \Longrightarrow \exists x[P(x) \Longrightarrow R(x)]$.

 c. "Sequence $\{a_n\}_{n=1}^{\infty}$ converges to L," written as: "$(\forall \epsilon)$

$$\left[\epsilon > 0 \Longrightarrow (\exists N \in Z^+)[(\forall n \in Z^+)(n > N \Longrightarrow |a_n - L| < \epsilon)]\right]."$$

d. Assume the mean value theorem is written as:

$$\forall f \left[C(f) \wedge D(f) \implies \exists\, c \left(c \text{ is in } (a,b) \wedge f'(c) = \frac{f(b) - f(a)}{b - a} \right) \right].$$

4.13 In Example 4.6b, why is z used at step 5?

4.14 In Example 4.3, supply the reasons left out.

4.15 Modify Example 4.6a, by replacing $\wedge$ with $\vee$. The new proposition is

$$(\exists x)[A(x) \vee B(x)] \implies (\exists x)[A(x)] \vee (\exists x)[B(x)].$$

Determine if this is true.

4.16 In the following problems rewrite the statements using the indicated abbreviations for the premises. Use the derivation method to then determine if the argument is valid. If invalid, try to explain why.

a. All students are clever. Some politicians are clever. Therefore, some politicians are students. $S(x)$, $C(x)$, and $P(x)$.

b. Some students are not imaginative. Some students attend class regularly. Therefore, some students who attend class regularly are not imaginative. $S(x)$, $I(x)$, $C(x)$.

c. Every member of the policy committee is either a democrat or a republican. Some members of the policy committee are wealthy. Burroughs is not a democrat, but he is wealthy. Therefore, if Burroughs is a member of the policy committee, he is a republican. $P(x)$, $D(x)$, $R(x)$, $W(x)$, b.

CHAPTER 5

INFORMAL PROOF
AND
THEOREM–PROVING TECHNIQUES

Previously, we delved rather deeply into the derivation method. We explored various ways, such as direct and indirect proofs of logical propositions, and we introduced proof techniques involving quantifiers. Now we touch upon ways to write a proof somewhat less formally than the columnar way done earlier.

Rather than write out the steps of a derivation, one may find that the essential ideas of such a derivation can be conveyed in narrative form. This may make it easier to read because informal proofs have wording that encourages the reader to follow the flow of ideas. In such a proof several steps may be combined when the result is essentially as understandable as when written out step–by–step. Additional consolidation is attained by omitting reasons for obvious steps. Nevertheless, such an informal proof must carry the most essential ingredients of a formal derivation and be easily understood; it should not require reading between the lines.

As an illustration, Example 2.6 is done informally.

EXAMPLE 5.1:   Verify $\left[(A \longrightarrow B) \wedge (B \longrightarrow C) \wedge A\right] \Longrightarrow C.$

PROOF: Since $A \longrightarrow B$ and $B \longrightarrow C$, $A \longrightarrow C$ by the transitive property of $\longrightarrow$. Thus since A is true by assumption and $A \longrightarrow C$, C follows by the law of detachment.

From this example it is clear that informal proofs may employ more words than their formal counterparts, but take fewer lines. They also coach the reader through the reasoning process by explanatory sentence structure. Typically they are easier to read and follow. This format is fairly typical for proofs in mathematics.

If quantifiers are involved, the technique of writing informal proofs is still much the same. The following proof illustrates the points made in the second paragraph above.

EXAMPLE 5.2:   Prove

$$\left[(\exists x)\left[P(x) \wedge Q(x)\right]\right] \Longrightarrow \left[\exists x\left[P(x)\right] \wedge \exists x\left[Q(x)\right]\right].$$

PROOF: Applying Rule E.S. to the premise $(\exists x)[P(x) \land Q(x)]$ yields $[P(y) \land Q(y)]$. Thus by simplification, $P(y)$ and, as well, $Q(y)$. Note y is free so by Rule E.G., $\exists x[P(x)]$. Again by Rule E.G. $\exists x[Q(x)]$. Then by conjuction $\exists x[P(x)] \land \exists x[Q(x)]$.

In the next subchapter, additional examples of informal proofs are given. At that point, techniques are considered which help structure the proof according to the type of theorem. As usual, practice is the best way to develop these skills. Even though the complexity of the material increases, your skill and ability to deal with the fundamentals and abstractions increases hand in hand. As we proceed, more and more abbreviations and combinations of obvious steps will be made. This consolidation of steps gives informal proofs advantages over formal derivations.

Finally, several points need to be emphasized before writing informal proofs.

- Although an informal proof may be a kind of abbreviation of a formal derivation, it must carry the essence of that derivation.

- Any step in an informal proof should be able to withstand the test of having justification in the same way as was done with a formal derivation.

- The reader should be led easily through the reasoning process by comments which explain from whence statements arise and why any less–than–obvious statements are true.

- More specifically, a proof should explain whether a given statement follows from the preceding statement by writing: *thus, therefore, consequently,* or *hence.* It should give a clue if a statement comes from an earlier theorem, an assumption, definition, or earlier statement in the proof by the inclusion of the words: *since, also, by assumption, from theorem such and such* or the like.

- If establishing an intermediate result is desirable, make absolutely sure that the reader will be able to follow what is being done. Unless you use explanatory language, he or she may think you are asserting something that has not been justified.

At the beginning of theorem–proving practice, it would be wise to put into each step, its reason and explanatory comments as mentioned previously. These guidelines may then be relaxed somewhat after progressing far enough to feel at ease with the procedures and after turning out reasonably problem–free derivations. In any case, it is helpful to follow the 1900–year–old advice of Quintilian, the Roman rhetorician: "<u>One</u>

should <u>not</u> <u>aim</u> at <u>being</u> <u>possible</u> <u>to</u> <u>understand,</u> <u>but</u> at <u>being</u> <u>impossible</u> <u>to</u> <u>misunderstand</u>."

Some Theorem–Proving Techniques:

In the development of a mathematical idea or in working with a body of mathematical facts, certain questions in the form of sentences arise. Perhaps it is simply a situation in which someone asks, "Is this true?" It may be that it is, but the way to be sure is to prove that it is. If that is not possible, an attempt should be made to disprove it. Such statements are called *conjectures* until their truth or falseness is established. After a conjecture has been proven, it is typically called a *theorem*. A very famous conjecture is the Goldbach conjecture which says: "Every even integer greater than 2 is the sum of two prime numbers." No one has succeeded in proving it is true, nor has anyone found a number that is not such a sum. In other words, the Goldbach conjecture is still unsolved after more than 250 years (see *Journey Through Genius* by William Dunham for more information). Another famous conjecture called, "Fermat's Last Theorem," has recently been proven. Many mathematicians worked on this conjecture for over 300 years, until finally in 1993 a very complex proof by Andrew Wiles was announced (see *Annals of Mathematics*, Vol. 141, pp. 443–551, May 1995).

In this section we begin to explore the process of proving mathematical assertions. Perhaps the most important thing (and sometimes the most difficult) is to know what to do or how to begin. In this section, we explore some of the most common forms in which mathematical assertions appear, how techniques for their proofs are developed, the usual means of disproving conjectures, or assertions, and common mistakes which lead to fallacious proofs.

The largest categories of theorems occur in one of two basic forms: $R \implies S$ or $R \iff S$. Each of these have special subcategories and some of the more important of these will be considered.

Perhaps the most common form is $P \implies S$, with one or two stated premises and one conclusion. A trivial example statement of this type is, "If x is a real number and $x < 4$ then $3x < 12$." The two main approaches to proving an assertion of the type $P \implies S$ are the direct and indirect methods. In a direct proof, the beginning point is the assumption that P is true. That is because the alternative, "P is false," automatically (or vacuously) makes $P \implies S$ true. The assumption that P is true may give lots to work with. In fact, everything defined or previously proven under the same assumptions may be used as unstated premises. By the methods considered earlier, the conclusion is then derived using previously proven results, tautologies, and pertinent definitions.

The actual path taken from P to S depends on skill, experience, and sometimes accident. As usual, the skill is derived by practice while experience comes from the knowledge of related results and methods which work on similar theorems. More specific suggestions for proving a theorem of this general form are given as we move forward.

Let us temporarily assume, for illustrative purposes, that certain appropriate arithmetic properties of addition, multiplication, and order hold for real numbers. Imagine that we are in the middle of a development of some mathematical facts and that certain facts have already been established. Since our purpose here is to illustrate how theorems are proven, not to develop a mathematical body of information, we enumerate some of the "prior facts" we need to use. In other words, assume we have the following "previously developed" facts available to use.

Assumed Properties:

A. Usual properties of addition and multiplication over the real number system, such as closure, commutative, associative, and distributive laws and properties pertaining to 0 and 1 for + and *times* and properties of inverses for + and *times*. Here R denotes the collection of real numbers.

B. Trichotomy law: For all $x, y \in R$, one and only one of the following is true. $(x < y)$ or $(x = y)$ or $(y < x)$.

C. For all x, y, z if $x, y, z \in R$ and $x < y$ and $z > 0$ then $xz < yz$.

D. For all x, y, z if $x, y, z \in R$ and $x < y$ then $x + z < y + z$.

E. For all x, y, z if $x, y, z \in R$ and $x < y$ and $y < z$ then $x < z$.

F. $0 < 1 < 2 < 3 < \cdots$ with no integer between any pair of consecutive integers in the list.

G. For all $x, y, z \in R$, $x < y < z$ means the same thing as $x < y$ and $y < z$.

H. The <u>absolute value</u> of x is defined by $|x| = \begin{cases} x, & \text{if } x \geq 0; \\ -x, & \text{if } x < 0. \end{cases}$

I. Properties of absolute value, such as the <u>triangle inequality</u>

$$(\forall x, y \in R)\big[\,|x + y| \leq |x| + |y|\,\big].$$

J. $\forall x, y, z$ if $z \geq 0$ then $|x - y| < z$ iff $y - z < x < y + z$.

Be aware that in a normal development of some area of mathematics we might not have been given an itemized list of assumed properties such as these. Instead, the properties gradually evolve from a few basic properties or axioms into a total body of information. The assumed properties above are just here to get us ready to investigate properties we will use to illustrate proof techniques. Many of these items will be revisited in Chapter 22 in a more complete and careful manner.

SAMPLE TYPE OF THEOREM: $P \Longrightarrow S$ with a direct proof.

THEOREM 5.1:   If a, b, and c are real numbers and $a < b \wedge c < 0$ then $ac > bc$.

Before proving this, let's think it through in order to devise a strategy. We know from Property C noted previously, that if we multiply both sides of $a < b$ by a positive number k then $ak < bk$. Can we multiply both sides of $a < b$ by a negative number and get the same inequality? Obviously not, since our theorem says the inequality gets reversed. To use Property C, we have to multiply by a positive number. Since $c < 0$, can we multiply both sides of $c < 0$ by -1 to get $-c > 0$? To do so, is to use the theorem we are proving; it is "begging the question." But there is another way: take $c < 0$, then using the Property D, we <u>add $-c$</u> to both sides of $c < 0$ to get $c + (-c) < 0 + (-c)$, which yields $0 < -c$. This is something we can work with. Now we can multiply both sides by the <u>positive</u> number $-c$, to get $a(-c) < b(-c)$, which is $-ac < -bc$. Can we now multiply this by -1 to get the result we want? NO!, because that is begging the question again. Instead, let's use the trick we used above. Add something to both sides to get rid of the minus signs. Now that we have devised a way through the proof, let us write it more carefully.

PROOF: Let a, b, and c be real numbers and suppose $a < b \wedge c < 0$ then, by adding $-c$ to both sides of $c < 0$ and using Property D, we get: $0 = c + (-c) < 0 + (-c) = -c$ so $-c > 0$.

Since we know $-c > 0$, we can use Property C (i.e., $x < y \wedge z > 0 \Longrightarrow xz < yz$) to get $a(-c) < b(-c)$. By Property A $-(ac) < -(bc)$. But then $-(ac) + (ac + bc) < -(bc) + (ac + bc)$, using Property D. By regrouping, using the associative and commutative laws of addition, (i.e., Property A) $0 + bc < 0 + ac$, which says $ac > bc$ as desired.

Notice how the proof assumed the premises were true, i.e., $a, b, c \in R$ and $a < b$ and $c < 0$. But notice also there are unstated premises. Those are the properties of inequalities we used, as well as the algebra of the reals we used for addition and multiplication, itemized under assumed properties. Finally, our proof ended with the conclusion, just as we ended with the conclusion Q in the derivation method.

Next, let us illustrate a direct proof by dealing with a different topic. Here the context is the integers. Consider the definitions of even and odd for integers.

DEFINITION: If $n \in Z$ then n is even iff $\exists (k \in Z)[n = 2k]$. If $n \in Z$ then n is odd iff $\exists (k \in Z)[n = 2k + 1]$.

EXAMPLE 5.3:   The integer 0 is even because $(\exists k \in Z)[0 = 2k]$ and -14 is even because $(\exists k \in Z)[-14 = 2k]$. Are the k's the same for these two examples? (Think about Rule E.S. What are the limitations in Rule E.S.?) Think about the sum and the product of even numbers. We are ready for another example theorem of the form $P \implies Q$.

THEOREM 5.2:   If $x, y \in Z$ and x and y are even then

a. $x + y$ is even.

b. xy is even.

PROOF: Suppose x and y are integers and x and y are even, then $\exists k \in Z$ such that $x = 2k$ and $\exists h \in Z$ such that $y = 2h$. (Why use different symbols h and k?) Substituting equals for equals and using the distributive law in Property A:

$$x + y = 2k + 2h = 2(k + h).$$

But $k + h \in Z$, so $\exists z \in Z$ such that $x + y = 2z$. (Exactly what is the z we asserted to exist?) Thus $x + y$ is even.

Likewise

$$xy = (2k)(2h) = 2(2kh).$$

Since $2kh \in Z$, $\exists z \in Z$ such that $xy = 2z$. That means xy is even.

Sometimes after a theorem is proven, another result follows immediately on its heals. Such a result may be called a *corollary*. We have a corollary to the theorem previously mentioned and numbered 5.2 to corollate with it. In the case $a = b$ the result in Part b says a^2 is even.

COROLLARY 5.2:   If a is an even integer, then a^2 is even.

Another special case of the general form $P \implies S$ that a theorem may take is $P \implies Q_1 \vee Q_2$ where S is the disjunction $Q_1 \vee Q_2$. An example statement of this form is, "If x is a real number and $x < 4$, then $x^2 < 16$ or $x \leq 0$." A proof of $P \Longrightarrow Q_1 \vee Q_2$ begins, as before, by assuming the premise(s). Since there is more than one outcome in the conclusion we can take advantage of a natural dichotomy. Either Q_1 is true or $\sim Q_1$ is true, by the law of excluded middle. If Q_1 is true, then certainly $Q_1 \vee Q_2$ is true, by the law of addition. Thus $P \Longrightarrow Q_1 \vee Q_2$ is true. On the other hand, if Q_1 is false, it is necessary to show that Q_2 must be true, so that $Q_1 \vee Q_2$ will be a valid consequence of P.

To illustrate how to handle this type of assertion in a direct proof, let us consider the example.

SAMPLE THEOREM FORM: $P \Longrightarrow Q_1 \vee Q_2$. By direct proof.

"THEOREM" 5.3:   If x is real and $x < 4$ then $x^2 < 16$ or $x \leq 0$.

Suppose x is real and $x < 4$. (Why don't we worry about what happens if $x \geq 4$?) We observe then that either $x \leq 0$ or $x \not\leq 0$. If $x \leq 0$, we are done; if $x \not\leq 0$ we must show that $x^2 < 16$. Let us see if we can do this more formally.

PROOF: Suppose x is real and $x < 4$. By the law of excluded middle, either $x \leq 0$ or $x \not\leq 0$. If $x \leq 0$, the conclusion is true. If $x \not\leq 0$, we need to show $x^2 < 16$. Suppose $x \not\leq 0$. Then from the trichotomy law, $0 < x$. By Assumed Property C, we can multiply both sides of the inequality $x < 4$ by the positive number x to get $x^2 = x \cdot x < 4 \cdot x = 4x$. Since $0 < 4$ (Which previous result is this?) and since $x < 4$, (Why?), $4x < 4 \cdot 4 = 16$. (What allows this?) Putting together the two inequalities, $x^2 < 4x$ and $4x < 16$, with the transitive property of $<$, we get $x^2 < 16$. In any case, the conclusion is either $x \leq 0$ or $x^2 < 16$. The proof is complete.

Another particular case of the more general form $P \Longrightarrow Q$ is $P_1 \vee P_2 \vee P_3 \Longrightarrow Q$. A direct proof of this is sometimes called the <u>method of exhaustion</u> or proof by cases, meaning that each premise is taken case by case, $P_1 \Longrightarrow Q$, $P_2 \Longrightarrow Q$, and $P_3 \Longrightarrow Q$. These are then put together using tautology 18a to get Q.

SAMPLE THEOREM FORM: $P_1 \vee P_2 \Longrightarrow Q$. By direct proof.

THEOREM 5.4:   If $x > 0$ or $x < 0$ then $x^2 > 0$.

We suppose the hypothesis. There are two cases: either $x > 0$ or $x < 0$. In the first case, since $x > 0$, we can multiply each member of $x > 0$ by x, using Property C to get $x \cdot x > x \cdot 0 = 0$. In the second case, $x < 0$, so when we multiply both sides by the negative number x, we get a reversal of the inequality using Theorem 5.1. So $x \cdot x > x \cdot 0$. Again, the result is $x^2 > 0$. Both cases lead to the same result. Now we will write the proof more formally.

PROOF: Suppose the hypotheses. By Theorem 5.1, multiplying both sides of $x < 0$ by the negative number x yields $x^2 = x \cdot x > x \cdot 0 = 0$. That is $x^2 > 0$. In the other case, $x > 0$. By Property C, multiplying both sides of $x > 0$ by x leaves the direction of the inequality unchanged. So $x^2 = x \cdot x > x \cdot 0 = 0$. Consequently, if $x < 0$ or $x > 0$ then $x^2 > 0$.

THEOREM 5.5:   For all real numbers x, $|x| \geq x$.

Recall that the definition of $|x|$ is given by: $|x| = \begin{cases} x, & \text{if } x \geq 0; \\ -x, & \text{if } x < 0. \end{cases}$

It seems that it will be necessary to divide this into cases in order to prove this theorem. In other words, we show the assertion is true first when $x \geq 0$, then when $x < 0$.

PROOF: Since we are to show this is true for all x, let x be an arbitrary real number. Then either $x \geq 0$ or $x < 0$. If $x \geq 0$ then $|x| = x$ by the definition, so $|x| \geq x$ follows. And if $x < 0$, then by the definition of $|x|$ and Theorem 5.1, $|x| = -x > 0$. By transitivity of $>$, since $0 > x$, and $|x| > 0$, we get $|x| \geq x$. Thus in any case $|x| \geq x$.

Many theorems are in the form of this sample theorem. One of the most famous turned out to be of this case. It is the so-called *Four Color Problem* that was first conjectured in 1852 by Francis Guthrie, a graduate student at University College, London. Essentially, the problem said that any map that can be drawn, subject to suitable limitations, would require at most four colors to color the different countries. Many mathematicians worked on it for about a hundred years. Finally, in the late 1970's a valid proof was published by Appel and Haken (see the *Bulletin of the American Mathematical Society*, 1976, **82**, 711–12). The proof had 1879 cases that had to be resolved. The authors were successful in showing that every map that could be drawn had to fit one of those cases. That proof required years of work to develop and test for validity.

It is important to note that in order to prove something by cases, one has to be sure that every case has been considered. For example, to prove that $n^2 \geq 0$ for all integers n by individual cases would require infinitely many steps: one for each case $n = 0$, $n = 1$, $n = -1$, $n = 2$, etc. Obviously that is not the way to proceed. To say $(17)^2 = 289 > 0$ and neglect the other cases would not suffice to show that $n^2 \geq 0$ for other n. Such reasoning is fallacious. As an illustration of the difficulty in proving by cases, see the next example.

EXAMPLE 5.4 Consider the conjecture: "$f(n) = n^2 - n + 41$ is a prime number for every natural number n." If someone were trying to establish this by cases, it might go as follows:

If $n = 1$, then $n^2 - n + 41 = 41$ is a prime number.
If $n = 2$, then $n^2 - n + 41 = 43$ is a prime number.
If $n = 3$, then $n^2 - n + 41 = 47$ is a prime number.
If $n = 4$, then $n^2 - n + 41 = 53$ is a prime number.
If $n = 5$, then $n^2 - n + 41 = 61$ is a prime number.

$$\vdots \qquad\qquad \vdots \qquad\qquad \vdots$$

If $n = 10$, then $n^2 - n + 41 = 131$ is a prime number.

$$\vdots \qquad\qquad \vdots \qquad\qquad \vdots$$

If $n = 40$, then $n^2 - n + 41 = 1601$ is a prime number.

In all 40 cases, the result is a prime number. At this point the fatigued theorem prover may want to conclude that since the "experiment" repeated for 40 different values always yielded the same conclusion, it must be true. But was every case considered? The conjecture says that $n^2 - n + 41$ is a prime for <u>every</u> natural number n. It certainly might be tempting to draw such a conclusion at this time. But it would be a fallacious conclusion since not every case was considered. In fact, if the value of $f(n)$ were calculated for $n = 41$, the result would be $n^2 - n + 41 = (41)^2 - 41 + 41 = (41)^2$ which is certainly NOT prime, since it is the product of two integers greater than 1.

At the risk of belaboring the obvious, a theorem cannot be proven true by cases unless every case is considered. However, there is a situation where an example pertains to proving something about a conjecture. This occurs if a conjecture is to be proven false. An example which shows that a conjecture is false is called a <u>counterexample</u>. Referring to the preceding example, a counterexample to, "If n is any natural number then $f(n)$ is a prime," is the number 41, since 41 makes $f(41)$ a non-prime number. That is because the number 41 makes the antecedent true and the consequent false (see *Table 1.5*).

Recall the Goldbach conjecture that states any even number greater than two is the sum of two primes. No one has been able to prove it nor has any one found a counterexample, so it is still a conjecture.

The last special case of the more general form $P \implies S$ that we con-

sider for direct proofs are the theorems whose conclusions are conditionals. That form reduces, as we have seen, to one of the forms previously
discussed.

THEOREM TYPE: $P \implies (R \implies Q)$.

PROOF FORM: Suppose P and further suppose R as a conditional
premise. Then deduce Q in a manner similar to any of the foregoing
methods, particularly $P \wedge R \implies Q$. You may conclude the proof by
saying, by the law of exportation: $P \implies (R \implies Q)$ (see Example 2.10).

Indirect Proof:

Sometimes a conjecture to be proven does not seem to lend itself well
to a direct proof. Maybe it is because there is no way that can be thought
of to approach it, or maybe there isn't enough obvious information to
proceed. In such a case it may be preferable to use an indirect method.
As mentioned in Chapter 4, one type of indirect proof of $P \implies Q$ is by
the contrapositive, i.e., prove $\sim Q \implies \sim P$. Here is an example.

We showed in Corollary 5.2 that if n is even, n^2 is even. Now let's
see if we can show the converse.

THEOREM 5.6:   If $n \in Z$ and n^2 is even then n is even.

If we were to try a direct approach by assuming n^2 is even we would
know there exists $k \in Z$ such that $\left[n^2 = 2k \right]$. This says 2 is a factor of
n^2, but it isn't clear that 2 is a factor of n, itself, from what we have
available. So we try something different. What if n were not even, then
n would be odd. Thus $n = 2k+1$ for some $k \in Z$. Then $n^2 = (2k+1)^2 =
4k^2+4k+1 = 2(2k^2+2k)+1$. Thus $\exists z \in Z$ such that $n^2 = 2z+1$ (Why?
Which z does it?), which says n^2 is odd. That says n^2 is not even. Let
us formalize the ideas into a proof.

PROOF: Since we are doing this proof by the contrapositive, we show
"n is not even" implies "n^2 is not even." So suppose n is not even, then
n is odd. Thus $\exists k \in Z$ such that $n = 2k + 1$. Then

$$n^2 = (2k + 1)^2 = 4k^2 + 4k + 1 = 2(2k^2 + 2k) + 1$$

which will be odd if $2k^2 + 2k \in Z$, which it is. So $\exists z \in Z$ such that
$n^2 = 2z + 1$, which is odd. Thus n^2 is not even. We have shown the
contrapositive of the given theorem. Hence Theorem 5.6 is valid.

Another way to approach the proof of a theorem indirectly is by
contradiction. Here is such a theorem.

THEOREM 5.7: $\sqrt{2}$ is irrational.

Before tackling this, we need to know exactly what rational and irrational mean. A real number r is <u>rational</u> iff $\exists m, n \in Z$ such that $n \neq 0$ and $r = \frac{m}{n}$. Any real number that cannot be so written is <u>irrational</u>. However, this definition of irrational does not seem to give us much to go on. We know more about rational numbers than irrational. Were we to assume $\sqrt{2}$ is rational, maybe we could find a contradiction. If $\sqrt{2}$ were rational that would mean $\sqrt{2} = \frac{p}{q}$. If p and q have any factors in common they can be divided out, so that the fraction is reduced. Let's see how it unfolds. As we proceed through the argument, we look for two contradictory statements. Notice how the proof begins with an alert to the reader that we are doing an indirect proof.

PROOF: Let us suppose the contrary. That means $\sqrt{2} = \frac{p}{q}$ for some integers p and q such that $q \neq 0$. If p and q have any factors in common, divide them out to get $\sqrt{2} = \frac{m}{n}$ with m and n having no factors in common except 1 and -1. Multiplying by n yields $n\sqrt{2} = m$. After squaring, we get $n^2 \cdot 2 = m^2$. This says m^2 is even. But by Theorem 5.6 that means m is even, i.e., $m = 2k$ for some integer k. Thus

$$2n^2 = m^2 = (2k)^2 = 2k(2k).$$

After dividing both sides by 2, we have $n^2 = 2k^2$. But this says n^2 is even. By Theorem 5.6 again, n is even. Thus 2 is a factor of m and 2 is a factor of n. But that says $\frac{m}{n}$ is NOT reduced. Thus $\frac{m}{n}$ IS reduced and it is NOT reduced (see tautology 23). This is a contradiction. Hence the assumption that $\sqrt{2}$ is rational is false. Consequently, $\sqrt{2}$ is irrational.

This proof is a classic example of an elementary proof by contradiction.

Another general form that theorems may have is $P \Longleftrightarrow Q$.

THEOREM TYPE: $P \Longleftrightarrow Q$.

PROOF FORM: The technique of proof is to take advantage of Tautology 5, $(P \Longleftrightarrow Q) \equiv (P \Longrightarrow Q) \land (Q \Longrightarrow P)$. Since we know several ways to do conditionals, this type of proof is generally no more difficult, except to remember that there are two parts to its proof.

Suppose we are in a development of properties of $n \times n$ matrices and we know the definition of the transpose A^T of the matrix A as well as the definition, "The square matrix A is <u>symmetric</u> iff $A^T = A$." Suppose also that we already know the "previous property" $(AB)^T = B^T A^T$ for all square matrices A and B. At this point it is not necessary to really know what these concepts are. We are just using them to illustrate a proof of a theorem whose conclusion is a biconditional.

THEOREM 5.8: If A and B are $n \times n$ symmetric matrices then $AB = BA$ if and only if AB is symmetric.

PROOF: Suppose A and B are $n \times n$ symmetric matrices. This means $A^T = A$ and $B^T = B$. There are two things to show:

a. If $AB = BA$ then AB is symmetric, and

b. If AB is symmetric then $AB = BA$.

[Actually these parts are each conditional proofs of the form $P \longrightarrow (Q \longrightarrow R)$ and $P \longrightarrow (R \longrightarrow Q)$ since there is an underlying assumption P that is the premise "A and B are symmetric matrices."]

a. Suppose $AB = BA$ then,

$$(AB)^T = B^T A^T \quad \text{Previously Proven Property}$$
$$= BA \quad \text{Since } A \text{ and } B \text{ are symmetric}$$
$$= AB \quad \text{Since we are assuming } AB = BA$$

Therefore $(AB)^T = AB$ Transitivity of $=$

Consequently AB is symmetric, so Part a is complete.

b. Suppose AB is symmetric. This means $(AB)^T = AB$. Then we show $AB = BA$. (We will start with AB and get a string of equalities and end with BA.)

$$AB = (AB)^T \quad \text{Assumption that AB is symmetric}$$
$$= B^T A^T \quad \text{Previously Proven Property}$$
$$= BA \quad \text{Since } A \text{ and } B \text{ are symmetric}$$

Therefore $AB = BA$ Transitivity of $=$

Thus Part b is established. Putting the two parts together, we have if A and B are symmetric matrices then $AB = BA$ iff AB is symmetric.

The final example illustrates that some experimentation and a few false starts may be a necessary part of getting the ideas together in a way that yields a nice simple proof.

THEOREM 5.9: If a and b are real numbers then $|a| - |b| \leq |a - b|$.

As usual let us think about what we might want to do to prove this. The assumption $a, b, c \in R$ seem at first not to tell us much, but there are unstated premises. Our assumption really tells us that the triangle inequality holds, as well as other things, such as were mentioned above. So $|x + y| \leq |x| + |y|$. Obviously this seems to be something to consider using. What if we considered $|a - b|$? By the triangle inequality $|a - b| = |a + (-b)| \leq |a| + |-b| = |a| + |b|$. Then we could subtract $|b|$ from both sides. That result is $|a - b| - |b| \leq |a|$. Although this is perfectly true, it doesn't seem to fit with what we want. But it does gives us an idea. $a = a - b + b$, so $|a| = |(a - b) + b| \leq |a - b| + |b|$. Now we can subtract $|b|$ from both sides and get what we want. Now let's write the proof.

PROOF: Let a, b, and c be real numbers. Then

$$|a| = |(a - b) + b| \leq |a - b| + |b|$$

by the triangle inequality. Then by subtracting $|b|$ from both sides we are led to the conclusion $|a| - |b| \leq |a - b|$.

Some Concluding Remarks:

This chapter contains a wide variety of forms in which theorems appear. Techniques for handling a proof by both direct and indirect methods are treated. This chapter contains a variety of approaches to tackling various types of proofs. These ideas will be referred to quite often in subsequent chapters. In addition, some examples of how to take known results and then apply them to new properties are also examined.

Let us summarize some specific ideas of this chapter. It is important in a proof to say enough to be sure the reader knows whether the student is asserting a statement P to be true or the student is saying he or she is going to show P is true. The distinction is important. Include wording that smooths the way for the reader. If P is being concluded from some earlier statement such as R, it is helpful to write "*since R.*" If is being concluded from the immediately preceeding statement, write "*thus P,*" "*so P is true,*" or "*consequently P* is true." If there is no wording between two mathematical assertions, such as P and Q, the interpretation the reader will logically make is that Q follows from P is meant. If that is not what is meant, be certain to put in some wording to make sure intentions are understood. And finally, if the concluding statement is written somewhere in the middle of the argument and it is not prefaced in some way to indicate it is something to be shown, then the argument appears to be begging the question. Perhaps the most important thing to remember is to write so that intentions are as clear as they can be made.

Exercises:

5.1 Identify the form in the problems below as being one of
$$P \Longrightarrow Q \ , \ P_1 \lor P_2 \Longrightarrow Q \ , \ P \Longrightarrow Q_1 \lor Q_2, \text{ etc.}$$

a. If $f'(x)$ exists at x_0 then f is continuous at x_0.

b. If p is a prime and p is a factor of ab then p is a factor of a or p is a factor of b.

c. If $|x| > a > 0$ then $x > a$ or $x < -a$.

d. If $|x| < a$ then $x < a$ and $x > -a$.

e. If f is continuous on $[a,b]$ and $a < c < b$ and f has a relative maximum at c then $f'(c) = 0$ or $f'(c)$ does not exist.

f. If $x < 1$ then $2x < 2$.

g. If $f(x) = x^2$ then $f'(x) = 2x$.

h. If m is even then m^2 is even.

i. If $0 < x$ and $x < 1$ then $x^2 < x$.

j. If $x < 5$ then $x^2 < 25$ or $x \leq 0$.

k. If $f(x) = x^n$, $f(x) = \sin x$, or $f(x) = e^x$ then $f'(x)$ exists at all x.

l. If x is a real number then $|x| = x$ if and only if $x \geq 0$.

m. If m^2 is odd then m is odd.

5.2 In each of Exercises 5.1a–m, write the negation of the given statement.

5.3 Parts a–m. In each of Exercises 5.1a–m, indicate the form that a direct proof would take. Do not try to prove the assertions, just indicate what is assumed and what is to be shown.

5.4 Prove the assertion in Exercise 5.1f directly with the assumption that the x's are integers.

5.5 Prove the assertion in Exercise 5.1h indirectly by the contrapositive with the assumption that the m's are integers.

5.6 Prove the assertion in Exercise 5.1 Part i directly with the assumption that the x's are real numbers.

5.7 Prove the assertion in Exercise 5.1j directly with the assumption that the x's are real numbers.

5.8 Prove the assertion in Part l of Exercise 5.1.

5.9 Prove the assertion in Exercise 5.1m with the assumption that the m's are integers.

5.10 Prove or disprove the conjecture, If x is a rational number then $x^2 \geq x$.

5.11 Prove or disprove the conjecture, If a, b, c are integers and $ab = ac$ then $b = c$.

5.12 Parts a–m. Indicate the form an indirect proof would take for each of the parts of Exercise 5.1a–m. Be sure to indicate which type of proof (contrapositive or contradiction) you are outlining. Do not try to prove the assertions, just indicate what is assumed and what is to be shown.

5.13 a. $|x| = |-x|$.
   b. If $b > 0$ then $|x| < b$ iff $-b < x < b$.
   c. If a and b are real numbers then $\big||a| - |b|\big| \leq |a - b|$. Hint, see Theorem 5.9.

5.14 Prove $\forall x[x^2 \geq 0]$.

CHAPTER 6

ON THEOREM PROVING AND WRITING PROOFS

Introduction:

In the preceding chapter we explored some proof techniques. We looked at various ways, such as direct and indirect proofs of mathematical statements. In this chapter we learn more about writing proofs to make them more readable as well as how to tackle specific difficulties that arise in proving theorems.

One of the first things that some people ask about proving an elementary theorem is, "Why bother with this; it is obviously true." A good response to such a question is that it is easier to learn how to prove easy theorems than hard theorems and our purpose here is to learn techniques, not move back the frontiers of mathematics quite yet. A more compelling answer is that sometimes there is something subtle going on in an "easy" theorem that would be missed if a proof were not attempted. To be more specific, we briefly allude to some history of set theory. About 1900 A.D., various intuitive ideas in set theory were considered to be apparently true, and not requiring proofs, that is, until Bertrand Russell and others pointed out certain paradoxes. The paradoxes and other shortcomings made a reorganization of the subject necessary to be sure that these shortcomings did not cause set theory to have internal contradictions (see R.L. Wilder, *Introduction to the Foundations of Mathematics*). In view of this, it is wise for us to be as sure as we can that an assertion that seems plausible, not be false.

Not all theorems are known. There are no books that have all theorems. Mathematics has been, and is still being, created by the human mind. Most often, the creative process of developing mathematics occurs when someone is studying an area of mathematics and sees some relationship between ideas. A conjecture is formulated and then a proof is attempted. The idea may have been thought of before and perhaps even a proof was devised. When a mathematician does this, it may be suitable for publication. Students, on the other hand, are usually guided by a professor or text through a previously developed area of mathematics in an apprenticeship. Sometimes the student is asked to draw some conclusion from given information and so he or she is being asked to formulate a conjecture. Whether it is a known theorem or a conjecture, the student needs to know how to proceed with devising his or her own proof and how to write it so it will be readable. That is the focus of this section.

Strategy and Analysis:

We have seen that there may be several ways to approach the proof of a theorem. There are various direct proofs depending on the structure of the theorem and there are indirect proofs. The actual final product is shaped by necessity and experience as well as taste or even accident. Some approaches to proving a particular theorem may be considered superior for aesthetic reasons because of the clever way in which the result is established or for economic reasons due to the direct way in which the result is established. In any case, for the novice it may seem bewildering to be confronted with a theorem to prove.

Before the student actually begins to compose the proof, it would probably be constructive to brainstorm the relationships between the hypothesis and the conclusion. The student may want to refer to Chapter 5 for some ideas about what general procedure might be suitable, based on the logical structure of the theorem statement. Since some theorems are stated in a form more complicated than a simple, "If P then Q," it is necessary to know what the assumptions and the conclusions are.

Organizing and Writing the Proof:

If the theorem is of the form $P \implies Q$, where P is a conjunction of premises, a good way to begin writing the proof (<u>after</u> putting in the heading *Proof*), is to write "*suppose P*" or "*if P*" and include all the hypotheses. This corresponds to Rule P of the derivation method and is a standard beginning in any direct proof. The next thing may be to restate the premise(s) in different words by using definitions and/or previous theorems. It helps the reader to see how the ideas progress if the change is accompanied by an explanation. This may be accomplished by writing, "*in other words* $\cdots$," before the rewording.

Next, it may be helpful to put into the argument, "*to show Q*," whatever Q is. It is beneficial to say exactly what is planned. It clearly focuses the work toward the objective.

The student may see a way, at this point, to get the desired conclusion Q from the premises P. If that is true, he or she should go ahead. If not, rewrite the conclusion Q using a logical equivalent of Q and prefacing what is written by a phrase like, "*in other words, show* $\cdots$." At this stage be very precise about what is to be done. Writing an exact statement of the objective in the argument is one of the most helpful things in learning to construct proofs. It is important to preface that rewritten objective with a phrase that indicates its status, otherwise the reader would interpret what has been said as a conclusion, when in fact, it would just be rephrasing the objective. Many times after doing these things, a path will be seen that goes from the assumptions all the way through to the conclusion. If the way through to the conclusion is still not seen, continue to use the premises, rules of logic, relevant definitions, and previously proven theorems to logically work along the path from the premises toward the rephrased conclusion and then the conclusion. While doing this, it is helpful to put in a brief commentary to clarify any step that does not follow from the immediately preceeding step.

If several (perhaps somewhat involved) steps have been performed in the deductive process, aiming at Q, it is sometimes helpful to summarize what has been done up to a certain point. This might take the form, *"thus, we have shown that* $\cdots$*."* The preceding suggestion is particularly appropriate at certain points inside a proof, if its conclusion clarifies a lengthy sequence of steps, for example when proving $P \Longleftrightarrow Q$. Since such a proof is frequently split into two parts, $P \Longrightarrow Q$ and $Q \Longrightarrow P$, it may be beneficial to summarize what has been done when each of the individual parts have been completed.

A longer proof may be divided into various parts, each with a particular objective. When each objective has been completed, so indicate. And when all the parts have been done and put together for the conclusion Q, it is good procedure to summarize what has been done. This might also take the form, *"thus we have shown that $P \Longrightarrow Q$, as intended."*

When all of the ideas and the thread of logical statements going from the premises to the conclusion have been done, rewrite what has been done to check that the ideas truly do follow logically from one to another, and do so in such a way that the reader will be coached step–by–step through that process. The process of rewriting and polishing the proof into a formal product is important; among other things it forces the student to make sure everything that may have been hurried through in the brainstorming stage was actually valid.

Don't expect to be able to sit down and write out the proof to a given assertion without a pause from the beginning to the end. There are times when this can be done in those kinds of theorems that have a "straightforward" proof. In these arguments, the idea for what to do as a next step is almost suggested by the preceeding step as long as the ultimate goal is kept in mind. Of course, this is more likely to occur after the student has had some experience with this process. Usually, however, one or more attempts may be made before finding something that works properly. It is easy to suggest ways to write an argument to make it easier for the reader to follow as we have tried to do. It is easy to indicate the various forms in which a theorem may appear as well as the general procedure which might be used to prove it. It is not easy to actually say how to creatively use the premises of any given proposition to obtain the ultimate conclusion. That is something the student must learn by doing and it depends heavily on the problem at hand.

Fortunately, these abilities improve with practice. The more practice, the more skillfully direct the flow of steps toward the objective will become. To repeat however, do not expect the proof to roll out from beginning to end without occasionally turning into a blind alley.

In the student's first few proofs, more details should be carefully included so that he or she become more aware of the nature of proving things and the way earlier facts are integrated into the current problem. Every step must have a specific reason which may be found at an earlier point either as a theorem, definition, logical principle or the like. In fact, a step should probably not even be written on paper unless it is justifiable in this fashion. There are occasions in which errors arise simply because of the faulty assumption that the step appears reasonable, and so there must be a law which supports it. Of course this is fallacious because some of what appears to be plausible may, in fact, be false as we have indicated before. Only after sufficient practice, should relevant steps be combined or eliminated. Later in this text, examples of arguments in which several steps are combined are illustrated.

As the student becomes more familiar with the processes and feels more at ease in proving theorems, he or she will see that the inclusion of every step may get to be tedious. This is very true; some proofs would be extraordinarily long if every minute detail were included, so what to include becomes a matter of judgment.

SUMMARIZING:

After writing the argument, go back over it once again to check it for validity. Ask the following questions.

- Are the premises correct?

- Does each assertion have a valid basis?

- Does the argument say exactly what was intended for it in an unambiguous fashion?

- Would another student, reading this for the first time, understand it?

- Are there any irrelevant steps in the proof? Sometimes in a first attempt at a proof, one may write down a logically justifiable step, but that step does not move him or her toward the goal.

- Is there good transitional language that explains what is being used to make the next assertion? These tend to help the reader follow the argument more easily.

Proving Theorems With Quantifiers:

We have seen how to do derivations with quantifiers and we have begun the process of writing informal proofs with non-quantified statements in order to accurately convey ideas to the audience. Now let us see how to construct proofs with quantified mathematical statements using the rules governing quantifiers.

In addition to the *assumed properties* mentioned in Chapter 5, suppose we have already proven that $(\forall x)[x^2 \geq 0]$. Here is a property we want to prove. $(\forall x)[(x^2 + 1)^2 \geq 1]$. How would it be done? It has a universal quantifier, so we will likely use Rule U.G. It is stated as a "property" instead of a theorem because it is contrived to illustrate a procedure, but is not a very useful result in its own right.

PROPERTY 6.1: $(\forall x)[(x^2 + 1)^2 \geq 1]$.

PROOF: The previous result says $(\forall x)[x^2 \geq 0]$, so by Rule U.S. $y^2 \geq 0$. Thus by Assumed Property D, $y^2 + 1 \geq 1$. Now since $1 > 0$ and $>$ is transitive, $y^2 + 1 > 0$. By assumed properties A and C, we get

$$(y^2 + 1)^2 = (y^2 + 1)(y^2 + 1) \geq (y^2 + 1) \cdot 1 = y^2 + 1.$$

What are the reasons for each connective $=$ and $\geq$?

We now know two things: $(y^2 + 1)^2 \geq y^2 + 1$ and $y^2 + 1 \geq 1$. Since $\geq$ is transitive, $(y^2 + 1)^2 \geq 1$.

Using Rule U.G. (since y was not free in any premise and was not freed by the use of Rule E.S.) we get $(\forall x)[(x^2 + 1)^2 \geq 1]$. Therefore we have proven the assertion.

Observe how the wording suggests how and why things are true in the argument above. Notice that the use of the word *since* brings in a result prior to the immediately preceding step. The word *thus* suggests we are using the prior step to conclude the next step.

Let us consider a familiar result from calculus concerning continuity of a function. In particular, let us prove the function $f(x) = 3x$ is continuous at $x_0 = 2$ using the $\epsilon - \delta$ definition of continuity. The first thing we need to have firmly in our grasp is the definition of continuity.

DEFINITION: If f is a function defined on an open interval containing x_0 then f is <u>continuous</u> at x_0 if and only if

$$(\forall \epsilon > 0)(\exists \delta > 0)\Big[\forall x\big(\text{ if } |x - x_0| < \delta \text{ then } |f(x) - f(x_0)| < \epsilon\big)\Big].$$

PROPERTY 6.2: If $f(x) = 3x$ then f is continuous at $x_0 = 2$.

Before starting the proof let us analyze what we have and what we want to get. We must show a certain quantified statement is true for all ϵ, so we will be using Rule U.G. To begin, we suppose $\epsilon > 0$ using the same symbol. Then we need to find a $\delta > 0$ that depends upon ϵ so that whenever $|x - 2| < \delta$ then necessarily $|3x - 3(2)| = |f(x) - f(x_0)| < \epsilon$. Now $|3x - 3(2)| = 3|x - 2|$, so we have a way to connect the inequalities concerning ϵ and δ. In fact, if we choose $\delta = \frac{\epsilon}{3}$ then

$$|f(x) - f(x_0)| = |3x - 6| = 3|x - 2| < 3\delta = 3 \cdot \frac{\epsilon}{3} = \epsilon$$

will hold for all x satisfying $|x - 2| < \delta = \frac{\epsilon}{3}$. That is the trick.

PROOF: Let $\epsilon > 0$. Choose $\delta = \frac{\epsilon}{3}$. Then δ exists and $\delta > 0$. Now the objective is to do the following.

SHOW: For all x, if $|x - 2| < \delta$ then $|3x - 6| < \epsilon$.

(Notice here that we have said exactly what we are going to do next, as suggested earlier. Since we are to establish a conditional, we suppose the antecedent and show the consequent.) To that end, suppose x is any real number satisfying $|x - 2| < \delta$. Then $|x - 2| < \frac{\epsilon}{3}$. Thus we get

$$|f(x) - f(x_0)| = |3x - 3(2)| = 3|x - 2| < 3 \cdot \frac{\epsilon}{3} = \epsilon.$$

We have shown for the arbitrarily selected $\epsilon > 0$ that there exists a $\delta > 0$ so that the conditional, "If $|x - 2| < \delta$ then $|3x - 6| < \epsilon$," holds for all x. That is

$$(\exists \delta > 0)\Big[(\forall x)\big(\text{ if } |x - 2| < \delta \text{ then } |3x - 6| < \epsilon\big)\Big].$$

Finally,

$$(\forall \epsilon > 0)(\exists \delta > 0)\Big[(\forall x)\big(\text{ if } |x - 2| < \delta \text{ then } |3x - 6| < \epsilon\big)\Big].$$

(which follows since ϵ was not free in a premise and was not freed by the use of Rule E.S.). In other words, $f(x) = 3x$ is continuous at $x = 2$ by the definition above so we have completed the proof.

In a subsequent mathematics course, if a student is asked to prove some quantified statement, he or she would usually not bother to mention the rules concerning quantifiers as we did in the preceeding proofs. Instead one might begin with, *Let ϵ be an arbitrary positive real number,* then show there is a positive number δ so that statement involving ϵ and δ, e.g., $P(\epsilon, \delta)$, is true. Then at the conclusion, observe that since the choice of the positive number ϵ was arbitrary, the assertion would be true for all positive real numbers. This is what will more likely be seen in real analysis and abstract algebra instead of a reference to the rules of quantifiers in Chapter 4.

Let us tackle a more difficult continuity problem.

PROPERTY 6.3: If $f(x) = x^2$ then f is continuous at $x_0 = 2$.

As before we must show a certain quantified statement is true for all ϵ. Let $\epsilon > 0$. Then we need to find a $\delta > 0$ that depends upon ϵ so that whenever $|x - 2| < \delta$ then necessarily $|x^2 - 2^2| < \epsilon$. Now $|x^2 - 4| = |(x-2)(x+2)|$ so we may have a way to connect the inequalities concerning ϵ and δ, but this time that connection isn't as clear. Note that if $|x + 2| \neq 0$ then maybe we can get $|x - 2| < \frac{\epsilon}{|x+2|}$, but unfortunately $|x + 2|$ varies. It is not like the 3 in the earlier example. Can we limit how much it varies by limiting δ somehow? After all, the size of $|x + 2|$ is inversely proportional to $\frac{\epsilon}{|x+2|}$. We need to limit δ anyway so that the size of $|x - 2|$ is limited and to keep $|x + 2| \neq 0$. Notice that $|x - 2| < \delta$ is equivalent to $-\delta < x - 2 < \delta$ which is equivalent to $2 - \delta < x < 2 + \delta$, then $4 - \delta < x + 2 < 4 + \delta$. So if we require $\delta \leq 1$, we won't have to worry about $|x + 2|$ becoming 0. Continuing the brainstorming, if δ were to satisfy $\delta \leq \frac{\epsilon}{|x+2|}$ then $|x - 2| < \delta \leq \frac{\epsilon}{|x+2|}$. Then the assumption

$|x - 2| < \delta$ would ensure $|x^2 - 4| = |x - 2||x + 2| < \epsilon$. The restriction $\delta \leq 1$ will guarantee

$$3 \leq 4 - \delta < x + 2 < 4 + \delta \leq 5.$$

Thus

$$\frac{\epsilon}{3} > \frac{\epsilon}{x + 2} > \frac{\epsilon}{5}.$$

Finally, we select δ to be the smaller of $\frac{\epsilon}{5}$ and 1. Thus, since $\delta \leq \frac{\epsilon}{5}$, we will get $|x - 2| < \frac{\epsilon}{5}$, and since $\delta \leq 1$ we also know that $\frac{\epsilon}{5} < \frac{\epsilon}{x+2}$. Putting these together we get

$$\left|x^2 - 4\right| = \left|x - 2\right|\left|x + 2\right| < \delta\left|x + 2\right| \leq \frac{\epsilon}{5}(x + 2) < \frac{\epsilon}{x + 2} \cdot (x + 2) = \epsilon.$$

That should do it. It is an exercise for you take what we have here and write it up in a manner similar to Property 6.2.

Recall that a rational number is a number of the form $\frac{p}{q}$ where p and q are integers and $q \neq 0$. Let's prove there is no smallest positive rational number. To make it easier, let Q^+ denote the set of all positive rational numbers. First, exactly what is it that we would have to show? If we write out symbolically exactly what we have to do, perhaps we will have an inspiration about how to proceed. If there were a smallest positive rational number r, what would r have to be? It would have to be an $r \in Q^+$ since it is positive, and it would have to have the property that $(\forall s \in Q^+)(r \leq s)$. Our goal is to show that no such r exists.

PROPERTY 6.4: $\sim \left(\exists r \in Q^+\right)\left[(\forall s \in Q^+)\left[r \leq s\right]\right].$

By our rules this is equivalent to

$$\left(\forall r \in Q^+\right)\left[\sim (\forall s \in Q^+)\left[r \leq s\right]\right].$$

Rewriting this we get

$$\left(\forall r \in Q^+\right)\left[(\exists s \in Q^+)\left[s < r\right]\right].$$

So to prove there is NO smallest positive rational number, it would suffice to take an arbitrary $r \in Q^+$ then find an $s \in Q^+$ so that $s < r$. Can we do this? Remember, the s depends upon the r we select! Let $r \in Q^+$ then $\frac{1}{2} \cdot r$ is in Q^+ and $\frac{1}{2} \cdot r < r$. (Why? Be precise!) So the s that must exist IN Q^+, may be taken as $\frac{1}{2} \cdot r$, because $\frac{1}{2} \cdot r \in Q^+$ and $\frac{1}{2} \cdot r < r$. So s DOES exist.

PROOF: Since

$$\sim \left(\exists r \in Q^+\right)\left[(\forall s \in Q^+)(r \leq s)\right] \equiv \left(\forall r \in Q^+\right)\left[(\exists s \in Q^+)(s < r)\right]$$

we need only to show, "For each $r \in Q^+$ there exists an $s \in Q^+$ such that $s < r$." To do this, let r be a positive rational number. Then from Assumed Property C, since $\frac{1}{2} > 0$ and $r > 0$, $\frac{1}{2} \cdot r > 0$ and again by the same property, since $\frac{1}{2} < 1$, we know that $\frac{1}{2} \cdot r < r$. Also $\frac{1}{2} \cdot r$ is in Q^+ since the product of positive rationals is a positive rational. Thus for each positive rational number r there exists s (what is s?) in Q^+ such that $s < r$. We have demonstrated what was to be shown.

Exercises:

In each of the following, write an informal proof according to the instructions and examples of this section.

6.1 Prove: $\forall x$ in $Z[$ if $x < 1$ then $2x < 2]$.

6.2 Prove: $\forall x$ in the reals $[$ if $x < 5$ then $x^2 < 25$ or $x < 0]$.

6.3 Prove: If $f(x) = x^n$ or $f(x) = \sin x$ or $f(x) = e^x$ then $f'(x)$ exists at all x.

6.4 Prove: $\forall m$ in $Z[$ if m^2 is odd then m is odd $]$.

6.5 Prove: $\forall x[$ If x is a real number then $|x| = x$ iff $x \geq 0]$.

6.6 Prove Property 6.3 informally using the ideas developed after its statement.

6.7 Prove or disprove:

$$(\forall x \in Z)\Big[(\forall y \in Z)[x < y \implies (\exists z \in Z)[x < z < y]]\Big].$$

6.8 Prove or disprove:

$$(\forall x \in Q)\Big[(\forall y \in Q)[x < y \implies (\exists z \in Q)[x < z < y]]\Big].$$

6.9 Prove or disprove: For all functions f if $\lim_{x \to 0^+} f(x) = \infty$ then the following improper integral does not exist.

$$\int_0^1 f(x)\, dx$$

6.10 Prove or disprove:

$$(\exists k \in N)(\forall n \in N)\left[\frac{1}{3n^2 + 6n} < \frac{1}{kn^2 - 4}\right].$$

CHAPTER 7

MATHEMATICAL INDUCTION

Introduction:

The process which takes known results such as axioms, definitions, and previously proven theorems, and follows the principles of logic to derive new results, is called *deductive reasoning*. This process has been employed in the preceding sections and in previous mathematics courses to prove various assertions. There is, however, another reasoning process used to justify an assertion by a kind of experimental method. In this process the experiment is performed numerous times, under carefully controlled conditions. If the same result is consistently found, then "inductively" the conclusion is, "under the prescribed experimental conditions, that result will always be found." Of course, such a conclusion cannot be 100% certain. This is one of the kinds of reasonings used by scientists to advance knowledge in their disciplines and is appropriate for their work. The experimental method is unsuitable, however, for proving mathematical assertions. The problem with using this type of induction in mathematics, as well as the sciences, is that the same result may have been obtained in each experiment due to an unknown or accidental influence. In fact, the example in Chapter 5 regarding the primeness of $n^2 - n + 41$ illustrates that this type of inductive reasoning is unsatisfactory for proving mathematical assertions.

On the other hand, inductive reasoning serves a very important role in mathematics. It may be used to test a given hypothesis. If a case can be found for which the assertion is false, the entire assertion, as stated, is invalid. If such a case cannot be found, an attempt to devise a deductive proof can be initiated. The deductive proof will then guarantee the validity of the assertion. If such a proof cannot be constructed, further inductive experimentation may be in order. Frequently, inductive procedures help the mathematician decide upon the proper hypotheses and/or conclusion to have in the assertion and may even suggest a means of proving the theorem. When all the particular examples fit the assertion, a deductive proof may be attempted.

The Principle of Mathematical Induction:

Let us turn to a specific example. We will be dealing with natural numbers, i.e., $N = \{1, 2, 3, 4, \ldots\}$, so we assume various algebraic properties have been established for N such as closure, associativity, and commutativity for $+$ and multiplication. Suppose we wish to find a formula for the sum $2 + 4 + 6 + \cdots + 2n$. This expression represents the sum of the even natural numbers up to and including the nth even natural number. Now $2n$ <u>is</u> the nth even natural number. This is interpreted as follows: when the first value of n is used, i.e., <u>$n = 1$</u>, $2n$ denotes the *first* even natural number. The first term is $2 \cdot 1 = 2$. The second even natural number is $2n = 2 \cdot 2 = 4$ with $n = 2$. The third even natural number is $2 \cdot 3 = 6$. Proceeding down the list, the *seventeenth* even natural number is $2 \cdot 17 = 34$. Here are five such sums.

$$2 = 2$$
$$2 + 4 = 6$$
$$2 + 4 + 6 = 12$$
$$2 + 4 + 6 + 8 = 20$$
$$2 + 4 + 6 + 8 + 10 = 30.$$

The sums we get: $2, 6, 12, 20, 30, \ldots$ do not at first have any obvious pattern. But let's try factoring the terms in the left and right sides of each equation in various ways. We get

$$2 \cdot 1 = 1 \cdot 2$$
$$2 \cdot 1 + 2 \cdot 2 = 6 = 2 \cdot 3$$
$$2 \cdot 1 + 2 \cdot 2 + 2 \cdot 3 = 12 = 2 \cdot 6 = 3 \cdot 4$$
$$2 \cdot 1 + 2 \cdot 2 + 2 \cdot 3 + 2 \cdot 4 = 20 = 2 \cdot 10 = 4 \cdot 5$$
$$2 \cdot 1 + 2 \cdot 2 + 2 \cdot 3 + 2 \cdot 4 + 2 \cdot 5 = 30 = 3 \cdot 10 = 5 \cdot 6.$$

The pattern is clearer now. Notice the "non-2" factor of the last term in the sum on the left is the same as the first factor on the right. That is, if the last term in the sum on the left is $2 \cdot n$, the right side is $n \cdot (n+1)$. This is true in each case above. It seems that we are ready to make a conjecture. Inductively, we conjecture

$$2 + 4 + 6 + 8 + \cdots + 2n = n(n+1).$$

It is clear that induction is a reasoning procedure from the particular to the general. In other words, a conclusion that results from viewing many particular cases is a conclusion that is drawn inductively. However, in view of some of the difficulties mentioned before, inductive reasoning must occur under very controlled conditions. You will recall a previous situation concerning Rule U.G. which reasoned from the particular to the general in a sound fashion. It occurred under strict conditions. In particular, the variable y must not have been free in any premise and must not have been freed by the use of Rule E.S. It is with no less care that we impose conditions on inductive reasoning as a valid means of verifying a conjecture. Here is the main tool to accomplish what we want.

PRINCIPLE OF MATHEMATICAL INDUCTION (P.M.I.): Suppose N denotes the natural numbers, and further suppose $P(n)$ is a predicate meaningful in the context N, and suppose

1. $P(1)$ is true, and
2. $\forall k \big[k \in N \wedge P(k) \implies P(k+1) \big]$. That is, "For all k, if $k \in N$ and if $P(k)$ is true then $P(k+1)$ is true."

   Then $\forall n \big[n \in N \implies P(n) \big]$.

Notice that the second premise is a conditional. The $P(k)$ of that premise is sometimes called the <u>induction hypothesis</u>, abbreviated as I.H. The assertion of P.M.I. says that after conditions 1 and 2 are established the conclusion is, "$P(n)$ is true for each natural number n." In other words, if $P(n)$ is a predicate, meaningful in the context of N then to use P.M.I. to prove $P(n)$ is true for all n, simply show conditions 1 and 2 hold. THEN, after invoking P.M.I., $P(n)$ is true for all n. In the previously discussed example, $2 + 4 + 6 + \cdots + 2k = k(k+1)$ is the induction hypothesis $P(k)$. After it has been established, the implication in condition 2 says, "If $P(k)$ is true at any k then it is true at the next value $k+1$."

The principle of mathematical induction, also called *the principle of finite induction*, will be taken as an axiom. There are other principles logically equivalent to P.M.I. One of these is called the *well ordering property of N*. In some texts the well ordering property is taken as an axiom and mathematical induction is proven to be logically equivalent, although we will not do that here. Another equivalent principle is called strong mathematical induction. It is described later in this chapter. Any of these principles may be assumed as axioms, not proven, just as the axioms of geometry are assumed about plane geometry.

Now a non–mathematical example is considered, which illustrates the effect of each part of the principle of mathematical induction.

EXAMPLE 7.1: Suppose one has an unknown, but fixed finite number n of dominoes. Perhaps there are 2 or 2 million. Also suppose each domino is numbered with a different number, from 1 for the first domino, to n for the last. In addition, suppose we know that no matter which domino k is given, if the kth domino is tipped over, the next, the $(k+1)^{\text{st}}$ domino, will also fall. This condition can be assured by the proper physical arrangement of each domino with the preceding one. This knowledge corresponds to knowing that condition 2 of P.M.I. holds. Now, further suppose that the first domino falls. This corresponds to knowing that condition 1 holds. Which dominoes will fall? All of them. This is because the first one falls by condition 1. Then by condition 2, the second domino falls. Again, by condition 2, the third domino falls since the second fell. Again, since the third one fell, the fourth will fall by condition 2, and so on. Thus all n dominoes will fall no matter how many there are.

Next, we set up for a proof by mathematical induction. Recall that in order to use mathematical induction to prove some predicate $P(n)$ is true for all natural numbers, the procedure is to prove that both conditions 1 and 2 of P.M.I. are true. Then the principle of mathematical induction says the predicate $P(n)$ is true for all n.

Let us return to the example concerning summing the first n even natural numbers.

EXAMPLE 7.2:   Prove. If $P(n)$ is the predicate

$$2 + 4 + 6 + \cdots + 2n = n(n+1)$$

then $\forall n[n \in N \Longrightarrow P(n)$ is true$]$. That is, prove

$$\forall n \in N[2 + 4 + 6 + \cdots + 2n = n(n+1)].$$

Before doing the proof, let us identify what we have and what we are to do. We are given the predicate $P(n)$. Notice that $P(n)$ really is a predicate, represented by an equation. As a predicate, we know, "If n is assigned a specific natural number value, then the equation $P(n)$ is either true or false." We must show $P(1)$ is true. That is, we must show that, the given equation holds in case $n = 1$. Next we must show that, "Whenever $P(k)$ is true then $P(k+1)$ MUST be true."

PROOF: Let $P(n)$ be the predicate given by $[2+4+\cdots+2n = n(n+1)]$.

CASE 1: When $n = 1$, the left side of the equation $P(1)$ is just one term of the sum, i.e., 2. The right side of $P(1)$ is $1(1+1)$, which is also 2. Since these are equal, $P(1)$ is true.

CASE 2: Suppose $k \in N$. The objective is to show, "If $P(k)$ is true then $P(k+1)$ is true." Since this is a conditional, we suppose the induction hypothesis $P(k)$ holds for some particular, but otherwise unspecified k in N. Then

$$2 + 4 + 6 + \cdots + 2k = k(k + 1) \tag{1}$$

is true. Now the objective is to show $P(k + 1)$ must be true. In other words, <u>show</u>

$$2 + 4 + 6 + \cdots + 2k + 2(k + 1) = (k + 1)\Big((k + 1) + 1\Big). \tag{2}$$

To do this, <u>start with the left member</u>. [Do NOT work with both members of the equation to be established such as (2).]

$$
\begin{aligned}
2 + 4 + \cdots + 2k + 2(k + 1) &= k(k + 1) + 2(k + 1) && \text{I.H. (How?)} \\
&= (k + 2)(k + 1) && \text{Right Distr. Law} \\
&= (k + 1)(k + 2) && \text{Commutative Law} \\
&= (k + 1)\big((k + 1) + 1\big) && \text{Associative Law}
\end{aligned}
$$

The first step is justified because it depends upon the induction hypothesis in equation (1), i.e., the first k terms are replaced by what I.H. says. Step two is the right distributive law and the others employ laws given in the assumed properties in Chapter 5. Thus we have shown that when $P(k)$ is true then $P(k+1)$ is true. Now by the principle of mathematical induction, $\forall n [n \in N \implies P(n)]$. This completes the proof.

A Word of Caution:

There is a temptation when doing proofs by mathematical induction to prove $P(k + 1)$ by using $P(k + 1)$. Of course, this is absurd when it is said like this, because it is begging the question. To prove $P(k + 1)$ assume $P(k)$ is true and prove $P(k+1)$. Whether the statement $P(k+1)$ is an equality, as above, or an inequality as we shall see, *begin with the left member of the equation (or inequality) and work step–by–step to the right member* or occasionally from the right member to the left. Most particularly, <u>do not</u> start with both members of $P(k + 1)$ and arrive at some identity. Look carefully at the proofs in this section to see various legitimate means that may be employed when doing such an induction proof. To belabor the point, notice that we did not work with both sides of the objective equation.

EXAMPLE 7.3: If $P(n)$ is the predicate $n \leq n^2$, prove $\forall n [n \in N \implies P(n)]$. That is, prove $\forall n [n \in N \implies n \leq n^2]$.

PROOF: Let $P(n)$ be the predicate $n \leq n^2$. We will use P.M.I., so we must verify conditions 1 and 2.

CASE 1:   Since $1 \leq 1^2$ is true then $P(1)$ is true.

CASE 2:   Suppose $k \in N$ is arbitrary then $k \geq 1$. The objective is to show $P(k) \implies P(k+1)$. In other words, the objective is to <u>show</u>

$$\text{If } k \leq k^2 \text{ then } (k+1) \leq (k+1)^2.$$

Toward that end, suppose $P(k)$, i.e., suppose, $k \leq k^2$, then by assumed properties of Chapter 5:

$$
\begin{aligned}
k + 1 &\leq k^2 + 1 & &\text{I.H. and Property D of Chapter 5} \\
&\leq k^2 + 2k + 1 & &\text{Since } k \geq 1 > 0 \text{ and Property C of Chapter 5} \\
&\leq (k+1)^2 & &\text{Algebraic Identity}
\end{aligned}
$$

Thus $(k+1) \leq (k+1)^2$. Since k was an arbitrary element of N, then $\forall k \big[k \in N \wedge P(k) \implies P(k+1) \big]$. Hence by the principle of mathematical induction, $\forall n \big[n \in N \implies n \leq n^2 \big]$.

Some additional relationships that are easily established by mathematical induction are given in the next property, but first a definition and some explanation of notation is in order. Some mathematical notation is defined "inductively" or "recursively." That is, a concept is given a starting value. Next, if any particular value is known, then a means for getting to the subsequent value is provided in the definition.

Here are examples of inductive definitions of *exponents, multiples, higher order derivatives*, and *summation notation*.

Let a be a number, then define $a^1 = a$, and for each $k \in N$ define $a^{k+1} = a^k a$. Then a^n has been defined for all natural numbers n. That is because we know what a^1 is, and we know how to get to the next exponent and the next and the next, etc. So

$$a^2 = a^{1+1} \overset{\text{def}}{=} a^1 a \overset{\text{def}}{=} aa \quad \text{and} \quad a^3 = a^{2+1} \overset{\text{def}}{=} a^2 a = (aa)a.$$

In a similar fashion define $1a = a$ and for each $k \in N$, define $(k+1)a = ka + a$. Thus

$$2a = (1+1)a \overset{\text{def}}{=} 1a + a = a + a$$

and

$$3a = (2+1)a \overset{\text{def}}{=} 2a + a = (a+a) + a.$$

Still another familiar concept occurs with higher order derivatives. In the appropriate setting define $f^{(1)}(x) = f'(x)$. For each $k \in N$ define

---

$f^{(k+1)}(x) = (f^{(k)})'(x)$. Thus we know what the first derivative is. Knowing that, we can get the second derivative and knowing that, get the third, etc. This defines the nth derivative of f for all natural numbers n.

Another notation that can be described through an inductive definition is *summation notation*. The expression $\sum_{i=1}^{n} a_i$ is defined as follows. First $\sum_{i=1}^{1} a_i = a_1$. And if $\sum_{i=1}^{k} a_i$ has been defined then $\sum_{i=1}^{k+1} a_i = \sum_{i=1}^{k} a_i + a_{k+1}$. Thus $\sum_{i=1}^{n} a_i$ has been defined for all natural numbers.

For example, $\sum_{i=1}^{3} a_i = \sum_{i=1}^{2} a_i + a_3$. Therefore more generally, $\sum_{i=1}^{n} a_i$ is interpreted as an abbreviation for $a_1 + a_2 + a_3 + \cdots + a_n$.

Now we pursue a series of properties using inductive definitions.

PROPERTY:
   a. If a is real and $a \geq 0$ then $a \leq na$ holds for all n in N.
   b. If a is real and $a \geq 1$ then $a \leq a^n$ holds for all n in N.

The justification is left to the exercises (see Exercises 7.4 and 7.5).

EXAMPLE 7.4: Prove $n < 2^n$ for all $n \in N$.

PROOF: Let $P(n)$ be the predicate $n < 2^n$.

CASE 1: Since $0 < 1$ then by Assumed Property D of Chapter 5, $1 = 0 + 1 < 1 + 1 = 2 = 2^1$, so $P(1)$ holds.

CASE 2: Suppose k is in N and $P(k)$ is true, then $k < 2^k$. This is the induction hypothesis. The objective is to show $P(k+1)$, i.e., <u>show</u> $(k+1) < 2^{k+1}$ We do this by beginning with the left member.

$$
\begin{aligned}
k+1 &< 2^k + 1 && \text{I.H. and Property D of Chapter 5}\\
&< 2^k + 2 && \text{Since } 1 < 2 \text{ using Property C of Chapter 5}\\
&\leq 2^k + 2^k && \text{Property b previously mentioned}\\
&= (2^k)2 && \text{Distributive law (How?)}\\
&= 2^{k+1} && \text{Recursive definition of exponents}
\end{aligned}
$$

Thus $P(k) \implies P(k+1)$ holds for all $k \in N$. By P.M.I., $n < 2^n$ holds for all $n \in N$. (Notice, we did NOT start with $(k+1) < 2^{k+1}$ and work with both sides; that would be fallacious argumentation.)

There is another type of mathematical induction that is equivalent to P.M.I. It is called *strong* or *complete induction*. It is similar to P.M.I. except that Part 2 is changed to, "If $P(m)$ is true for all $m \leq k$ then $P(k+1)$ is true." The induction hypothesis assumes that all prior cases to $P(k)$, including $P(k)$, are true, not just $P(k)$ is true. If we know $P(1)$ is true in addition to the conditional, "$P(m)$ is true for all $m \leq k$ implies $P(k+1)$ is true," then $P(n)$ is true for all n. Here is the statement.

STRONG MATHEMATICAL INDUCTION (S.M.I.): Suppose N denotes the natural numbers and further suppose $P(n)$ is a predicate, meaningful in the context N, and suppose:

1. $P(1)$ is true, and
2. $\forall k\big[(k \in N) \wedge \ \forall m[(m \in N) \wedge (m \leq k) \implies P(m)] \implies P(k+1)\big]$
   is true. That is, "If $P(m)$ is true for all $m \leq k$, where $m \in N$ and $k \in N$, then $P(k+1)$ is true."

   Then $\forall n\big[n \in N \implies P(n)\big]$.

There are occasions where strong induction is more useful in proving a theorem than P.M.I. Some of these instances occur in working with *factorials* and *Fibonacci sequences*, named after Leonardo of Pisa (also called Fibonacci, which means "son of Bonaccio") and who published *Liber Abaci* in 1202.

EXAMPLE 7.5: Suppose the sequence $a_1, a_2, a_3, \ldots, a_n, \ldots$ is given by $a_1 = 1$, $a_2 = 2$ and $a_{n+1} = 3a_n - 2a_{n-1}$ for all $n \geq 2$. Conjecture a general term a_n for the sequence and verify it by strong induction.

SOLUTION: In order to get an idea of what a general term might be, let us write out the first few terms of the sequence to see what we get. $a_1 = 1$, $a_2 = 2$, $a_3 = 4$, $a_4 = 8$, $a_5 = 16$, and $a_6 = 32$. Our conjecture is $P(n)$: $a_n = 2^{n-1}$. This works for each term above. If $k \in N$, the next step is to suppose $P(m)$ is true for each $m \leq k$. Then $a_m = 2^{m-1} \ \forall m \leq k$. The objective is to show $a_{k+1} = 2^{(k+1)-1}$.

$$
\begin{aligned}
a_{k+1} &= 3a_k - 2a_{k-1} &&\text{Definition} \\
&= 3\big(2^{k-1}\big) - 2\big(2^{(k-1)-1}\big) &&\text{I.H.} \\
&= 3\big(2^{k-1}\big) - 2^{k-1} &&\text{Property of exponents} \\
&= 2\big(2^{k-1}\big) &&\text{Distributive law (How?)}
\end{aligned}
$$

$$= 2^k \qquad \text{Recursive definition of exponents}$$

$$= 2^{(k+1)-1} \qquad \text{Assumed Properties of Chapter 5}$$

Thus $P(k+1)$ is true whenever $P(m)$ is true for all $m \leq k$. So by S.M.I. $P(n)$ is true for all n.

Let us look at another definition given inductively. We define the concept of $\underline{n\ \text{factorial}}$. The symbol for n factorial is $n!$. To define $n!$ for all integers n such that $n \geq 0$, we first define $0! = 1$. Now, if $k!$ has been defined then $(k+1)! = (k+1) \cdot k!$. For example, $1! = (0+1)! = (0+1) \cdot 0! = 1$, and $2! = (1+1)! = (1+1) \cdot 1! = 2$, and $3! = (2+1)! = (2+1) \cdot 2! = 3 \cdot 2! = 6$. Also, $4! = 4 \cdot 3! = 24$, and so on.

There are some predicates $P(n)$ that may not be true for all $n \in N$, but are true for all $n \in N$ such that $n \geq r$ where r is some fixed natural number. To prove $P(n)$ is true for all $n \geq r$ where r is some fixed natural number, simply show $P(r)$ is true; this gets the process started. Then show Part 2 of P.M.I. Then the assertion is true for all natural numbers greater than or equal to r. An example of such an assertion is $P(n)$: $n^2 \leq n!$. This predicate is not true for $n = 2$, or 3, but $\underline{\text{is}}$ true for $n = 1$ and for $n = 4, 5, 6, \ldots$ As an exercise, determine if $n^2 \leq n!$ for $n = 1, 2, 3$, and 4 by direct calculation.

EXAMPLE 7.6: Prove $n^2 \leq n!$ for all $n \in N$ such that $n \geq 4$.

PROOF: Let $P(n)$ be the assertion $n^2 \leq n!$. To show $P(n)$ is true for all $n \geq 4$ we show first

CASE 1: $P(4)$ is true. $4^2 = 16 \leq 24 = 4!$, so $P(4)$ is true

CASE 2: Suppose $P(k)$ is true for some $k \geq 4$. That means $k^2 \leq k!$. Our objective is to show $P(k+1)$ has to be true. That is, we must show $(k+1)^2 \leq (k+1)!$. [Note that we DON'T work with both sides of this latter inequality, rather we start with the left member $(k+1)^2$ and work toward the right if possible with a string of inequalities. You fill in the reasons for each equality or inequality.]

$$(k+1)^2 = k^2 + 2k + 1 \leq k! + 2k + 1 \leq k! + k^2 + k^2$$
$$\leq k! + k! + k! = 3k! \leq (k+1)k! = (k+1)!.$$

Exercises:

7.1 Suppose we define t_n recursively for all $n \in N$ by $t_1 = 1$, $t_2 = 1$, and $\forall k \in N$, $t_k = t_{k-1} + t_{k-2}$. Find $t_1, t_2, t_3, t_4, t_5, t_6$. These are the first few terms in the *Fibonacci sequence*.

7.2 Prove: $n + 1 \leq \frac{3}{2}n$ for all $n \geq 2$.

7.3 Prove: $1 + 2 + 3 + 4 + \cdots + n = \frac{n(n+1)}{2}$ for all $n \in N$.

7.4 Prove: If a is real and $a \geq 0$ then $a \leq na$ for all $n \in N$.

7.5 Prove: If a is real and $a \geq 1$ then $a \leq a^n$ for all $n \in N$.

7.6 Calculate the number $(2n-1)$ for all $n = 1, 2, 3, 4, 5, \ldots, 17$. What is the 17th odd natural number?

    a. Calculate the sum $s_n = 1+3+5+\cdots+(2n-1)$ for $n = 1, 2, 3, 4$. That is, find s_1, s_2, s_3, and s_4.

    b. Conjecture a value for $s_n = 1 + 3 + 5 + \cdots + (2n - 1)$ in the form of a function of an arbitrary n in N.

    c. Prove: $\sum_{i=1}^{n}(2i - 1) = n^2$ for all $n \in N$.

    d. What is the sum of the first 100 odd positive integers?

7.7 For $n = 1, 2, 3, 4$ find $\sum_{i=1}^{4} \frac{1}{i(i+1)} = \frac{1}{1 \cdot 2} + \frac{1}{2 \cdot 3} + \frac{1}{3 \cdot 4} + \cdots + \frac{1}{n(n+1)}$. Look at the sums you get, then try to conjecture a single term expression to represent the sum as a function of n. Next

$$\text{Prove: } \sum_{i=1}^{n} \frac{1}{i(i + 1)} = \frac{n}{n + 1} \text{ for all } n \text{ in } N.$$

7.8 Prove: If $r \neq 1$ then $1 + r + r^2 + r^3 + \cdots + r^{n-1} = \frac{1-r^n}{1-r}$ for all n in N by mathematical induction.

7.9 Prove: $1 \cdot 2 + 2 \cdot 3 + 3 \cdot 4 + \cdots + n(n + 1) = \frac{n}{3}(n + 1)(n + 2)$ for all n in N.

7.10 Prove: $1^2 + 2^2 + 3^2 + \cdots + n^2 = \frac{n}{6}(n + 1)(2n + 1)$.

7.11 Conjecture a simple formula (a one–term expression) for the sum given by: $B_n = 1(1!) + 2(2!) + 3(3!) + \cdots + n(n!)$. Then prove it by mathematical induction.

7.12 Prove the generalized DeMorgan's laws for propositions P_1, P_2, $P_3, \ldots, P_n$.

    a. $\sim(P_1 \vee P_2 \vee \cdots \vee P_n) \equiv \; \sim P_1 \wedge \; \sim P_2 \wedge \; \sim P_3 \wedge \cdots \wedge \sim P_n$.

    b. $\sim(P_1 \wedge P_2 \wedge \cdots \wedge P_n) \equiv \; \sim P_1 \vee \; \sim P_2 \vee \; \sim P_3 \vee \cdots \vee \sim P_n$.

7.13 Prove the generalized distributive laws for propositions P, R_1, R_2, $R_3, \ldots, R_n$.

    a. $P \wedge (R_1 \vee R_2 \vee R_3 \vee \cdots \vee R_n) \equiv (P \wedge R_1) \vee (P \wedge R_2) \vee \cdots \vee (P \wedge R_n)$.

    b. $P \vee (R_1 \wedge R_2 \wedge R_3 \wedge \cdots \wedge R_n) \equiv (P \vee R_1) \wedge (P \vee R_2) \wedge \cdots \wedge (P \vee R_n)$.

7.14 If $f(x) = xe^x$, find $f'(x)$, $f''(x)$, $f^{(3)}(x)$, ... Find a predicate $P(n)$ which you can use as a formula for $f^{(n)}(x)$. Then prove it by P.M.I.

7.15 Prove: If $x \geq -1$ then $(1 + x)^n \geq (1 + nx)$ $\forall n$ in N.

7.16 Prove: $3^{n+1} > 1 + 2^{n+1}$ for all n in N.

7.17 Prove: $\sum_{i=1}^{n} i^3 = \left[\frac{n(n+1)}{2}\right]^2$ for all natural numbers n.

7.18 Suppose the sequence $a_1, a_2, a_3, \ldots, a_n, \ldots$ is given by $a_1 = 1$, $a_2 = 3$, and for all $n \geq 2$ $a_{n+1} = 3a_n - 2a_{n-1}$. Conjecture a general term a_n for the sequence and verify it by strong induction.

7.19 Prove: If $D_x(f(x)) = \frac{d}{dx}f(x)$ and $D_x^2(f(x)) = D_x(D_x(f(x))) = \frac{d}{dx}(\frac{d}{dx}f(x))$ and more generally, $D_x^{n+1}(f(x)) = D_x(D_x^n(f(x)))$ then $D_x^n(x^n) = n!$ $\forall n$.

7.20 Using $\lim_{x \to a} x = a$ and other fundamental limit properties, prove $\lim_{x \to a} x^n = a^n$ by induction.

7.21 What is wrong with this proof of the "theorem"

All horses have the same color.

"Proof:" Let S_k denote the statement, "All horses in any set of k horses have the same color."

CASE 1: If $k = 1$, this is true since all horses in a set of one horse certainly have the same color.

CASE 2: Suppose S_k is true for some $k \geq 1$. We must show it is true for $k + 1$. So suppose we have a set of $k + 1$ horses. Take the subset of k horses obtained by removing the $k + 1^{st}$ horse. Then, since the resulting set has k horses, they are all the same color. Now put that $k + 1^{st}$ horse back in and take the first one out. The resulting set has k horses so they are all the same color again. So the $k + 1^{st}$ horse has the same color as the first one. Thus the set of all $k + 1$ horses have the same color. Hence by P.M.I. all horses have the same color.

SECTION II

SETS

CHAPTER 8

SETS AND SET OPERATIONS

Introduction:

The focus of this study is now directed toward some of the most fundamental notions of present day mathematics–the notions of sets and operations with sets. The material covered in the preceding section has laid the groundwork for this study by developing the tools of logic needed in the remainder of this text. In fact, many of the results in this chapter closely parallel the corresponding results in Chapter 1. After a foundation is formed with sets, their applications in several areas, such as functions and cardinal numbers, are pursued to provide concepts useful in other areas of mathematics.

Historical Overview:

Though the idea of a set had been tacitly assumed in mathematics for many years, it was not until the late 1800s that serious consideration was given to putting the foundations for set theory in good order. The mathematician, Georg Cantor (1845–1918), while working on some problems in a trigonometric series, accidently stumbled upon some results concerning infinite sets that induced a revolution in mathematical thought. The results, though not well received at first by all mathematicians, eventually dispelled the confidence that sets were totally understood. By 1900, there were research efforts begun by several mathematicians which attempted to establish a firmer footing for set theory and interrelate it to allied areas in mathematics. But just as the new interest in set theory began to gain wider appeal, some additional stumbling blocks in the form of paradoxes arose, indicating that the intuitive notions about sets were contradictory. To meet this challenge, whole new approaches were initiated by the mathematicians, Russell, Zermelo, and others. These approaches, which centered around the establishment of various axiomatic frameworks, were designed to eliminate the paradoxes and provide a really logical and totally reliable basis for set theory. The objective was to formulate an axiomatic theory reminiscent of *Euclid's Elements,* but more abstract and less open to challenge on logical grounds.

These new axiomatic structures, though successful at eliminating the paradoxes and having certain necessary mathematical traits, made set theory substantially more complex and regretfully fell short of achieving all of the objectives or features desired. Indeed, many of the notions of intuitive set theory and, in fact, some sets in intuitive set theory, had no place in some of these axiomatic systems. Many mathematicians were, and are, unwilling to accept this state of affairs. Thus research in set theory continues.

Since the early 1900s interest in set theory has remained quite high. Largely fueled by the presence of deep philosophical questions and unresolved problems, research interest has continued in the realm of set theory. From time to time, one of these questions is settled or partly settled, causing a kind of subdued mathematical rejoicing. Then, as often as not, a new problem, perhaps arising from the dust of one of the earlier solutions, presents itself to the mathematical world to spark new debate. With this seemingly endless resupplying of problems, the study of set theory continues today to meet the new challenges as well as tackle the older unsolved problems.

In this text, a less axiomatic more intuitive approach is followed. Deeper questions and subtleties are overlooked in the interest of covering a wider variety of topics and making the material accessible to the student at an earlier stage of his or her mathematical development.

Sets, Elements, and Subsets:

In geometry there are the undefined terms *point* and *line*. In set theory there are also *undefined terms* whose presence is required for the following reason. Without undefined terms, one would have one of two situations arise: a sequence of infinitely many terms, each used to define another, or a circle of terms, each defining another. The first situation is impossible and the second is unsatisfactory, especially if one does not know any of the words in the circle.

Technically, almost any interpretation may be attached to an undefined term as long as it is consistent with the axioms. Since we are not rigorously following an axiomatic system, other means such as examples will be provided below to narrow the scope of interpretation of undefined terms.

The first undefined notion is *set*. The student probably already has a concept of what a set is or should be. Some of the collections of objects we will refer to as sets and probably some of the things thought of as sets are not sets in the most rigorous sense. However, we will refer to these as sets, overlooking certain subtle mathematical objections. Until we get further into the subject, you may just think of a set as a collection of objects.

One of the most common and simplest ways to represent sets is by capital letters, and at times lower case letters, such as A, B, x, or S.

The second undefined notion is that of being *an element of*. The symbol for the phrase, "is an element of," is $\in$. The complete sentence, "$x \in S$," is read EXACTLY as: "x is an element of S," "x is in S," or "x is a member of S." The symbol $\in$ is NOT used to replace only the noun *element;* it replaces the phrase "is an element of." As an abbreviation, $a, b \in S$ will mean $a \in S \wedge b \in S$.

If S is a known set (i.e., it is fixed, but arbitrary) and a is fixed, $a \in S$ is a statement, so $a \in S$ is either TRUE or FALSE. In the latter case we write $\underline{a \notin S}$ as the notation.

DEFINITION: If S is a set and $a \in S$ then a is called an <u>element</u>.

An object is entitled to be called an element if there is a set that it is an element of. Once again, be certain that the symbol $\in$ is not used to call something an element by replacing the word "element" by the symbol $\in$.

Another common notation for certain sets is the *enumeration notation*. This notation enumerates the elements of the set inside of braces $\{\ ,\ ,\ \}$ with commas separating the individual elements. For example, the set S whose elements are precisely: 2, 4, 6, and 8 is denoted by $\{2, 4, 6, 8\}$ or $S = \{2, 4, 6, 8\}$. In this example, the statement

$$x \in \{2, 4, 6, 8\} \text{ means } (x = 2) \vee (x = 4) \vee (x = 6) \vee (x = 8).$$

Remember, use braces to enclose the set and commas to separate individual elements.

A bit of caution needs to be exercised when enumerating the elements of larger sets. The use of ellipses is acceptable if it is clear which elements are being replaced by the ellipsis. For example, $\{1, 2, 3, 6, \ldots\}$ might be $\{1, 2, 3, 6, 18, 108, 864, \ldots\}$ or $\{1, 2, 3, 6, 12, 24, 48, \ldots\}$. In the former, each number after the third is the product of the previous two and in the latter, each number after the first is the sum of all of the previous numbers. To avoid possible ambiguity, the inclusion of a general term in the enumeration, when practical, would be ideal. For example $\{1, 3, \ldots, (2n - 1), \ldots\}$ is the set of odd natural numbers. Note the presence of the general term leaves no doubt about the nature of the set. Lacking this option, a simple explanatory comment, in conjunction with the set, proves helpful to the reader.

DEFINITION: (*Subset*) Let A and B be sets. <u>A is a subset of B</u>, (denoted $A \subseteq B$) iff $\forall x[x \in A \implies x \in B]$.

Note in the definition of $\subseteq$, the universal quantifier and the conditional $\implies$ are used. Also note the containment notation $\subseteq$ is read as "is a subset of," or "is contained in." That is, when reading a statement such as $A \subseteq B$, simply say, "A is a subset of B," replacing the symbol $\subseteq$ by "is a subset of." Once again, the symbol $\subseteq$ does NOT refer to the noun "subset;" it replaces the words "is a subset of" in a sentence. So it is IMPROPER to write, "Let A be a $\subseteq$ of T."

The next theorem describes two basic properties possessed by the subset relation. Its proof further illustrates some of the techniques that may be employed in an informal proof and it illustrates how and why we developed the techniques in Section 1.

THEOREM 8.1: Let A, B, and C be sets. Then

   a. $A \subseteq A$

   b. $A \subseteq B \wedge B \subseteq C \implies A \subseteq C$.

PROOF PART a: By observing the definition of $\subseteq$ it is clear that in order to show $A \subseteq A$, one must show $\forall x[x \in A \implies x \in A]$. To accomplish this use Exercise 2.8a, which says $P \implies P$. But then $y \in A \implies y \in A$ is a tautology. Thus by Rule U.G. $\forall x[x \in A \implies x \in A]$. Therefore $A \subseteq A$.

PART b: Suppose $A \subseteq B$ and $B \subseteq C$, then from the definition of $\subseteq$ we know that $\forall x[x \in A \implies x \in B]$ and $\forall x[x \in B \implies x \in C]$. By Exercise 4.7, $\forall x[x \in A \implies x \in C]$. Again, by the definition of $\subseteq$, $A \subseteq C$.

The definition for set equality is quite similar to the definition of the subset relation.

DEFINITION: (*Set Equality*) Let A and B be sets. A is equal to B if and only if $\forall x[x \in A \iff x \in B]$.

By observing the definitions of "is a subset of" for sets and "set equality," it is clear how heavily the concepts of sets are depending on the logic developed in Section 1.

AXIOM OF EXTENT: Let A be a set. If $x \in A$ and $x = y$ then $y \in A$.

EXAMPLE 8.1: Let $S = \{a, b\}$ and $T = \{a, b, c\}$. We show how to prove that $S \subseteq T$.

PROOF: Let $y \in S$. Since $S = \{a, b\}$, $(y = a) \vee (y = b)$. If $y = a$, then since $a \in T$, $y \in T$ by the axiom of extent. Similarly, if $y = b$ then $y \in T$

by the axiom of extent. So, by using Rule 18a from *Table 2.5*, $y \in T$ in any case. Now we know $y \in S \Longrightarrow y \in T$. Thus $\forall x[x \in S \Longrightarrow x \in T]$ by Rule U.G. Consequently, $S \subseteq T$ by the definition of $\subseteq$.

EXAMPLE 8.2: If $S = \{a, b, c\}$ and $T = \{b, c, a\}$ then $S = T$.

A step toward the justification for the conclusion may go as follows.

$$(y \in S) \overset{\text{why}}{\equiv} (y = a \lor y = b \lor y = c)$$

$$\overset{\text{why}}{\equiv} (y = b \lor y = c \lor y = a) \overset{\text{why}}{\equiv} (y \in T).$$

The first and last of the equivalences are true because of the axiom of extent and the definition of set equality. (Why is the middle equivalence true?) Hence, by the transitive property of $\Longleftrightarrow$, $(y \in S) \Longleftrightarrow (y \in T)$. Thus $\forall x[x \in S \Longleftrightarrow x \in T]$. Consequently, $S = T$ by the definition of set equality.

THEOREM 8.2: If A, B, and C are sets then

    a. $A = A$.

    b. If $A = B$ then $B = A$.

    c. If $A = B$ and $B = C$ then $A = C$.

    d. $A = B$ if and only if $A \subseteq B$ and $B \subseteq A$.

PROOF: It is fairly easy to see that Parts a and c are similar to parts of Theorem 8.1. They follow from tautologies $P \Longleftrightarrow P$, (see Exercise 2.8) and $((P \Longleftrightarrow Q) \land (Q \Longleftrightarrow R)) \Longrightarrow (P \Longleftrightarrow R)$, (see Tautology 25). Part b follows from the tautology $(P \Longleftrightarrow Q) \Longrightarrow (Q \Longleftrightarrow P)$. We will establish Part d using Rule 5 of *Table 2.5* and the strategies mentioned in the sample proof of the biconditional in the paragraph preceding the Summary in Chapter 5.

PART d: Because this is an "iff," there are two main cases to prove.

CASE 1: $(A = B) \Longrightarrow (A \subseteq B) \land (B \subseteq A)$, and

CASE 2: $(A \subseteq B) \land (B \subseteq A) \Longrightarrow (A = B)$.

CASE 1: Suppose $A = B$, then $\forall x[x \in A \Longleftrightarrow x \in B]$. By Rule U.S. $y \in A \Longleftrightarrow y \in B$ and hence, by Rule 5 and simplification, $y \in A \Longrightarrow y \in B$ and $y \in B \Longrightarrow y \in A$. This exploits what we have from the assumption $A = B$. Keep in mind that there are two things to be shown in this case: $A \subseteq B$ and $B \subseteq A$. To get the first of these two containments, apply Rule U.G. to the implication: $y \in A \Longrightarrow y \in B$. The result is $\forall x[x \in A \Longrightarrow x \in B]$. Hence $A \subseteq B$. By analogy, one may take the other implication: $y \in B \Longrightarrow y \in A$, to obtain $B \subseteq A$. Thus we have shown $(A = B) \Longrightarrow (A \subseteq B \wedge B \subseteq A)$.

CASE 2:   The converse part is left as an exercise (see Exercise 8.14).

After cases 1 and 2 are established, a second appeal to Rule 5 yields:

$$(A = B) \equiv (A \subseteq B) \wedge (B \subseteq A).$$

Theorem 8.2d, which is the set theoretic analogue of Rule 5, is used numerous times in the ensuing development for establishing that two sets are equal, and consequently, is a very important property.

At this point it is convenient to indicate the hierarchy of application of set theoretic and logical connectives. Set theoretic connectives must be executed before logical connectives. Thus in the statement in Case 1 of the proof of Theorem 8.2d, one may replace the expression $(A = B) \Longrightarrow (A \subseteq B) \wedge (B \subseteq A)$ with $A = B \Longrightarrow A \subseteq B \wedge B \subseteq A$.

DEFINITION: (*Proper Subset*) Let A and B be sets. Then A is a proper subset of B if and only if $A \subseteq B$ and $A \neq B$.

A common notation for "proper subset" is $\subset$. In particular, $A \subset B$ means, "A is a proper subset of B." Formally we would write logical equivalents to "$A \subset B$" as follows

$$A \subset B \equiv A \subseteq B \wedge A \neq B$$
$$\equiv (\forall x)[x \in A \Longrightarrow x \in B] \wedge (\exists x)[x \in B \wedge x \notin A].$$

An example of two sets that exhibit the proper subset relationship may be found in Example 8.1.

Set Builder Notation:

If set theory were limited to the means we have considered so far for describing sets, mathematics would be severely limited. Fortunately there is a very powerful and useful notation for describing sets. It is called set builder notation.

If $P(x)$ is a predicate and U is a known set, then we say P(x) is meaningful in the context of U if and only if $P(x)$ and U are so related that no matter which a is selected from U, either $P(a)$ is true or $P(a)$ is false.

As an example, suppose U is the set N of natural numbers and $P(x)$ is the predicate, "x is a prime number." Then $P(x)$ <u>is</u> meaningful in the context of U, because for each natural number x, either x is prime or x is not prime. On the other hand, if $Q(x)$ is the predicate "x is red," $Q(x)$ is not meaningful in the context N.

Now suppose $P(x)$ is a predicate and U is a set such that $P(x)$ is meaningful in the context U. Then one may be interested in those objects $x \in U$ such that $P(x)$. Is this collection of objects a set? The next axiom says the answer is YES.

AXIOM OF SPECIFICATION: Suppose U is a set and $P(x)$ is a predicate meaningful in the context of U. Then there is a set S whose elements consist of precisely those elements $x \in U$ such that $P(x)$ is true. In other words, if $(\forall x)[x \in S \longleftrightarrow (x \in U \land P(x) \text{ is true})]$ is a tautology then S is a set.

The notation for the set described above in the axiom of specification is the set

$$S = \{x \mid x \in U \land P(x)\}.$$

A more brief version of the same notation is $\{x \in U \mid P(x)\}$. The set notation given here is called <u>set</u> <u>builder</u> <u>notation</u> <u>and</u> <u>is</u> <u>read</u> <u>as</u> *the set of* <u>all</u> *x such that $x \in U$ and $P(x)$.* It is called set builder notation because the "plans" for building the set are given. Another way to help interpret the notation is that the set described is like a special club. The membership requirements are given by $x \in U$ and $P(x)$ and appear between the $\mid$ and the $\}$. This is no ordinary club though, because if an element meets the membership requirements then that element is in the club and if an element does NOT meet the membership requirements then that element is NOT in the club. This follows because the set S consists of ALL OF those elements x such that $x \in U$ and $P(x)$. To reemphasize a point from the axiom of specification

An element $y \in \{x \mid x \in U \land P(x)\}$ iff $y \in U$ and $P(y)$ is true.

Many problems that students have later in this course or in other mathematics courses centers on their not fully realizing this simple, but powerful equivalence.

Another important and useful fact that results through set builder notation is the logical equivalence

$$\left[\{x \in U \mid P(x)\} = \{x \in U \mid Q(x)\}\right] \equiv \left[(\forall x \in U)[P(x) \Longleftrightarrow Q(x)\right].$$

In other words, the given sets are equal if and only if their membership requirements are logically equivalent. Let us look at an example. Let U be the set of real numbers, $P(x)$ the predicate $x^2 = 1$, and $Q(x)$ the predicate $|x| = 1$. Then

$$\{x \mid x \in U \land x^2 = 1\} = \{x \mid x \in U \land |x| = 1\}$$

because

$$\left(x^2 = 1 \text{ iff } |x| = 1\right) \text{ is a tautology.}$$

In fact we can do even better than this. If $S = \{x \mid x \in U \land P(x)\}$ and $T = \{x \mid x \in U \land Q(x)\}$ then $S \subseteq T$ iff $(\forall x \in U)[P(x) \implies Q(x)]$. Thus, for example, $\{x \mid x \text{ is real } \land (x^2 = 1)\} \subseteq \{x \mid x \text{ is real } \land x^3 = x\}$. This is because if $x^2 = 1$ then $x^2 \cdot x = 1 \cdot x$, so $x^3 = x$. Note that the converse of this implication is false, so you should not expect the given sets to be equal.

Set builder notation allows one to consider many diverse types of sets. Some sets, such as the following set,

$$\{x \mid x \text{ is real and } 0 < x < 1\} = \text{ "open interval" } (0, 1),$$

CANNOT be given by enumeration, as we shall see later. And still another example that set builder notation allows us to describe is the set $\mathcal{P} = \{A \mid A \subseteq S\}$ which we would call the set of all subsets of S, and is in fact a set of sets as we shall see.

Another very important set that we can describe by set builder notation is the empty or null set. From the axiom of specification, the set given by $\{x \mid x \text{ is real and } x \neq x\}$ is a set, but it has no elements whatsoever. This is because the predicate "x is real and $x \neq x$ is false for all $x \in U$."

DEFINITION: (*Empty Set*) The set having no elements is called the empty or null set. It is denoted by the Greek letter $\emptyset$ or by $\{\ \}$.

Notice the presence of the article "the" in referring to "*the* empty set." This is meant to imply there is only one such set, an easily verifiable fact (see Exercise 8.12).

AXIOM OF PAIRING: If A and B are sets then there exists a set C such that $C = \{A, B\}$.

This axiom provides a way of making a new set C from two given sets A and B. In addition, it allows for the construction of some very important sets, not the least of which are singletons and ordered pairs, as we shall see later.

If a is a set then by the axiom of pairing, $\{a, a\}$ is a set, but $\{a, a\} = \{a\}$ by the axiom of extent (see Exercise 8.23). So we have the notion of a <u>singleton</u> set $\{a\}$. In general, any set having exactly one element is called a singleton set. Note that the singleton set $\{a\}$ is NOT the same as the element a.

EXAMPLE 8.3:   Here is a variety of sets described in different ways, illustrating a variety of different characteristics.

a. Assume $S = \{a, b, c\}$ is the set consisting of the first three letters of the alphabet and $T = \{1, 2\}$ consisting of the first two natural numbers. Then $C = \{S, T\} = \{\{a, b, c\}, \{1, 2\}\}$ is a set by the axiom of pairing. Notice that C has two different elements, (<u>not five</u>). The two elements of C are S which is $\{a, b, c\}$ and $T = \{1, 2\}$. The elements a, b, and c are elements of S, not C and 1 and 2 are elements of T, not C.

b. Since $\emptyset$ is a set, $\{\emptyset, \emptyset\} = \{\emptyset\}$ is a set. Is $\{\emptyset\}$ empty? In other words, is $\{\emptyset\} = \emptyset$? Why?

c. Since $\emptyset$ is a set, $\{\emptyset, \{\emptyset\}\}$ is a set with two distinct elements, $\emptyset$ and $\{\emptyset\}$. They are distinct because the answer to Example 8.3b is NO.

d. $\{x \mid x$ is a set $\wedge\ x \neq x\} = \emptyset = \{x \mid x$ is a real no. $\wedge\ x = x + 1\}$.

e. If $S = \{x \mid x$ is a real number and $0 < x < x^2 < 2\}$ then $1.003 \in S$, $1.1 \in S$, $\pi/3 \in S$ and $1.4 \in S$. But $0.1 \notin S$, $0 \notin S$, $\frac{1}{2} \notin S$, $1.5 \notin S$ and $-3 \notin S$. As an exercise justify these assertions (see Exercise 8.15). Find other elements in S and other elements, not in S?

f. $\{s \mid s$ is an integer and $-3 < s < -1\} = \{-2\}$ and consequently is a singleton.

g. $\{x \mid x$ is a real number and $(x - 1)(x - 1) = (x - 1)(2x - 3)\} = \{1, 2\}$ is a pair called a <u>doubleton</u>. Why are the two sets equal? (See Exercise 8.22.)

h. $2 \in \{2, 3\}$ and $2 \in \{2\}$, but $2 \notin \{3\}$ and $2 \neq \{2\}$. More generally, $a \in \{a\}$ and $a \neq \{a\}$ for any a.

i. The set $\{x \mid x$ is a real number and $x^2 - 7x + 5 = x - 2\}$ and the set $\{z \mid z$ is a real number and $z^2 - 7z + 5 = z - 2\}$ are equal sets. More generally, if $P(x)$ is a predicate then $\{y \mid y \in U \wedge P(y)\} = \{z \mid z \in U \wedge P(z)\}$.

THEOREM 8.3: If S is any set then $\emptyset \subseteq S$.

PROOF: Let S be a set. Now $y \in \emptyset$ is a contradiction, so $y \in \emptyset \Longrightarrow y \in S$ is vacuously true. Thus $\forall x[x \in \emptyset \Longrightarrow x \in S]$. Consequently, $\emptyset \subseteq S$ by the definition of subset.

EXAMPLE 8.4: If $S = \{x \mid x$ is real and $(x^2 - x - 6 = 0)\}$ then $3 \in S$ and $-2 \in S$ and all other real numbers are not in S. Why?

AXIOM OF POWER SET: If S is any set then there exists a set, denoted by $\mathcal{P}(S)$, such that $A \in \mathcal{P}(S)$ if and only if $A \subseteq S$. In other words, $\mathcal{P}(S) = \{A \mid A \subseteq S\}$.

EXAMPLE 8.5: The power set of the set $\{a, b, c\}$ is the set given by

$$\mathcal{P}(\{a,b,c\}) = \{\emptyset, \{a\}, \{b\}, \{c\}, \{a,b\}, \{a,c\}, \{b,c\}, \{a,b,c\}\}$$

because each element of $\mathcal{P}(\{a,b,c\})$ is a subset of $\{a,b,c\}$ and conversely.

We have been careful to emphasize that the axiom of specification guarantees the existence of a set consisting of elements selected from a known set and satisfying some predicate AND we were careful in the discussion of, "$P(x)$ is meaningful in the context of U," NOT to say U is, "the universal set." Many mathematicians before Bertrand Russell (1872–1969) had used the concept freely. It would be convenient in set builder notation to have a set containing everything, so that way nothing was overlooked when seeking $\{x \mid P(x)\}$. Russell showed that it is impossible to consider as a set, the collection U of all sets, because that leads to a "logical paradox." In other words, Russell showed there is NO set of all sets. Here is Russell's paradox.

Russell's Paradox:
A set may contain certain sets as elements, as seen in Example 8.5. For example, the set $\{a, b\}$ is an element of the set $\mathcal{P}(\{a,b,c\})$. Since this is possible, we suppose there is a "universal set" U which is the set of all sets. Since all sets are in U then U, in particular, is also an element of U. Likewise, there may be other sets which are elements of themselves. Define a set T to be a *regular* set if $T \notin T$. For example, the "universal set" we discussed is not a regular set. Also, even though T may have sets as elements inside it, if T is regular then T itself is not an element of T. If T is not regular then T is called *irregular*. Thus every set is either regular or irregular.

Now let $S = \{T \mid T \in U \wedge T \notin T\}$. In other words, S is the set of all regular sets. Because U is the set of all sets, $S \in U$. So either S is regular or S is irregular. In case S is regular, we have $S \in S$, since S contains all regular sets. Thus S satisfies all membership requirements of S. That is $S \in U$ AND $S \notin S$. Thus, "S is regular implies S is irregular" is established. On the other hand, if S were irregular then $S \notin \{T \mid T \in U \wedge T \notin T\}$. The failure of S to belong in S, means that S does NOT meet the membership requirements. So either $S \notin U$, which cannot be, or S does not meet the requirement $S \notin S$. But this means $S \in S$. Thus, "S is irregular implies S is regular." So we have shown that if a universal set exists then the set S is regular iff S is irregular, which is obviously a contradiction. So a universal set does not exist.

This paradox was one of the things mentioned at the beginning of this section which caused such vexation and therefore induced a much more careful development of set theory as an axiomatic system. Prior concepts of sets would allow very large collections of things like U to be considered to be sets. To avoid this difficulty, one can let U be a set just large enough to contain all the elements needed in a given discussion. For example, if one is considering various sets of real numbers, the set E_1 of all real numbers would make a satisfactory U. The set U so selected is called the <u>universe of discourse</u>.

To aid in referring quickly to certain sets of numbers, we adopt the symbols given here for the indicated number system.

N is the set of natural numbers
Z is the set of integers
Q is the set of rational numbers
E_1 is the set of real numbers
C is the set of complex numbers

If you would like to know more about these sets of numbers now, consult Chapters 18–22 in this text. We end this section with a few more examples.

EXAMPLE 8.6: In this example we introduce a variation on the set builder notation and additional example sets.

a. For example, $\{1, 8, 27, \ldots\} = \{1^3, 2^3, 3^3 \ldots\} = \{n^3 \mid n \in N\}$. If we wrote this set by using the originally described set builder notation, we would get $\{y \mid \exists n \, (n \in N \wedge y = n^3)\}$. Obviously the latter description is more obscure than the former. As a result, if you are careful, you may modify the set builder notation as we did in this example and as we do in some of the following situations.

   b. $\{2n-1 \mid n \in N\} = \{2 \cdot 1 - 1, 2 \cdot 2 - 1, 2 \cdot 3 - 1, \dots\} = \{1, 3, 5, 7, \dots\}$

   c. $\{n \mid \sqrt[3]{n} \in N\} = \{1, 8, 27, \dots\}$

   d. $\{x \in E_1 \mid x^2 < -\frac{1}{2}x\}$ is the open interval $(-\frac{1}{2}, 0)$. Why?

Exercises:

8.1 Referring to Example 8.2, explain why

$$(y = a \lor y = b \lor y = c) \equiv (y = b \lor y = c \lor y = a).$$

8.2 Referring to Example 8.2, explain why $y \in S$ iff $y \in T$.

8.3 Referring to Theorem 8.2,

   a. Write out an informal proof of Part a, i.e., prove $A = A$.

   b. Verify the tautology $(P \iff Q) \equiv (Q \iff P)$. Then prove Part b informally.

8.4 Using set builder notation describe each of the following sets.

   a. $A = \{1, 4, 9, 16, 25, \dots\}$.

   b. $B = \{2, 3, 5, 7, 11, \dots\}$.

   c. $C = \{\frac{1}{2}, \frac{2}{3}, \frac{3}{4}, \frac{4}{5}, \dots\}$.

   d. $D = \{3, 7, 11, 15, 19, \dots\}$.

   e. E = set of all real numbers whose square is between 0 and 1.

   f. $F = \{0, 3, 8, 15, 24, \dots\}$.

   g. G = set of all real numbers which exceed their square.

8.5 Using the enumeration method, write out the elements of the following sets. (Ellipses are acceptable, subject to non-ambiguity.)

   a. $A = \{x \mid x \in N \land 12 < x < 17\}$.

   b. $B = \{y \mid y \in E_1 \land y^2 - 7y + 6 = 0\}$.

   c. $C = \{x \mid x \in N \land x^2 < 10\}$.

   d. $D = \{z \mid z \in Z \land z^2 - 2 = 0\}$.

   e. $E = \{z \mid z \in E_1 \land z^2 - 2 = 0\}$.

   f. $F = \{x \mid x \in E_1 \land x^3 = 2x\}$.

   g. $G = \{x \mid x \in Z \land x^2 - 2x \text{ is divisible by } 2\}$.

   h. $H = \{x \mid x \in Z \land x^2 - 3x \text{ is divisible by } 2\}$.

8.6 Is $\emptyset = \{\emptyset\}$? Why? Is $\{\emptyset\} = \{\{\emptyset\}\}$?

8.7 a. By negating the right side of the definition of $\subseteq$ write out the definition of $A \not\subseteq B$.

  b. By negating the definition of set equality, write out a condition for $\neq$, for sets.

8.8 $(A \subseteq B \wedge B \subseteq C) \Longrightarrow A \subseteq C$ and $(A = B \wedge B = C) \Longrightarrow (A = C)$ are facts established in the foregoing section. Answer the following questions.

  a. Does $A \subset B \wedge B \subset C$ imply $A \subset C$?

  b. Does $A \not\subseteq B \wedge B \not\subseteq C$ imply $A \not\subseteq C$?

  c. Does $A \in B \wedge B \in C$ imply $A \in C$?

  d. Does $A \subseteq B \wedge B \not\subseteq C$ imply $A \not\subseteq C$?

  e. Does $A \in B \wedge B \not\subseteq C$ imply $A \notin C$?

  f. Does $\{x \mid x \in E_1 \wedge 2x = 6\} = 3$?

  g. Does $\{\{x\} \mid x \in E_1 \wedge 2x = 6\} = \{\{3\}\}$?

  h. Is $\{3\} \in \{x \mid x \in E_1 \wedge x^2 = 9\}$?

  i. Is $\{-3\} \in \{\{x\} \mid x \in E_1 \wedge x^2 = 9\}$?

  j. Does $y \in \{x \mid x \in E_1 \wedge x^2 = 1\} \Longrightarrow y = 1$?

  k. If $y \in \{x \mid x \in E_1 \wedge x^2 = 4\}$ is $y \in \{x \mid x \in E_1 \wedge |x| = 2\}$?

  l. Does $y \in \{x \mid x \in E_1 \wedge x^2 = 1\} \Longrightarrow y \in \{-1, 1\}$?

8.9 How many elements are in $\emptyset$, $\{\emptyset\}$, $\{\{1\}, \{1, 2\}\}$, or $\{\{1, 2\}\}$?

8.10 Are $\{a, b, c\}$ and $\{a, c, b\}$ equal sets?

8.11 Are $\{a, b, c\}$ and $\{a, c, a, b\}$ equal sets?

8.12 Prove there is only one empty set. Hint, show if E is an empty set then $E = \emptyset$.

8.13 Prove Theorem 8.2c, i.e., If $A = B$ and $B = C$ then $A = C$.

8.14 Complete the proof of Theorem 8.2d, by establishing the converse part. That is, show $(A \subseteq B) \wedge (B \subseteq A) \Longrightarrow A = B$.

8.15 Justify the assertions made in Example 8.3e.

8.16 Negate the "right side" of the definition of *proper* subset. That is, write an equivalent to $\sim (A \subset B)$.

8.17 Does $\emptyset$ have any proper subsets? Why?

8.18 Write out $\sim (\forall x)[x \in A \Longrightarrow x \in B]$ without using $\sim$. What symbol can you use with A and B to describe this?

8.19 Write out the negation of $A = B$ using the equivalence $(A = B) \equiv (A \subseteq B \wedge B \subseteq A)$. That is, $A \neq B$ iff ____.

8.20 Let $S = \{a, b, c\}$, $T = \{b, c, d, e\}$, and let $U = \{a, b, c, d, \ldots, z\}$ be the universe of discourse.

   a. Find A if $A = \{x \in U \mid x \in S \wedge x \in T\}$.

   b. Find B if $B = \{x \in U \mid x \in S \vee x \in T\}$.

   c. Find C if $C = \{x \in U \mid x \in T \wedge x \notin S\}$.

   d. Find D if $D = \{x \in U \mid x \in S \Longrightarrow x \in T\}$.

8.21 Complete the following equivalences by stating conditions on the elements a, b, and c for which the condition is true. For example,

$$\big[\{b, c\} \subseteq \{a\}\big] \text{ is true if } \underline{\big[(b = a) \wedge (c = a)\big]}.$$

   a. $\big[\{a\} = \{b, c\}\big] \equiv$ _____

   b. $\big[a \in \{b, c\}\big] \equiv$ _____

   c. $\big[\{a\} \subseteq \{b, c\}\big] \equiv$ _____

   d. $\big[\{a\} \in \{b, c\}\big] \equiv$ _____

   e. $\big[\{\{a\}\} = \{b, c\}\big] \equiv$ _____

   f. $\big[\{a\} \supseteq \{b, c\}\big] \equiv$ _____

   g. $\big[\{b, c\} \in \{a\}\big] \equiv$ _____

8.22 Why are the given sets in Example 8.3g equal? More specifically, why are the two sets
$\{x \mid x \text{ is a real number and } (x - 1)(x - 1) = (x - 1)(2x - 3)\}$ and $\{1, 2\}$ equal?

8.23 Prove $\{a, a\} = \{a\}$.

8.24 If $S = \{1, 2, 3, 4\}$ then find $\mathcal{P}(S)$.

8.25 If $n \in N$ or $n = 0$ and if S has n elements, how many elements does $\mathcal{P}(S)$ have? Verify by proof.

CHAPTER 9

SET UNION, INTERSECTION, AND COMPLEMENT

Introduction:

Our study of sets thus far has been somewhat limited in that it chiefly has been concerned with given sets and how they may or may not relate to one another with respect to the relations $\subseteq$, $\in$, and $=$. In this section the study will be expanded to explore the manner in which new sets can be formed from given sets. This will be done using the notions of union of sets, intersection of sets, and their complementation, and it will focus on related properties.

Before getting to the details of unions, intersections, and complements, a few suggestions are made to help you with your theorem proving. While the hints are directed more toward proofs of set theoretic concepts, there are general principles which are applicable in other areas of mathematics as well. In addition, comments are provided which address common areas of difficulty in doing and writing arguments.

A number of the theorems that are considered in the text as well as in the exercises may be so simple that they are considered to be obvious and not requiring a proof. But learning to apply the ideas presented here on simpler things builds skills that help with certain important manipulations. It may not be as likely that these skills would be learned by working on more difficult, less obvious theorems.

Throughout this text a variety of illustrative techniques are used for proving a theorem. The proof that is selected for a given theorem is determined not only by what is appropriate, but also to illustrate a particular technique. As a consequence, there are direct, indirect, and conditional proofs, as well as a variety of ideas, within each of these categories.

In the proofs of some of the following theorems, a bit more detail is given to illustrate what care must be taken, at least mentally if not on the paper. Many errors arise simply because of the sloppy assumption that the step appears reasonable and so there must be a law which supports it. Certainly this is a faulty reasoning procedure, because some of what appears to be plausible may in fact be false. Care should be exercised to be sure that every step is legitimate and that nothing is overlooked. After adequate practice, steps may be combined or eliminated. Example proofs, in which the combining of steps occurs to varying degree, are also given later to illustrate how this may be done. It may also be beneficial to consult Chapters 5 and 6.

Union, Intersection, and Relative Complement:

Let us now turn our attention to the main thrust of this section: the study of the concepts mentioned above of *unions, intersections*, and *complementation* of sets.

In the remainder of this section, it is assumed, unless otherwise noted, that there is a universe of discourse. It will be denoted by U and it is further assumed that U contains as subsets, each set in a given discussion. In other words, if a theorem asserts something regarding sets A, B, and C, it is assumed that there is a set U such that $A \subseteq U$, $B \subseteq U$, and $C \subseteq U$. Thus in particular, $\forall x[x \in A \implies x \in U]$.

DEFINITION: Let A and B be sets and let U a universe of discourse.

- The <u>union</u> of A and B, denoted by $A \cup B$, is given by

$$A \cup B = \{x \mid x \in A \vee x \in B\}.$$

- The <u>intersection</u> of A and B, denoted by $A \cap B$, is given by

$$A \cap B = \{x \mid x \in A \wedge x \in B\}.$$

- The <u>relative complement of B in A</u>, denoted by $A - B$ is

$$A - B = \{x \mid x \in A \wedge x \notin B\}.$$

It should be noted that the predicate part (i.e., the "membership requirements") of the set builder notation descriptions of $A \cup B$ and $A \cap B$ use disjunction $\vee$, and conjunction $\wedge$, respectively. Confusion on this point should be minimal since there is a strong similarity between the set theoretic symbols $\cup$ and $\cap$ and their logical counterparts $\vee$ and $\wedge$. However, in spite of this, difficulties may arise.

A common mistake is to say $A \cup B$ is the set of all elements that are in A and B, but such a statement is clearly FALSE. $A \cup B$ is the set of all elements x that are in either A <u>or</u> B. It is $A \cap B$ that is the set of all elements that are in A and B.

To aid in verbally communicating these sets, let $A \cup B$ and $A \cap B$ be read (or said) as <u>A union B</u> and <u>A intersection B</u>, respectively. The relative complement of B in A may be read as <u>A minus B</u>.

EXAMPLE 9.1: Let $A = \{a, b, c\}$ and $B = \{b, c, d, e\}$ then:

   a. $A \cup B = \{x \mid x \in A \vee x \in B\} = \{a, b, c, d, e\}$

   b. $A \cap B = \{x \mid x \in A \wedge x \in B\} = \{b, c\}$

c. $A - B = \{x \mid x \in A \wedge x \notin B\} = \{a\}$

d. $B - A = \{x \mid x \in B \wedge x \notin A\} = \{d, e\}$

In each of these examples as well as in any use of set builder notation, it is extremely important to understand and employ the full capabilities of the notation. Recall that if $P(x)$ is a predicate, then $\{x \mid x \in U \wedge P(x)\}$ is the set of ALL x in U such that $P(x)$ is true. It means ALL. If an element $y \in U$ makes P true [i.e., $P(y)$ is true] then y IS IN the set, and if $y \notin U$ or y does NOT make P true, [i.e., $P(y)$ is false] then y is NOT IN the set. In other words

$$y \in \{x \mid x \in U \wedge P(x)\} \text{ if and only if } y \in U \wedge P(y) \text{ is true.}$$

This equivalence was employed in Example 9.1a above. The $P(x)$ in Example 9.1a is $x \in A \vee x \in B$. Since $a \in A$ then $a \in A \vee a \in B$ is true, by the law of addition. Thus, $P(a)$ is true, hence $a \in A \cup B$. Likewise b, c, d, e are also elements of $A \cup B$. However, since $h \in A \vee h \in B$ is false, $P(h)$ is false, so $h \notin A \cup B$. Likewise, $f, b, 1, 2, 17$, and π are NOT in $A \cup B$.

The $P(x)$ in Part b of Example 9.1 above is $x \in A \wedge x \in B$. Since $a \in A \wedge a \in B$ is false, $P(a)$ is false. Thus $a \notin A \cap B$. Similarly, $P(d)$, $P(e)$, $P(f)$, $P(17)$, and $P(\pi)$ are all false, so $d, e, f, 17, \pi$ are not in $A \cap B$. Now $P(b)$ is true, so $b \in A \cap B$. Similarly $c \in A \cap B$. Since there are no other elements in $A \cap B$, $A \cap B = \{b, c\}$. You may want to do this kind of reasoning process in the other parts of Example 9.1.

In a similar use of set builder notation, since $\emptyset = \{x \mid x \neq x\}$, $x \in \emptyset$ is a contradiction for all x. On the other hand, since U contains all elements needed in a given discussion, $x \in U$ is a tautology for all x.

As is customary after the introduction of some new concepts, properties of these concepts are sought and verified. Some of the obvious questions regarding union and intersection are, do the commutative, associative, and distributive laws hold for set connectives? How do sets connected by the *set connectives* $\cap$, $\cup$, and $-$ compare by the relation "is equal to" or the relation "is a subset of" to other sets such as the empty set or the universe of discourse? By exploring the similarity with logical connectives, conjectures about these relationships can be formulated. The results of these conjectures which can be answered in the affirmative are stated in the theorems below. Some proofs are included in detail and others are left to the exercises.

It is appropriate to ask exactly how union and intersection interrelate with other notations. In particular, what is the hierarchy of operations in an expression involving $\in$, $\cup$, $\cap$, or involving $\subseteq$ and $\cup$ and $\cap$? The answer is, "By standard usage $\cup$ and $\cap$ are executed before $\in$, $=$, $\subset$, and $\subseteq$." For example,

$x \in A \cup B$ means $x \in (A \cup B)$ and $C \subseteq A \cup B$ means $C \subseteq (A \cup B)$.

But, just as with $\wedge$ and $\vee$, we make no order preference for the execution of the symbols $\cap$ and $\cup$. In other words, parentheses <u>must</u> <u>be</u> <u>used</u> to specify the order of operation when these two set connectives are used in the same expression. The expression $A \cap B \cup C$ is ambiguous, just as the corresponding sentence with logical connectives.

In the following theorems, A, B, C are sets and U is the universe of discourse.

THEOREM 9.1:   Let A, B, and C be sets then

   a. $A \cup B = B \cup A$ b. $A \cap B = B \cap A$

The second part is left to the exercises. In order to show $A \cup B = B \cup A$ it is required that the statement $\forall x[x \in A \cup B \iff x \in B \cup A]$ be established. To achieve that, we will apply Rule U.G. to the statement $y \in A \cup B \iff y \in B \cup A$. So the objective is to establish $y \in A \cup B \iff y \in B \cup A$. Also note in the proof below how we use the set connectives and logical connectives in proper syntax.

PROOF: Let $y \in U$. If $y \in A \cup B$ then $y \in A \vee y \in B$. By the commutative property of $\vee$, $y \in B \vee y \in A$. Hence, $y \in B \cup A$. Thus we have shown $y \in A \cup B \implies y \in B \cup A$. (Have we finished? NO. In fact, we are not quite halfway.) Similarly, if $y \in B \cup A$, then $y \in B \vee y \in A$. By commutativity of $\vee$, $y \in A \vee y \in B$, thus $y \in A \cup B$. Therefore $y \in B \cup A \implies y \in A \cup B$. By combining the two implications established above and by using Rule 5, we have $y \in A \cup B \iff y \in B \cup A$. Then by Rule U.G. $\forall x[x \in A \cup B \iff x \in B \cup A]$. Finally, by the definition of set equality, $A \cup B = B \cup A$ as desired.

THEOREM 9.2:   Let A, B, and C be sets. Then

   a. $A \cup (B \cup C) = (A \cup B) \cup C$ b. $A \cap (B \cap C) = (A \cap B) \cap C$

The first part is left to the exercises. We will show the second part, but we will use a more refined method to establish it than we used in the proof of Theorem 9.1. This is quite similar to the "algebra of statements" emphasized in Theorems 2.1, 2.2, and 2.5. The type of proof used here illustrates another technique that may be used. This technique is easy to misuse so be careful. In order to prove $A \cap (B \cap C) = (A \cap B) \cap C$ we will have to show $\forall x[x \in A \cap (B \cap C) \Longleftrightarrow x \in (A \cap B) \cap C]$. Before doing that we will show $y \in A \cap (B \cap C) \Longleftrightarrow y \in (A \cap B) \cap C$.

PROOF: Let A, B, and C be sets and let $y \in U$.

$$
\begin{aligned}
y \in A \cap (B \cap C) &\Longleftrightarrow y \in A \land y \in (B \cap C) && \text{Definition of } \cap \\
&\Longleftrightarrow y \in A \land (y \in B \land y \in C) && \text{Definition of } \cap \\
&\Longleftrightarrow (y \in A \land y \in B) \land y \in C && \text{Associative for } \land \\
&\Longleftrightarrow (y \in A \cap B) \land y \in C && \text{Definition of } \cap \\
&\Longleftrightarrow y \in (A \cap B) \cap C. && \text{Definition of } \cap
\end{aligned}
$$

Applying transitivity of $\Longleftrightarrow$ to the sequence of steps above, we get $y \in A \cap (B \cap C) \Longleftrightarrow y \in (A \cap B) \cap C$. Since y was arbitrarily selected, we know $\forall x[x \in (A \cap B) \cap C \Longleftrightarrow x \in A \cap (B \cap C)]$. Finally, by the definition of set equality, $A \cap (B \cap C) = (A \cap B) \cap C$.

When using the method employed above of writing one equivalence after another, care must be taken to *be sure that the member to the left of each $\Longleftrightarrow$ is indeed logically equivalent to the member on the right.* The work proceeds from left to right and thus the reasoning may be only from left to right, even though the symbol $\Longleftrightarrow$ means that the reasoning must go both ways.

THEOREM 9.3: Let A, B, and C be sets.

    a. $A \cup (B \cap C) = (A \cup B) \cap (A \cup C)$

    b. $A \cap (B \cup C) = (A \cap B) \cup (A \cap C)$

We will show that union distributes over intersection. Part b is left to the exercises. We follow the style of argument used in Theorem 9.2, except some additional consolidation shortens the proof a bit.

PROOF: Let A, B, and C be sets and let $y \in U$.

$$
\begin{aligned}
y \in A \cup (B \cap C) &\Longleftrightarrow y \in A \lor y \in (B \cap C) && \text{Definition } \cup \\
&\Longleftrightarrow y \in A \lor (y \in B \land y \in C) && \text{Definition } \cap \\
&\Longleftrightarrow (y \in A \lor y \in B) \land (y \in A \lor y \in C) && \text{Rule 12b} \\
&\Longleftrightarrow (y \in A \cup B) \land (y \in A \cup C) && \text{Definition } \cup \\
&\Longleftrightarrow y \in (A \cup B) \cap (A \cup C) && \text{Definition } \cap
\end{aligned}
$$

Thus $\forall x[x \in A \cup (B \cap C) \Longleftrightarrow x \in (A \cup B) \cap (A \cup C)]$. (Why? What step or steps did we skip to get the preceeding sentence?) Consequently, $A \cup (B \cap C) = (A \cap B) \cup (A \cap C)$, by definition of set equality.

THEOREM 9.4: Let A be any set, then

    a. $A \cup \emptyset = A$ b. $A \cap \emptyset = \emptyset$

The proof is an exercise.

THEOREM 9.5: If A is any set then

    a. $A \cup U = U$ b. $A \cap U = A$

Part b is left to the exercises. We do Part a by showing two containments: $A \cup U \subseteq U$ and $U \subseteq A \cup U$. Then we will employ Rule 24a, to get the desired result. Keep in mind that $A \subseteq U$ and $U \subseteq U$, so $y \in A \implies y \in U$ (Why?) and $y \in U \implies y \in U$ (Why?).

PROOF: Let A, B, and C be sets and let $y \in U$.

$$y \in A \cup U \implies y \in A \lor y \in U \qquad \text{Definition of } \cup$$
$$\implies y \in U \lor y \in U \qquad \text{Rule 24a}$$
$$\implies y \in U. \qquad \text{Exercise 2.8c}$$

Thus $A \cup U \subseteq U$. (Why?) Conversely, by the law of addition, $y \in U \implies y \in A \lor y \in U$. Hence, by Rule U.G. and the definition of $\subseteq$, $U \subseteq A \cup U$. Thus $A \cup U = U$.

THEOREM 9.6: If A is any set then

    a. $A \cup (U - A) = U$ b. $A \cap (U - A) = \emptyset$

Part a is an exercise and Part b is done by contradiction. This proof differs in style from the ones given above. It is in a more narrative informal form.

PROOF: PART b: Let A be a set and suppose on the contrary that the set $A \cap (U - A)$ is not empty. Then, from the definition of empty set, $\exists x [x \in A \cap (U - A)]$. Let that x be y. Then y satisfies $y \in A \cap (U - A)$. Then $y \in A \land y \in (U - A)$. But this says $y \in A \land (y \in U \land y \notin A)$. By simplification, $y \in A \land y \notin A$. This is a contradiction, so $A \cap (U - A) \neq \emptyset$ is false. Therefore $A \cap (U - A) = \emptyset$.

The indirect technique employed here for showing $A \cap (U - A) = \emptyset$ by supposing the set to be nonempty, is frequently a successful way to show a set to be empty.

THEOREM 9.7: If A is any set then

    a. $A \cup A = A$ b. $A \cap A = A$

The proof of this is left to the exercises. There are names attached to these properties. We say any set A is *idempotent* with respect to $\cup$ and to $\cap$ since $A \cup A = A$ and $A \cap A = A$, respectively.

THEOREM 9.8: If A and B are any sets then

    a. $A \cup (B - A) = A \cup B$ b. $A \cap (B - A) = \emptyset$

Part b is an exercise. Part a uses Rule 27 as a crucial step.

PROOF: Let A and B be sets and $y \in U$. Then

$$
\begin{aligned}
y \in A \cup (B - A) &\Longleftrightarrow y \in A \lor y \in (B - A) &&\text{Definition of } \cup \\
&\Longleftrightarrow y \in A \lor (y \in B \land y \notin A) &&\text{Why?} \\
&\Longleftrightarrow y \in A \lor (y \in B \land \sim (y \in A)) &&\text{Why?} \\
&\Longleftrightarrow y \in A \lor y \in B &&\text{Rule 27} \\
&\Longleftrightarrow y \in A \cup B. &&\text{Why?}
\end{aligned}
$$

Then by Rule U.G. and the definition of set equality, $A \cup (B-A) = A \cup B$.

THEOREM 9.9: If A and B are any sets and $A \subseteq B$ then $B-(B-A) = A$.

To do this, suppose A and B are sets and $A \subseteq B$. (Do we worry about what happens if $A \not\subseteq B$? Why?) To establish the indicated equality, show $B - (B - A) \subseteq A$ and $A \subseteq B - (B - A)$.

PROOF: Suppose A and B are sets and $A \subseteq B$ and let $y \in U$. If $y \in B - (B - A)$ then $y \in B \land y \notin (B - A)$. Thus

$$
(y \in B) \land \left(\sim (y \in (B - A)) \right).
$$

Therefore

$$
(y \in B) \land \left(\sim (y \in B \land y \notin A) \right)
$$

by the definition of $B - A$. Therefore, by DeMorgan's law and double negation, $(y \in B) \land (y \notin B \lor y \in A)$. Therefore

$$
(y \in B \land y \notin B) \lor (y \in B \land y \in A).
$$

Why? What principle? It follows that $y \in B \land y \in A$. (Why?) Then $y \in A$ (Why?). Consequently $B - (B - A) \subseteq A$.

Conversely, if $y \in A$ then since $A \subseteq B$, $y \in B$. Since $y \in A$ then, by the law of addition, $y \notin B \lor y \in A$. By the law of double negation, $y \notin B \lor (\sim (\sim y \in A))$. Then, from the definition of $\in$ (really $\notin$), $\sim (y \in B) \lor (\sim (y \notin A))$. Consequently, by DeMorgan's law $\sim (y \in$

$B \wedge y \notin A$)). Since ($y \in B$) and since $\sim (y \in B \wedge y \notin A)$, then ($y \in B) \wedge \sim (y \in B \wedge y \notin A)$. Thus ($y \in B) \wedge (\sim (y \in (B-A)))$. Therefore, by the definition of set difference, $y \in B-(B-A)$. Recall that we assumed that $y \in A$ above, and we concluded, in that case, $y \in B - (B - A)$. By Rule U.G. and the definition of $\subseteq$, $A \subseteq B - (B - A)$. Consequently, we have shown that $A \subseteq B \Longrightarrow [(B - (B - A)) \subseteq A$ and $A \subseteq B - (B - A)]$. This means that $A \subseteq B \Longrightarrow B - (B - A) = A$. (Why?)

Further Hints Regarding the Writing of Proofs:

After working through the argument, checking it carefully for errors, and making any necessary changes that improve its readability or accuracy, try reading it over again. Read what you have written from the reader's perspective, devoid of unwritten assumptions or insights. Is it clear? Does it convey to the reader in an understandable and logical fashion what has been done? Reading this for the first time would what is being done be understood? If the answer is "no" to any of these questions, then a reworking of the proof may be in order.

In the case that the proof does not really satisfy the questions above, the fault may be with the way the ideas are presented rather than the underlying logic. As we have said before, a feature that is very beneficial to the reader of a mathematical work is the presence of symbols or explanatory wording describing the relationship of one mathematical statement to another. The reader needs to know whence statements arise. Don't just write it down as an unsubstantiated assertion. If a statement follows logically from the previous one, say, *thus, therefore,* or *consequently,* to name a few possible words. If writing a statement Q that is a logical consequence of a statement R from several sentences above, say, "*Since R is true, then Q.*" This is helpful to the reader in that he or she knows how the statement Q fits in relation to other statements in the proof. Without the use of "Since ..." the reader is tempted to think that Q follows from the immediately preceding statement. Attention to such details can greatly improve the readability of a proof.

Another common mistake that can be alleviated with a little effort is the mixing or using of logical and set theoretic symbols in the wrong syntax. *Logical connectives* like $\wedge$, $\vee$, $\sim$, $\Longrightarrow$, $\Longleftrightarrow$, or $\equiv$ separate (or go between) statements or predicates while *set connectives* (or *set theoretic connectives*) $\cup$, $\cap$, $\subseteq$, $\in$, $-$, $\subset$, and $=$ all separate (or go between) sets. It may help to remember that A SET IS NOT A STATEMENT AND A STATEMENT IS NOT A SET.

Two examples of proper uses of statements, sets and logical and set theoretic connectives are

$$(A \subseteq B) \iff (\forall x \in U)\big[x \in A \implies x \in B\big].$$

$$\Big[\{x \in U \mid P(x)\} = \{x \in U \mid Q(x)\}\Big] \equiv \Big[\forall x[P(x) \iff Q(x)]\Big].$$

In these examples, note that the members of the connectives $\wedge$, $\vee$, $\iff$, $\implies$, and $\equiv$, etc. are statements (or predicates) while the members of $\subseteq$, $=$, and $\in$ are sets. Placing set connectives between statements or predicates or placing logical connectives between sets or elements is not only improper syntax, it is difficult or impossible to interpret the resulting sentence.

A test you may use to uncover some poor notation or form, such as abuses committed in parts of Exercise 9.1, is to read the work for grammatical construction; putting in the *exact* English equivalent for each mathematical symbol. For example, you would read $A \subseteq B$ as, "A is a subset of B" and $(\forall x \in U)[x \in A \implies x \in B]$ as, "for all x in U, if x is an element of A then x is an element of B." If such a translation does not yield reasonable grammar or make sense, there is a strong liklihood that nonsense has been written.

There are ways that are sometimes used to relax the notation and improve its readability. The statement $\exists x[P(x)]$ is usually read as, "There exists an x such that $P(x)$." A symbol that sometimes replaces the words <u>such that</u> in an existentially quantified statement is $\ni$. The new symbolism is $\exists x \ni P(x)$ and is read as, "There exists an x such that $P(x)$."

Likewise $\forall x[x \in A \implies P(x)]$ which is read as, "For all x, if x is an element of A then $P(x)$ is true," may be changed to $(\forall x \in A)[P(x)]$. This is read as, "For all x in A, $P(x)$ is true."

These alternatives ways to write and verbalize quantified statements will be used more extensively later in this text.

Finally, don't be too discouraged if your proof does not roll out from beginning to end with professional quality. It is unrealistic to expect to be able to immediately write such proofs, since mathematicians frequently write and rewrite and polish their first drafts or initial sketches of proofs. On the other hand you should take the time to assimilate these ideas so that your writing of mathematics improves.

Exercises:

9.1 Answer the following questions true if the set theoretic and logical connectives and quantifiers are used in the proper syntax. Answer false otherwise. Do not answer based on the truth of the sentence, but on whether it is put together in such a way that it makes sense, is not ambiguous, and would have some truth value. Let $\mathcal{L}$ be the set of all lines in a given plane and $\mathcal{P}$ be the set of all points in that plane. Let $L_1\|L_2$ mean L_1 is parallel to L_2.

a. Let $L_1, L_2 \in \mathcal{L}$, then $L_1\|L_2 \iff [L_1 \cap L_2 = \emptyset]$.

b. Let $L_1, L_2 \in \mathcal{L}$. $\sim[L_1\|L_2] \iff \exists x[x \in \mathcal{P} \wedge x \in L_1 \wedge x \in L_2]$.

c. Let $L_1, L_2 \in \mathcal{L}$. $[L_1\|L_2] \iff \sim[x \in L_1 \wedge x \in L_2]$.

d. $\forall L_1 \in \mathcal{L}[\forall x \in \mathcal{P}(x \notin L_1 \implies \exists! L_2 \ni (x \in L_2 \wedge (L_1 \cap L_2 = \emptyset)))]$.

e. Let $P(x)$ and $Q(x)$ be predicates. Then

$$\{x \mid P(x) \wedge Q(x)\} \subseteq \{x \mid P(x) \vee Q(x)\}.$$

f. Let $P(x)$ be a predicate and S a set. Then $\{P(x) \mid x \in S\} = P(x)$.

g. Let $P(x)$ be a predicate and S a set. Then $S \implies \forall x P(x)$.

h. Let S and T be sets then $\{x \mid x \in S \wedge x \in T\} = S \wedge T$.

i. Let S and T be sets then $\exists x \ni (x \in S \vee x \in T) \iff S \cup T$.

j. Let S and T be sets then $y \notin (S \cap T) \implies \sim[S \cap T]$.

k. Let S be a set. $(S \neq \emptyset) \implies \exists x \ni [x \in S]$.

l. Let S be a set. $(S = \emptyset) \implies \sim x \in S$.

m. Let x be an $\in S$ then $S \neq \emptyset$.

n. $x \in A \cup B \cap C \implies x \in A \vee x \in B \cap C$.

o. Let S and T be sets then $S \cap T \subseteq \{x \mid x \in S \implies x \in T\}$. Are these two sets actually equal?

9.2   a. Let $A = \{1, 2, 3\}$, $B = \{1, 2, 3, 4, 5\}$, and $C = \{2, 3, 4\}$. Compute the following sets: $A \cup B$, $A \cap B$, $A - B$, $B - A$, $C - A$, and $\{x \mid x \in B \wedge (x \in C \implies x \in A)\}$.

     b. Using $A = \{1, 2, 3\}$, $B = \{2, 3, 4, 5\}$, and $U = \{1, 2, 3, \cdots, 10\}$ compute $U - (A \cup B)$, $(U - A) \cap (U - B)$.

     c. Using A and B as previously mentioned, compare the two sets you determined in Part b.

9.3   Prove Theorem 9.1b, i.e., $A \cap B = B \cap A$ for all sets A and B.

9.4   Prove Theorem 9.2a, i.e., prove $A \cup (B \cup C) = (A \cup B) \cup C$ for all sets A, B, and C.

9.5   Prove Theorem 9.3b, i.e., prove $A \cap (B \cup C) = (A \cap B) \cup (A \cap C)$.

9.6   a. Prove Theorem 9.4a. That is, prove $A \cup \emptyset = A$ for all sets A.

  b. Prove Theorem 9.4b, i.e., prove $A \cap \emptyset = \emptyset$ for all sets A.

9.7 Prove Theorem 9.5b, i.e., prove $A \cap U = A$ for all sets A.

9.8 Prove Theorem 9.6a, i.e., prove $A \cup (U - A) = U$ for all sets A.

9.9 a. Prove Theorem 9.7a, i.e., prove $A \cup A = A$ for all sets A.

  b. Prove Theorem 9.7b, i.e., prove $A \cap A = A$ for all sets A.

9.10 Prove Theorem 9.8b, i.e., prove $A \cap (B - A) = \emptyset$ for all sets A and B.

9.11 a. Prove $A \subseteq A \cup B$ for all sets A and B. See Theorem 9.10a.

  b. Prove $A \cap B \subseteq A$ for all sets A and B. See Theorem 9.10b.

Now we continue this study by considering some additional theorems. In this part, we introduce a new technique that works in some arguments very nicely, but does not work in all. Its convenient use is limited to certain equalities and containments. On other theorems the methods of an earlier section may be applied more easily. The new technique requires the same degree of caution mentioned at the end of the proof of Theorem 9.2 in order to assure valid results. In addition, it employs the ideas introduced in the *set builder* notation section, Chapter 8. For example, if $(\forall x)[P(x) \iff Q(x)]$ then $\{x \mid x \in U \land P(x)\} = \{x \mid x \in U \land Q(x)\}$. To distinguish it from the methods of the earlier part of Chapter 9, we will call this technique, *proof by set builder notation*, while we will call the former methods, *proof by logical identities* or *by algebra of statements*.

As before, suppose A, B, and C are sets and U is a suitable universe of discourse. To introduce the technique, let us return to a former theorem, Theorem 9.3a.

Note the close parallel to the earlier proof of this theorem. The reasons are the same here as in the earlier proof.

ALTERNATE PROOF: Let A, B, and C be sets. Then

$$
\begin{aligned}
A \cup (B \cap C) &= \{x \mid x \in A \lor x \in B \cap C\} \\
&= \{x \mid x \in A \lor (x \in B \land x \subset C)\} \\
&= \{x \mid (x \in A \lor x \in B) \land (x \in A \lor x \in C)\} \\
&= \{x \mid x \in A \cup B \land x \in A \cup C\} \\
&= (A \cup B) \cap (A \cup C).
\end{aligned}
$$

Next let us resume our development of theorems about sets, which spell out relationships between set theoretic connectives. Some of these will be proven in a manner similar to the proof above, while others will be left as exercises to practice theorem proving, using all techniques developed thus far.

THEOREM 9.10:   Let A and B be sets, then

   a. $A \subseteq A \cup B$ b. $A \cap B \subseteq A$

PROOF: We do Part a. Let A and B be sets and suppose $y \in U$. By the law of addition $y \in A \Longrightarrow y \in A \vee y \in B$ so

$$A = \{x \mid x \in A\} \subseteq \{x \mid x \in A \vee x \in B\} = A \cup B.$$

Part b is an exercise.

THEOREM 9.11:   If A and B are sets and $A \subseteq B$ then $B = A \cup B$.

PROOF: Suppose A and B are sets and $A \subseteq B$ then it is true that $\forall x (x \in A \Longrightarrow x \in B)$. Therefore

$$
\begin{aligned}
A \cup B &= \{x \mid x \in A \vee x \in B\} &&\text{Definition of } \cup \\
&\subseteq \{x \mid x \in B \vee x \in B\} &&\text{Rule 24a} \\
&= \{x \mid x \in B\} &&\text{Exercise 2.8} \\
&= B. &&\text{Notation}
\end{aligned}
$$

Hence $A \cup B \subseteq B$. (Why is the second step $\subseteq$ and not $=$?)

   On the other hand

$$
\begin{aligned}
B &= \{x \mid x \in B\} &&\text{Notation} \\
&\subseteq \{x \mid x \in A \vee x \in B\} &&\text{Law of addition} \\
&= A \cup B. &&\text{Definition of } \cup
\end{aligned}
$$

Thus $B \subseteq A \cup B$. Consequently, by Theorem 8.2d, $B = A \cup B$.

THEOREM 9.12:   If A and B are sets and $A \subseteq B$ then $A \cap B = A$.

   The proof of this theorem as well as the next three theorems, are left to the exercises. In those exercises any method discussed may be used.

THEOREM 9.13:   If A, B, and C are sets and $A \subseteq C$ and $B \subseteq C$ then $A \cup B \subseteq C$.

THEOREM 9.14:   If A, B, and C are sets and $A \subseteq B$ and $A \subseteq C$ then $A \subseteq B \cap C$.

THEOREM 9.15:   If A, B, and C are sets and $A \subseteq C$, $B \subseteq C$, $A \cup B = C$, and $A \cap B = \emptyset$ then $A = C - B$.

Now the new concept of *set complementation* is introduced. We know what relative complement means: $B - A$ is the relative complement of A in B, i.e., $\{x \mid x \in B \wedge x \notin A\}$. This idea is related to that concept.

DEFINITION (COMPLEMENTS): If A is a set and U a universe of discourse then the <u>complement</u> of A, denoted by A', is given by $A' = U - A$. An alternate notation for the complement of A is sometimes denoted by $\overline{A}$. Both concepts represent the set $\{x \in U \mid x \notin A\}$.

THEOREM 9.16: Let A and B be sets then

   a. $(A \cup B)' = A' \cap B'$ b. $(A \cap B)' = A' \cup B'$

PROOF: Part b is an exercise. To show Part a, suppose $y \in U$ then

$$
\begin{aligned}
y \in (A \cup B)' &\Longleftrightarrow y \in U - (A \cup B) \\
&\Longleftrightarrow y \in U \wedge y \notin (A \cup B) \\
&\Longleftrightarrow y \in U \wedge \sim (y \in (A \cup B)) \\
&\Longleftrightarrow y \in U \wedge (\sim (y \in A \vee y \in B)) \\
&\Longleftrightarrow y \in U \wedge (y \notin A \wedge y \notin B) \\
&\Longleftrightarrow y \in U \wedge y \in U \wedge y \notin A \wedge y \notin B \\
&\Longleftrightarrow y \in U \wedge y \notin A \wedge y \in U \wedge y \notin B \\
&\Longleftrightarrow y \in (U - A) \wedge y \in (U - B) \\
&\Longleftrightarrow y \in A' \wedge y \in B' \\
&\Longleftrightarrow y \in A' \cap B'
\end{aligned}
$$

Thus $\forall x [x \in (A \cup B)' \Longleftrightarrow x \in A' \cap B']$. Consequently, $(A \cup B)' = A' \cap B'$.

Next we redo the proof using set builder notation. Note the very close parallel to the proof previously mentioned. At any rate, the earlier admonition that care be exercised to be sure that each step is actually an equality, rather than a containment, is still in order. It is easy to repetitiously, but erroneously, put in an equality mark or a biconditional when something else is correct. Test your work by going through the string of $=$'s or $\Longleftrightarrow$'s in reverse to see if they follow logically in that direction.

ALTERNATE PROOF:

$$
\begin{array}{lll}
(A \cup B)' = \{x \mid x \in U - (A \cup B)\} & \text{Def. Complement} \\
\quad = \{x \mid x \in U \wedge x \notin (A \cup B)\} & \text{Rel. Complement} \\
\quad = \{x \mid x \in U \wedge \sim (x \in A \cup B)\} & \text{Definition } \notin \\
\quad = \{x \mid x \in U \wedge \sim (x \in A \vee x \in B)\} & \text{Why?} \\
\quad = \{x \mid x \in U \wedge (x \notin A \wedge x \notin B)\} & \text{Why}^2? \\
\quad = \{x \mid x \in U \wedge x \in U \wedge x \notin A \wedge x \notin B\} & \text{Exercise 2.8} \\
\quad = \{x \mid x \in U \wedge x \notin A \wedge x \in U \wedge x \notin B\} & \text{Rule 10a}
\end{array}
$$

$$= \{x \mid x \in (U - A) \wedge x \in (U - B)\} \qquad \text{Def. Complement}$$
$$= \{x \mid x \in A' \wedge x \in B'\} \qquad\qquad\quad \text{Why?}$$
$$= A' \cap B' \qquad\qquad\qquad\qquad\qquad\quad \text{Why?}$$

THEOREM 9.17: Let A be any set then

    a. $A - \emptyset = A$

    b. $\emptyset - A = \emptyset$

    c. $A - A = \emptyset$

The proofs of these assertions are left as exercises.

THEOREM 9.18: a. $\emptyset' = U$ b. $U' = \emptyset$

PROOF: We establish Part a, deferring Part b to the exercises. By the definition of complement, $\emptyset' = U - \emptyset$. But, by Theorem 9.17, $U - \emptyset = U$, so $\emptyset' = U$ by Theorem 8.2c.

THEOREM 9.19: Let A be a set.

    a. $A \cup A' = U$ b. $A \cap A' = \emptyset$.

PROOF: We establish Part b leaving Part a as an exercise.

By use of Theorem 9.6, $A \cap A' = A \cap (U - A) = \emptyset$.

The proofs to Theorems 9.18 and 9.19 illustrate that it is not necessary to go all the way back to first considerations to prove a theorem. Frequently, after several basic theorems have been established, these theorems can be put to use in proving certain other results. Take advantage of this situation whenever it occurs. Not having to "reinvent the wheel," saves time and reinforces understanding of the interrelationships that exist between the various theorems in a body of mathematical information.

A property illustrated in Theorem 9.19b is the property of *disjointness*. Sets A and B are <u>disjoint</u> if and only if $A \cap B = \emptyset$. There are exercises on this concept (see Exercise 9.21).

THEOREM 9.20: If A is any set then $(A')' = A$.

The proof is left to the exercises.

Exercises:

Prove the following by appropriate methods.

    9.12 a. Prove Theorem 9.11. That is, prove: If $A \subseteq B$ then $B = A \cup B$.

            b. Prove Theorem 9.12. That is, prove: If $A \subseteq B$ then $A \cap B = A$.

9.13 Prove Theorem 9.13. That is, prove: If $A \subseteq C$ and $B \subseteq C$ then $A \cup B \subseteq C$.

9.14 Prove Theorem 9.14. That is, prove: If $A \subseteq B$ and $A \subseteq C$ then $A \subseteq B \cap C$.

9.15 Prove Theorem 9.15. That is, prove: If $A \subseteq C$, $B \subseteq C$, $A \cup B = C$ and $A \cap B = \emptyset$ then $A = C - B$.

9.16 Prove Theorem 9.16b. That is, prove: $(A \cap B)' = A' \cup B'$. To what logical principle is this theorem analogous?

9.17 a. Prove Theorem 9.17a. That is, prove: $A - \emptyset = A$.

     b. Prove Theorem 9.17b. That is, prove: $\emptyset - A = \emptyset$.

     c. Prove Theorem 9.17c. That is, prove: $A - A = \emptyset$.

9.18 Prove Theorem 9.18b. That is, prove: $U' = \emptyset$.

9.19 Prove Theorem 9.19a. That is, prove: $A \cup A' = U$.

9.20 Prove Theorem 9.20. That is, prove: $(A')' = A$. To what rule of logic does this correspond?

9.21 Prove: If $A \subseteq C$, $B \subseteq D$ and C and D are disjoint then A and B are disjoint.

9.22 What is wrong with these actual "proofs" submitted by students of the proposition a. $A \subseteq U$ and the proposition b. $A \cup A = A$?

     a. "Proof:" $A \subseteq U$ if and only if $\forall x [x \in A \implies x \in U]$. By Rule U.S., $y \in A \implies y \in U$. Thus by Rule U.G. $A \subseteq U$.

     b. "Proof:" Show $\forall x (x \in A \cup A \iff x \in A)$. By Rule U.S. $y \in A \cup A$ iff $y \in A$. But by Exercise 2.20, $y \in A \vee y \in A \iff y \in A$, so $A \cup A = A$.

9.23 Prove: If A, B, and C are sets then $A - (B \cap C) = (A - B) \cup (A - C)$. To what rule of logic is this most analogous?

9.24 Prove: If A and B are sets then $A \subseteq B \implies B' \subseteq A'$. To what rule of logic is this analogous?

9.25 Prove the two *absorption* laws.

     a. $\forall A, B[A \cup (A \cap B) = A]$.

     b. $\forall A, B[A \cap (A \cup B) = A]$.

9.26 Prove: If A and B are sets and $C = B - A$ then $A \cup B = A \cup C$ and $A \cap C = \emptyset$.

9.27 What is wrong with the following "proofs" of Exercise 9.21? That is, of, "If $A \subseteq C$ and $B \subseteq D$ and C and D are disjoint then A and B are disjoint."

a. "Proof:" Assume $\forall x \{x \mid x \in A \longrightarrow x \in C\}$, $\forall x \{x \mid x \in B \longrightarrow x \in D\}$, $C \cup D = \emptyset$, etc. The "proof" goes on. Assess the part given so far. Then go to Parts b, c, and d for other attempts.

b. "Proof:"

$$
\begin{array}{ll}
A \subseteq C \equiv x \in A \longrightarrow c \in C & \text{Definition of } \subseteq \\
B \subseteq D \equiv x \in B \longrightarrow x \in D & \text{Definition of } \subseteq \\
C \cap D = \emptyset & \text{given} \\
A \cap B = C \cap D & \text{substitute 1,2} \\
 = \emptyset. & \text{substitute 3}
\end{array}
$$

c. "Proof:"

$$
\begin{aligned}
A \subseteq C &= \{x \mid x \in A \longrightarrow x \in C\} \\
B \subseteq D &= \{x \mid x \in B \longrightarrow x \in D\} \\
A \cap B &= \{x \mid x \in A \longrightarrow x \in C \wedge x \in B \longrightarrow x \in D\} \\
&\subseteq \{x \mid x \in C \wedge x \in D\} \\
&= C \cap D.
\end{aligned}
$$

But $C \cap D = \emptyset$ because they are disjoint so $A \cap B = \emptyset$ and they are therefore disjoint.

d. "Proof:" Let $y \in U$. If C and D are disjoint then $C \cap D = \emptyset$.

$$
\begin{array}{ll}
A \cap B = \{y \mid y \in A \wedge y \in B & \text{Definition of } \cap \\
 \subseteq \{y \mid y \in A \wedge y \in C \wedge y \in B \wedge y \in D\} & \text{Law of Addition} \\
 = \{y \mid y \in C \wedge y \in D\} & \text{Simplification} \\
 = C \cap D & \text{Notation}
\end{array}
$$

9.28   If A and B are any sets, is $B - (B - A) = A$? Explain.

9.29   Prove the empty set is disjoint from any set A.

CHAPTER 10

GENERALIZED UNION AND INTERSECTION

In mathematics there is a need to work with unions and intersections of more than just two sets. To handle these situations there are generalized unions and intersections. Before saying what these generalized concepts are, some terminology and notation regarding sets of sets is considered.

A set of sets is frequently called a family or collection of sets. For example, suppose we wish to discuss a family of sets consisting of A_1, A_2, etc. up to A_n, that is the family $\{A_1, A_2, A_3, \ldots, A_n\}$. This family could be denoted by

$$S_1 = \left\{ A_i \ \middle| \ i \in N \text{ and } 1 \leq i \leq n \right\}$$

that is $S_1 = \{A_1, A_2, A_3, \ldots, A_n\}$. With this idea for describing a finite set of sets, it is easy to generalize the concept to a certain infinite family S_2 of sets $S_2 = \left\{ A_i \mid i \in N \right\} = \left\{ A_1, A_2, A_3, \ldots, A_n, \ldots \right\}$. Once again, the power of set builder notation triumphs. The sets S_1 and S_2 may be described more precisely with set builder notation than by enumeration. In fact, we shall see later in this text that some sets simply cannot be described by enumeration, so a means of referring to such a family of sets is needed. One means of referring to such a set of sets is through what is called indexed families of sets.

The set N in S_2 above is called an index set. It indexes the sets A_i. This means each set A_i is identified by (or paired with) its subscript i in N. Indexing may be generalized from N by letting any set, perhaps symbolized by the Greek letter Λ, be an index set. Then each set $\lambda \in \Lambda$ will serve as a variable subscript, i.e., index for the A's in the same way as the i from N was the variable subscript in A_i in S_2. The family S of sets we wish to describe is $S = \{A_\lambda \mid \lambda \in \Lambda\}$. It is called the set of A_λ's indexed by Λ. Further, it is agreed that if $\alpha, \beta \in \Lambda$ then $\alpha = \beta$ implies $A_\alpha = A_\beta$, and conversely if $A_\alpha = A_\beta$ then $\alpha = \beta$. So there is a pairing of sets A_λ in S with the λ's in Λ.

To help make the concept of indexing a family of sets as clear as it should be, we draw an analogy to a "real world" situation with which the student should be familiar. The parallel is close enough to solidify understanding of the concept, but as with most analogies, it isn't perfect.

124   *Introductory Concepts for Abstract Mathematics*

EXAMPLE 10.1:   Suppose a school library uses a Library of Congress,
(L.O.C.) identification system for its holdings. And suppose

$$S = \{A \mid A \text{ is a book in your library}\}$$
$$\Lambda = \{\lambda \mid \lambda \text{ is an L.O.C. number of a book in } S\}$$
$$A_\lambda = \text{ the book whose index is } \lambda.$$

First, the Λ in this example is the set of L.O.C. numbers which index
books in your library. Each $\lambda \in \Lambda$ is a number in the L.O.C. system
which refers to a specific element (book) in S assuming every book has
an L.O.C. number. Furthermore, if $\alpha, \beta \in \Lambda$ then $\alpha = \beta$ if and only
if $A_\alpha = A_\beta$ assuming there are no exceptional cases where books are
misnumbered or there are several volumes of the same book in the library.
In this analogy the indices are symbols with rational numbers and letters
of the alphabet, whereas in the general case indices may be any kinds of
elements. Also, in this example, A_λ is really not a set; it is a book. Just
as in the general case though, the books may be identified by the indices
λ (L.O.C. numbers) in a similar way that λ refers to the A_λ in the general
case. In the general case, as well as in this example, the index λ is not
the same thing as the set A_λ, indexed by λ, though they are connected
or paired together in a certain sense.

This example illustrates some of the indexing ideas, but it is not very
helpful with what is coming next–unions and intersections of indexed
families of sets. Do not try to extend this analogy beyond what we have
already done.

Definitions for unions and intersections of families of sets indexed by
some index set Λ are now possible.

DEFINITION: Let $S = \{A_\lambda \mid \lambda \in \Lambda\}$ be a family of sets indexed by Λ.
The <u>generalized</u> <u>union</u> of this family of sets is denoted by and
defined as follows

$$\bigcup_{\lambda \in \Lambda} A_\lambda = \left\{x \mid \exists \lambda \ni (\lambda \in \Lambda \wedge x \in A_\lambda)\right\}.$$

Since $\{A \mid A \in S\} = \{A_\lambda \mid \lambda \in \Lambda\}$, an equivalent description is
given by

$$\bigcup_{\lambda \in \Lambda} A_\lambda = \bigcup_{A \in S} A = \left\{x \mid \exists A \ni (A \in S \wedge x \in A)\right\}.$$

An even shorter notation for $\bigcup_{\lambda \in \Lambda} A_\lambda$ is

$$\bigcup S = \bigcup_{\lambda \in \Lambda} A_\lambda.$$

Why is the existential quantifier involved in the first part of the definition of the union of a collection of sets? The answer is, that is really what union is for several sets. Suppose $S = \{A_1, A_2\}$. To say an element x is in the union $A_1 \cup A_2$ means there is a set $A \in S$ with $x \in A$. In other words, for x to belong to $A_1 \cup A_2$ there must be at least one of the sets, either A_1 or A_2, to which that x belongs. Note the use of the words "there is." That is why the existential quantifier is in the definition.

The second characterization above, of the generalized union, permits its description without an indexing set Λ. Instead of Λ, the union is indexed on S. So $\bigcup_{A \in S} A$ is the union of all sets A with A in S. Here we do not involve a different index set Λ; we simply use the family S to do our "indexing."

Using notation introduced earlier, the generalized union may be written as $\{x \mid \exists A \in S \ni x \in A\}$. The membership requirement is read as, "there is an A in S such that $x \in A$" where " $\ni$ " is read as "such that."

EXAMPLE 10.2: Let $\Lambda = \{\alpha, \beta, \gamma\}$, $A_\alpha = \{1, 2, 3, 4\}$, $A_\beta = \{2, 3, 4\}$, and $A_\gamma = \{2, 4, 6\}$. Then

$$S = \{A_\alpha, A_\beta, A_\gamma\} = \{A_\lambda \mid \lambda \in \Lambda\}$$

is the set of sets indexed by Λ. An element x belongs to the union $\bigcup_{\lambda \in \Lambda} A_\lambda$ if there is a $\lambda \in \Lambda$ so that $x \in A_\lambda$. Let's see what's in the union. Since $1 \in A_\alpha$, there exists $\lambda \in \Lambda$, (namely $\lambda = \alpha$) with $1 \in A_\lambda$. So 1 satisfies the membership requirements for $\bigcup_{\lambda \in \Lambda} A_\lambda$. That puts 1 in the union. Similarly 2 is in the union because there is some set A_λ, where λ is one of the elements of $\Lambda = \{\alpha, \beta, \gamma\}$ and that A_λ contains the element 2. Also there is some set A_λ that contains 3, so 3 is in the union. Likewise, 4 and 6 belong to the union, but 5 does not belong to the union, because there is no set A_λ where $\lambda \in \Lambda$ that has 5 in it. In other words, 5 does not satisfy the membership requirements. There are no other reasonable candidates for membership in $\bigcup S$. Consequently,

$$\bigcup_{\lambda \in \Lambda} A_\lambda = \left\{x \mid \exists \lambda (\lambda \in \Lambda \wedge x \in A_\lambda)\right\} = \left\{1, 2, 3, 4, 6\right\}.$$

DEFINITION: Let $S = \{A_\lambda \mid \lambda \in \Lambda\}$ be a set of sets indexed by Λ. The <u>generalized</u> <u>intersection</u> of this family of sets is denoted by and defined as

$$\bigcap_{\lambda \in \Lambda} A_\lambda = \left\{ x \;\middle|\; \forall A(A \in S \implies x \in A) \right\}.$$

Alternate and equivalent descriptions are given by

$$\bigcap_{A \in S} A = \bigcap_{\lambda \in \Lambda} A_\lambda = \left\{ x \;\middle|\; \forall A(A \in S \implies x \in A) \right\}$$

$$= \left\{ x \;\middle|\; \forall A \in S, x \in A \right\}.$$

An even shorter notation for $\bigcap_{\lambda \in \Lambda} A_\lambda$ is $\bigcap S$, i.e.,

$$\bigcap S = \bigcap_{\lambda \in \Lambda} A_\lambda.$$

In words, an element x belongs to the intersection of a collection S of sets A, if x belongs to every A in that collection S, i.e., if $x \in A$ for every $A \in S$ then $x \in \bigcap_{A \in S} A$. So a characterization of what is in the intersection of a family of sets is given by using the universal quantifier and $\implies$. Thus $x \in \bigcap_{A \in S} A$ if and only if $\forall A[A \in S \implies x \in A]$.

Some equivalent ways to write the generalized intersection are expressed in the equations

$$\bigcap S = \bigcap_{\lambda \in \Lambda} A_\lambda = \left\{ x \;\middle|\; \forall \lambda \in \Lambda, x \in A_\lambda \right\}.$$

EXAMPLE 10.3:   Let Λ, A_α, A_β, A_γ, and S be as in Example 10.2. That is, let $\Lambda = \{\alpha, \beta, \gamma\}$, $A_\alpha = \{1, 2, 3, 4\}$, $A_\beta = \{2, 3, 4\}$, and $A_\gamma = \{2, 4, 6\}$. Now because $1 \notin A_\beta$ and $\beta \in \Lambda$, it is false that $\forall \lambda[\lambda \in \Lambda \implies 1 \in A_\lambda]$. Consequently, $1 \notin \bigcap_{\lambda \in \Lambda} A_\lambda$. Since $2 \in A_\alpha$ and $2 \in A_\beta$ and $2 \in A_\gamma$ and $\Lambda = \{\alpha, \beta, \gamma\}$, then $2 \in \bigcap_{\lambda \in \Lambda} A_\lambda$ is true, because $(\forall \lambda \in \Lambda)[\lambda \in \Lambda \implies 2 \in A_\lambda]$. Similarly $4 \in \bigcap_{\lambda \in \Lambda} A_\lambda$. All other elements such as $3, 5, 6, 17$ are not in $\bigcap_{\lambda \in \Lambda} A_\lambda$ since they fail the membership requirements. Thus

$$\bigcap_{\lambda \in \Lambda} A_\lambda = \left\{ x \;\middle|\; \forall \lambda[\lambda \in \Lambda \implies x \in A_\lambda] \right\} = \left\{ 2, 4 \right\}.$$

The property that $A \cap B \subseteq A$ from Theorem 9.10 has a generalization for intersections of families of sets.

THEOREM 10.1:   If C is a nonempty set of sets and $A \in C$ then $\bigcap C \subseteq A$.

PROOF: Let C be a nonempty set of sets and $A \in C$. In order to show $\bigcap C \subseteq A$, it is sufficient to show $\forall x[x \in \bigcap C \Longrightarrow x \in A]$. If $\bigcap C$ is empty, this implication is vacuously true. If $\bigcap C \neq \emptyset$, then let $y \in \bigcap C$. Now by definition, $\bigcap C = \{x \mid \forall C[C \in C \Longrightarrow x \in C]\}$. Since y is in $\bigcap C$ then $\forall C[C \in C \Longrightarrow y \in C]$. By Rule U.S. (freeing A), we get $A \in C \Longrightarrow y \in A$. And since $A \in C$ by assumption, then we get by detachment that $y \in A$. Thus the implication $y \in \bigcap C \Longrightarrow y \in A$ is established. By Rule U.G. $\forall x[x \in \bigcap C \Longrightarrow x \in A]$. But then by the definition of set containment, $\bigcap C \subseteq A$ as required.

The generalized union, $\bigcup_{\lambda \in \Lambda} A_\lambda$, and intersection $\bigcap_{\lambda \in \Lambda} A_\lambda$ are indeed generalizations of union and intersection. For if $S = \{A, B\}$ then

$$\bigcup S = \{x \mid \exists S \ni [S \in S \wedge x \in S]\} \overset{1}{=} A \cup B.$$

(See Exercise 10.10. The student will be asked to verify that equality 1 holds.) This example uses the form of union that does not employ an index set per se.

To illustrate that intersection of a family of two sets is just their normal intersection, we employ a scheme which does use the index set. Suppose $A = A_1$, $B = A_2$ and $\Lambda = \{1, 2\}$, so that $S = \{A_\lambda \mid \lambda \in \Lambda\}$ then $\bigcap S = \{x \mid \forall \lambda[\lambda \in \Lambda \Longrightarrow x \in A_\lambda]\} = \{x \mid x \in A_1 \wedge x \in A_2\} = A_1 \cap A_2$. Thus, $A \cap B = \bigcap\{A, B\}$.

EXAMPLE 10.4:   Let $\Lambda = Z$ and for each $\lambda \in \Lambda$ let A_λ be the open interval $(\lambda - 1, \lambda)$ between $\lambda - 1$ and λ. In other words, A_λ is given by $A_\lambda = \{x \mid x \in E_1 \wedge (\lambda - 1 < x < \lambda)\}$, so $\bigcup_{\lambda \in \Lambda} A_\lambda$ is given by: $\{x \mid \exists \lambda \in \Lambda \ni x \in A_\lambda\}$ which is $\{x \mid x \in E_1 \wedge x$ is not an integer $\}$. (Why?) Consequently, $\bigcup_{\lambda \in \Lambda} A_\lambda = E_1 - Z$.

EXAMPLE 10.5:   Let $\Lambda = Q^+$ be the set of positive rational numbers and for each $\lambda \in \Lambda$, let $A_\lambda = (-\lambda, \lambda)$. Note A_λ is the open interval between the rational numbers $-\lambda$ and λ, so it is symmetric about 0. Then $\bigcap_{\lambda \in \Lambda} A_\lambda = \{x \mid \forall \lambda \in \Lambda, x \in A_\lambda\} = \{x \mid (\forall \lambda \in Q^+)[x \in (-\lambda, \lambda)]\}$. Your intuition should tell you there is only one number, the number 0, in this intersection of open sets. More justification for this is supplied in Section V where properties of the real numbers are pursued. But to clarify this as much as we can now, consider any non–zero real number x. If $0 < x$ then a rational number λ_0 can be found, so that $0 < \lambda_0 < x$. (What is a good choice for λ_0?) Then $x \notin (-\lambda_0, \lambda_0) = A_{\lambda_0}$. Therefore $x \notin A_{\lambda_0}$. If $x < 0$ then again, a positive rational number λ_0 can be found so that $x < -\lambda_0 < 0$. Again, $x \notin A_{\lambda_0}$. So any non–zero x is certainly not in all of the A_λ's. Also, it is clear that 0 <u>is</u> in the intersection. The only possible conclusion is that $\bigcap_{\lambda \in \Lambda} A_\lambda = \{0\}$.

THEOREM 10.2: Suppose Λ is an index set and for each $\lambda \in \Lambda$, both A_λ and B_λ are sets and $A_\lambda \subseteq B_\lambda$. Then

$$\text{a. } \bigcup_{\lambda \in \Lambda} A_\lambda \subseteq \bigcup_{\lambda \in \Lambda} B_\lambda \qquad\qquad \text{b. } \bigcap_{\lambda \in \Lambda} A_\lambda \subseteq \bigcap_{\lambda \in \Lambda} B_\lambda.$$

Is this believable? Let us look at a special case. If $A_1 \subseteq B_1$ and $A_2 \subseteq B_2$ then according to Part a, $A_1 \cup A_2 \subseteq B_1 \cup B_2$, which is certainly believable. Also $A_1 \cap A_2 \subseteq B_1 \cap B_2$. Let us see if we can prove Part a of the theorem.

PROOF PART a: Suppose all the hypotheses, including $A_\lambda \subseteq B_\lambda$, for all $\lambda \in \Lambda$.

If $\bigcup_{\lambda \in \Lambda} A_\lambda = \emptyset$, the assertion is automatically true by Theorem 8.3. So suppose $\bigcup_{\lambda \in \Lambda} A_\lambda \neq \emptyset$. Since the union is not empty we can let $y \in \bigcup_{\lambda \in \Lambda} A_\lambda$. Since $\bigcup_{\lambda \in \Lambda} A_\lambda$ is defined by $\{x \mid \exists \lambda \ni x \in A_\lambda\}$ then $y \in \{x \mid \exists \lambda \ni x \in A_\lambda\}$. So $\exists \lambda \ni y \in A_\lambda$, because y has to meet the membership requirements of the generalized union. Now by Rule E.S., $y \in A_{\lambda_0}$ for some $\lambda_0 \in \Lambda$. (Here Rule E.S. freed λ_0.) But $A_{\lambda_0} \subseteq B_{\lambda_0}$, so $y \in B_{\lambda_0}$. By Rule E.G. $\exists \lambda \in \Lambda \ni y \in B_\lambda$. But then $y \in \{x \mid \exists \lambda \in \Lambda \ni x \in B_\lambda\}$ which by definition is $\bigcup_{\lambda \in \Lambda} B_\lambda$. Let's summarize what we have shown. We showed if $y \in \bigcup_{\lambda \in \Lambda} A_\lambda$ then $y \in \bigcup_{\lambda \in \Lambda} B_\lambda$. By Rule U.G. we get $\bigcup_{\lambda \in \Lambda} A_\lambda \subseteq \bigcup_{\lambda \in \Lambda} B_\lambda$.

Part b is for the student to do.

Notice the reliance on the fundamentals developed in the section on quantifiers. Also notice that we relied heavily on the definitions and on set builder notation and what it means to belong to a set that is described by set builder notation. We used the definition of subset and took advantage of the law of excluded middle which says a set is either empty or nonempty.

EXAMPLE 10.6: Let S be the empty set. What are $\bigcup S$ and $\bigcap S$?

SOLUTION: Since $\bigcup S$ is defined by

$$\left\{ x \mid \exists A \ni (A \in S \wedge x \in A) \right\},$$

what elements y are in this set, if S is empty? In order to belong to this set, an element y must make the statement $\exists A \ni [A \in S \wedge y \in A]$ true. But this can never happen because $A \in S$ is false for all sets A. Thus there is NO y meeting the membership requirement: $A \in S \wedge y \in A$. So $\bigcup S = \emptyset$.

Filled with such success, let's go to the next part of the question. In order for y to belong to $\bigcap S$, y must be one of the x's satisfying $\forall A (A \in S \implies x \in A_\lambda)$. In other words, $\forall A (A \in S \implies y \in A_\lambda)$. What y's meet this condition? The answer is left up to the student (see Exercise 10.7).

AXIOM OF UNIONS: If S is a set of sets then $\bigcup S$, i.e., $\bigcup_{A \in S} A$ is a set.

This axiom guarantees, among other things, that if A and B are sets and $S = \{A, B\}$ then $\bigcup S = A \cup B$ is a set.

Exercises:

10.1 Let $\Lambda = \{\alpha, \beta, \gamma\}$, $A_\alpha = \{a, b, c\}$, $A_\beta = \{c, d, e, f\}$, $A_\gamma = \{a, c, g\}$, and $S = \{A_\lambda \mid \lambda \in \Lambda\}$ then find the following:

    a. $\bigcup S$, i.e., $\bigcup_{\lambda \in \Lambda} A_\lambda$.

    b. $\bigcap S$, i.e., $\bigcap_{\lambda \in \Lambda} A_\lambda$.

10.2 Let $\Lambda = Z$ and $A_\lambda = (\lambda, \lambda+1] = \{x \mid x \in E_1 \wedge (\lambda < x \leq \lambda+1)\}$ for all $\lambda \in \Lambda$. Find the following.

    a. $\bigcup S$, i.e., $\bigcup_{\lambda \in \Lambda} A_\lambda$.

    b. $\bigcap S$, i.e., $\bigcap_{\lambda \in \Lambda} A_\lambda$.

10.3 Let Λ be the indicated index set and $A_\lambda = (-\lambda, \lambda)$ be the open interval $\{x \mid x \in E_1 \wedge -\lambda < x < \lambda\}$. Find each of the following sets. Whenever possible, answer in interval form.

    a. $\bigcup_{\lambda \in \Lambda} A_\lambda$ provided $\Lambda = \{1, 2, 3\}$.

    b. $\bigcap_{\lambda \in \Lambda} A_\lambda$ provided $\Lambda = \{1, 2, 3\}$.

    c. $\bigcup_{\lambda \in \Lambda} A_\lambda$ provided $\Lambda = Z^+$ the positive integers.

    d. $\bigcap_{\lambda \in \Lambda} A_\lambda$ provided $\Lambda = Z^+$ the positive integers.

    e. $\bigcup_{\lambda \in \Lambda} A_\lambda$ provided $\Lambda = Q^+$ the positive rational numbers.

    f. $\bigcap_{\lambda \in \Lambda} A_\lambda$ provided $\Lambda = Q^+$ the positive rational numbers.

    g. $\bigcup_{\lambda \in \Lambda} A_\lambda$ provided $\Lambda = \{a, b, c\}$ provided $0 < a < b < c$.

    h. $\bigcap_{\lambda \in \Lambda} A_\lambda$ provided $\Lambda = \{a, b, c\}$ provided $0 < a < b < c$.

10.4 Prove: If A is a set and C is a set of sets and if the statement $\forall C[C \in C \implies C \subseteq A]$ is true then $\bigcup_{C \in C} C \subseteq A$. Notice that this property generalizes the property $B \subseteq A$ and $C \subseteq A \implies B \cup C \subseteq A$.

10.5 Prove: If A is a set and C is a set of sets and if the statement $\forall C[C \in C \implies A \subseteq C]$ is true then $A \subseteq \bigcap_{C \in C} C$. What property, similar to the one mentioned in Exercise 10.4, does this generalize?

10.6 Prove: If C is a nonempty collection of sets and if $A \in C$ then $A \subseteq \bigcup C$. What property about two sets A and B does this generalize?

10.7 Suppose Λ is an index set and $\forall \lambda \in \Lambda, A_\lambda \neq \emptyset$. See if you can determine what $\bigcap_{\lambda \in \Lambda} A_\lambda$ is, if $\Lambda = \emptyset$.

10.8 Prove: If C is a nonempty collection of sets then $\bigcap C \subseteq \bigcup C$. Why is C assumed to be nonempty? What property about two sets A and B does this generalize?

10.9 Prove: If C is a nonempty collection of sets and if A is a set then

    a. $A \cap \left(\bigcup_{C \in C} C \right) = \bigcup_{C \in C} (A \cap C)$.

    b. $A \cup \left(\bigcap_{C \in C} C \right) = \bigcap_{C \in C} (A \cup C)$.

    c. What properties about three sets A, B, and C do these properties generalize? Find it in this text.

10.10 Work out the details in the assertions that if $S = \{A, B\}$ then $\bigcup S = A \cup B$ and $\bigcap S = A \cap B$.

10.11 In a branch of mathematics called *Topology*, we find some important applications to generalized unions and intersections. If S is a non–empty set, then a topology on S is a set $\mathcal{T}$ of subsets of S satisfying the following conditions:

    i. $\emptyset \in \mathcal{T}$

    ii. If $\mathcal{T}_1 \subseteq \mathcal{T}$ then $\bigcup \mathcal{T}_1 \in \mathcal{T}$

    iii. If $\mathcal{T}_1 \subseteq \mathcal{T}$ and $\mathcal{T}_1$ has a finite number of elements then $\bigcap \mathcal{T}_1$ is an element of $\mathcal{T}$.

a. Show that if $S = \{a, b, c\}$ and $T = \left\{ \emptyset, \{a, b, c\}, \{a\}, \{b, c\} \right\}$

then T is a topology on S.

b. Show that if $S = \{a, b, c, d\}$ and $T = \left\{ \emptyset, \{a, b, c, d\} \right\}$ then T

is a topology on S.

c. Show that if $S = \{a, b, c\}$ and T is given by

$$T = \left\{ \emptyset, \{a\}, \{b\}, \{c\}, \{a, b\}, \{a, c\}, \{b, c\}, \{a, b, c\} \right\}$$

then T is a topology on S.

10.12   Use the principle of mathematical induction to establish generalizations of the properties in Exercise 9.23. Let $K = \{1, 2, 3, \ldots, n\}$
and let $\bigcup_{i=1}^{n} A_i$ denote $\bigcup_{i \in K} A_i$ and $\bigcap_{i=1}^{n} A_i$ denote $\bigcap_{i \in K} A_i$.

    a. Show that $U - \bigcup_{i=1}^{n} A_i = \bigcap_{i=1}^{n} (U - A_i)$.

    b. Show that $U - \bigcap_{i=1}^{n} A_i = \bigcup_{i=1}^{n} (U - A_i)$.

10.13   Establish the finite generalizations of the distributive laws by induction. Let $K = \{1, 2, 3, \ldots, n\}$. Let $\bigcup_{i=1}^{n} A_i$ denote $\bigcup_{i \in K} A_i$,
and $\bigcap_{i=1}^{n} A_i$ denote $\bigcap_{i \in K} A_i$. Use P.M.I. to:

    a. Show that $A \cap \left(\bigcup_{i=1}^{n} B_i \right) = \bigcup_{i=1}^{n} (A \cap B_i)$.

    b. Show that $A \cup \left(\bigcap_{i=1}^{n} B_i \right) = \bigcap_{i=1}^{n} (A \cup B_i)$.

10.14   Referring to Exercise 10.5, what is wrong with this "statement"
of the assertion and its corresponding "proof?" That is, what is
wrong with the following "theorem?"
"Theorem:" If $\forall C [C \in \mathcal{C} \Longrightarrow A \subseteq C] \Longrightarrow A \subseteq \bigcap_{C \in \mathcal{C}} C$.
"Proof:" Show $\forall x (x \in A \Longrightarrow x \in \bigcap_{C \in \mathcal{C}} C)$. By Rule U.S.
$y \in A \Longrightarrow y \in \bigcap_{C \in \mathcal{C}} C$, so by Rule U.G. $\forall x (x \in A \Longrightarrow x \in \bigcap_{C \in \mathcal{C}} C)$.

SECTION III

FUNCTIONS AND RELATIONS

CHAPTER 11

CARTESIAN PRODUCTS

Introduction:

In mathematical as well as in everyday experiences, pairs of objects are compared to one another by an abstract device called a binary relation, although in this text it will simply be called a relation. For example, the objects may be pairs of persons while the device could be the relation, "is the same age as," or, "is a parent of." In the first case, a specific pair of persons a and b is distinguished from other pairs when it is really true that a is the same age as b. If the sentence, "a is the same age as b," is true, we say the pair consisting of a and b belongs to the relation, "is the same age as." Of course, in the second relation, the pair would be so distinguished if, "a is a parent of b," and the relation could be called the, "is the parent of," relation.

More mathematical examples of such a "pairing" are: "b is an element of B," or "A is a subset of C," where the objects being compared are elements and sets or sets and sets. In the first of the latter two mathematical examples, the pair b and B would be distinguished from other pairs, if indeed $b \in B$. That is: if $b \in B$ is true, the pair b and B would belong to the relation: "is an element of." If $b \notin B$ then b and B would not belong to the relation. In the second relation, if $A \subseteq C$ is true, then the pair A and C would belong to that relation. If $A \subseteq C$ is false, then the pair A and C would not belong to the relation.

Several things need to be noted about relations. Perhaps the most obvious is that the comparison involves exactly two objects at a time. In addition, the two objects involved in the comparison are not necessarily interchangeable. This is certainly clear in the relation, "is a parent of," because if a is the parent of b then b is not the parent of a. In particular, it is clear that the pairs which are subject to the relations $\in$, $\subset$, and $\subseteq$ are not necessarily interchangeable, e.g., if $A \subset B$ then $B \not\subset A$. So the pair must be "ordered." Another point worthy of noting is that the two objects being compared do not have to come from the same set. For example, b and B need not come from the same set for the relation $\in$.

Another important point is that the result of applying a relation to a specific pair of objects results in a statement. It is similar to applying a test to a pair of objects. If the test results in a true statement, we keep the pair. In other words, the result of applying a relation to a pair of objects is <u>not</u> a set and <u>not</u> an element, but IS a statement. For example,

"$a \in A$" is a statement. It is the result of applying the relation, "is an element of," to the pair a and A. The sentence, "John Doe is a parent of Johnny Doe," is a statement.

If this discussion regarding the concept of relations seems a bit vague, have patience. We are about to make it much more manageable. What is lacking is precision with terms and concepts. Once again, sets can come to the rescue to provide that necessary precision. The topics of ordered pairs and Cartesian products are the means of gaining the required precision.

Cartesian Products:

A procedure employed by some authors to make the concept of an ordered pair precise enough is through its definition in terms of a set. This is the procedure chosen here, except that the major result is stated, but no proof is given, because it has several cases and is not aesthetically pleasing.

DEFINITION: Let x and y be elements of a set S. Then the <u>ordered</u> <u>pair</u> (x,y) is defined by $(x,y) = \{\{x\}, \{x,y\}\}$. The element x is called the <u>first</u> <u>coordinate</u> and y is called the <u>second</u> <u>coordinate</u>.

The most important feature that is needed from this definition is stated in the following theorem. Feel free to devise your own proof using the definition above.

THEOREM 11.1: If S is a set and $a, b, c, d \in S$ then $(a, b) = (c, d)$ if and only if $a = c$ and $b = d$.

DEFINITION: Let S and T be sets. Then the <u>Cartesian</u> <u>product</u> of S and T is given by

$$S \times T = \Big\{ (s,t) \;\Big|\; s \in S \wedge t \in T \Big\}.$$

EXAMPLE 11.1: The term *Cartesian product* comes from Rene Descartes, the father of analytic geometry. The subject of analytic geometry is concerned with the Cartesian plane which is given by: $E_1 \times E_1 = \{(x,y) \mid x,y \in E_1\}$ where E_1 is the set of all real numbers. The set $E_1 \times E_1$ is interpreted geometrically as the entire coordinate plane with the first mentioned set associated with the horizontal axis and the second associated with the vertical axis of that plane.

Figure 11.1

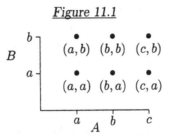

EXAMPLE 11.2:   The Cartesian product of the sets $A = \{a, b, c\}$ and $B = \{a, b\}$ is

$$A \times B = \{(a, a), (a, b), (b, a), (b, b), (c, a), (c, b)\}$$

which may be thought of as a *grid* of 6 discrete points each denoted by • in *Figure 11.1*.

Each of the six points of $A \times B$ occur at the intersection of the invisible vertical lines through a, b, or c on the horizontal axis with the invisible horizontal lines through a or b on the vertical axis. Notice that in a row, the second coordinates are all equal, and in a column the first coordinates are equal. This is similar to the coordinate plane of analytic geometry, except the points of this Cartesian product $A \times B$ are discrete.

EXAMPLE 11.3:   We claim that $A \times \emptyset = \emptyset$ and $\emptyset \times B = \emptyset$.

SOLUTION: We will show $A \times \emptyset = \emptyset$ indirectly. If $A \times \emptyset \neq \emptyset$, then there exists a y such that $y \in A \times \emptyset$. But $y \in A \times \emptyset$ says $y = (a, b)$ where $a \in A$ and $b \in \emptyset$. By simplification $b \in \emptyset$ which is a contradiction. Thus $A \times \emptyset = \emptyset$. Similarly $\emptyset \times B = \emptyset$.

THEOREM 11.2:   If A, B, and C are sets then $A \subseteq B$ implies $A \times C \subseteq B \times C$.

PROOF: Suppose A, B, and C are sets and $A \subseteq B$. If C is empty then $A \times C$ and $B \times C$ are both empty from Example 11.3, so $A \times C = B \times C$. If $A = \emptyset$ then $A \times C = \emptyset$, so $A \times C \subseteq B \times C$ by Theorem 8.5. If A and C are not empty then further suppose $y \in A \times C$. Then, by the definition of Cartesian product, there is an $a \in A$ and $c \in C$ such that $y = (a, c)$. Since $a \in A$ and $A \subseteq B$, $a \in B$, so $y = (a, c) \in B \times C$. Consequently, $A \times C \subseteq B \times C$. So in any of the three cases $A \times C \subseteq B \times C$.

In this proof, notice when A and C were non–empty we began with $y \in A \times C$; we did NOT start with something else. That is because we are showing $A \times C \subseteq B \times C$. To do this the definition of $\subseteq$ requires us to show every element of $A \times C$ is in $B \times C$. Also, in this proof and in the preceeding example, see how a number of steps have been eliminated. In particular, there was no mention of Rules C.P. and U.G. or the definition of $\subseteq$. In a straightforward proof like this, these things are so clear that these steps may be left out.

THEOREM 11.3:   If A and B are sets then $A \times B = B \times A$ if and only if $A = B$, or $A = \emptyset$, or $B = \emptyset$.

Part of the proof was established in Example 11.3 and the rest is left as an exercise.

The next two theorems are examples of the types of things that are true of the interrelationships of Cartesian products with other set theoretic operations, such as union and intersection. Caution should be observed however, since some statements that are superficially similar to Theorems 11.4 and 11.5, are actually false. The proofs of Theorems 11.4 and 11.5 are exercises.

THEOREM 11.4:   If A, B, and C are sets then

$$A \times (B \cup C) = (A \times B) \cup (A \times C).$$

THEOREM 11.5:   If A, B, and C are sets then

$$A \times (B \cap C) = (A \times B) \cap (A \times C).$$

To summarize a bit, we have begun the groundwork which will enable us to introduce some concepts in future sections. What we have done here is to pin down exactly what an ordered pair is and what Cartesian products are. With these concepts we will be able to deal with relations, functions, and equivalence relations in the subsequent sections.

Exercises:

11.1   If $A = \{a, b, c\}$ and $B = \{a, b, c, d\}$ then find $A \times B$ and $B \times A$.

11.2   Is $A \times (B \times C) = (A \times B) \times C$? Explain. (Be Careful.)

11.3   Sketch the following subsets of $E_1 \times E_1$. Use shading if necessary.

   a. $\{(x, y) \mid (x, y \in E_1) \land (y \geq x^2) \land (y \leq x + 2)\}$.

   b. $\{(x, y) \mid (x, y \in E_1) \land (y = x^2)\}$.

   c. $\{(x, y) \mid (x, y \in E_1) \land (2x + 3y = 1)\}$.

11.4 Prove Theorem 11.3. That is, $A \times B = B \times A$ iff $A = B$ or $A = \emptyset$ or $B = \emptyset$.

11.5   a. Is $A \cup (B \times C) = (A \cup B) \times (A \cup C)$? Explain. (Be Careful.)

      b. Is $A \cap (B \times C) = (A \cap B) \times (A \cap C)$? Explain. (Be Careful.)

11.6 Prove Theorem 11.4.

11.7 Prove Theorem 11.5.

11.8 Prove: If $A, B_1, B_2, \ldots, B_n$ are sets then

$$A \times \left(\bigcup_{i=1}^{n} B_i \right) = \bigcup_{i=1}^{n} (A \times B_i).$$

(See Theorem 11.4, inductive definitions, and Exercise 10.13.)

11.9 Prove: If $A \neq \emptyset$ and $B \neq \emptyset$ then $A \times B \neq \emptyset$.

CHAPTER 12

RELATIONS

As mentioned in the introductory remarks to Chapter 11, a matter of frequent interest in mathematical as well as in day to day experience is a comparison between objects a pair at a time. The two objects are selected, one from one set and one from perhaps another set. A test is applied or a comparison made between pairs of objects. The comparison is some relation, for example, "is the same age as," for persons or, "is less than," for real numbers. We want to distinguish when a given pair compares favorably with respect to the relation as opposed to not comparing favorably. Since we are dealing with pairs of things and since those two things might not be interchangeable, we need to use ordered pairs. For example, if P is a certain set of persons and N the set of natural numbers, we may be concerned with the set of pairs $\{(n,p) \mid (n,p) \in N \times P$ and n is the age of $p\}$. This set of ordered pairs could be considered the *age relation* for the collection P, meaning that we include in the set, exactly those ordered pairs (n,p) for which the relation, "is the age of," applied to (n,p) is true. What we have is essentially an "age table" with the correct ages for the persons in P. For example, if Mary Jane Doe is 19 years old and she is in P, then $(19, \text{Mary Jane Doe})$ is in the age relation above. On the other hand, if John Doe is 23 years old, then $(22, \text{John Doe})$ is <u>not</u> in the age relation. If you are 20 and your name is M and you are in P, then $(20, M)$ is in the age relation.

Some additional relations are:

$$\{(x,y) \mid x,y \in P \text{ and } x \text{ is the parent of } y\}$$

and

$$\{(x,y) \mid x,y \in P \text{ and } x \text{ is the same age as } y\}.$$

These are each subsets of $P \times P$ and describe the "is the parent of relation" and "is the same age as" relation.

If one were considering the set T of all triangles in a certain plane, a possible relation for consideration is the relation, "is similar to." It is the set $\{(x,y) \mid x,y \in T \text{ and } x \text{ is similar to } y\}$. This relation is a subset of $T \times T$. The relation $\{(x,y) \mid x \in E_1 \wedge y \in T \wedge x \text{ is the area of } y\}$ is clearly a subset of $E_1 \times T$ and is the relation, "is the area of," for triangles in T.

The examples cited previously illustrate the defining feature of relations, namely that relations are subsets of Cartesian products.

DEFINITION: A <u>relation</u> <u>between</u> <u>sets</u> A and B is a subset r of $A \times B$. A <u>relation</u> <u>in</u> A is a subset of $A \times A$. If r is a relation between A and B and $(x,y) \in r$ then x is said to be related by the relation r to y. Even more simply, we could say <u>x is r related to y</u> iff $(x,y) \in r$.

NOTATION: If r is a relation between A and B i.e., $r \subseteq A \times B$, then $(x,y) \in r$ <u>if and only if</u> $x\,r\,y$. This equivalence is very useful because it provides a way of <u>translating</u> between the mathematical sentence such as $(x,y) \in r$ and an English sentence such as, "x is the same age as y," where the relation r is the relation, "is the same age as." You will note that the phrase, "is the same age as," may replace r in $x\,r\,y$.

If r is the relation, "is congruent to," for triangles, this important feature allows a direct translation of "$x\,r\,y$" as "x is congruent to y" by replacing the relation r by, "is congruent to." Of course, we have a special symbol for this relation in geometry. It is $\cong$, so $x \cong y$ is also read as, "x is congruent to y." <u>Merely</u> <u>substitute</u> <u>the</u> <u>describing</u> <u>words</u> <u>for</u> <u>the</u> <u>relation</u> <u>r when</u> <u>reading it</u>. Further examples of this translating capability are to follow.

EXAMPLE 12.1:   If T is the set of all triangles in a plane and $r = \{(x,y) \mid (x \in E_1) \wedge (y \in T) \wedge (x \text{ is the area of } y) \}$ then for a certain triangle b, if b has area a, then $(a,b) \in r$. If b does not have area a then $(a,b) \notin r$. In the former case we would write $a\,r\,b$ and say, "a is r related to b," or more preferably as, "a is the area of b." If it is false that "a is the area of b" then write $\sim (a\,r\,b)$.

EXAMPLE 12.2:   Let $r = \{(x,y) \mid (x,y \in E_1) \wedge (x < y)\}$ then r is the "is less than" relation in E_1. Since $(2,3) \in r$ we could write $2\,r\,3$ and read it as, "2 is less than 3." It is even possible to replace r by $<$ to get: $< = \{(x,y) \mid x,y \in E_1 \wedge x \text{ is less than } y\}$. But we will not normally do this, because it seems somehow strange. To continue, there are some familiar properties this relation has. One of them is $(x,y) \in r \wedge (y,z) \in r \implies (x,z) \in r$. This the so-called *transitive property* of $<$.

EXAMPLE 12.3:   If T is the set of triangles in a given plane and r is the relation, "is congruent to," more briefly symbolized by $\cong$, then we could write $r = \{(x,y) \mid (x,y \in T) \wedge (x \cong y)\}$. Some familiar properties of $\cong$, that are expressible with this kind of notation are the *reflexive* property: $(x, x) \in r \; \forall x \in T$, the *symmetric* property: $[(x, y) \in r \implies (y, x) \in r]$ $\forall x, y \in T$, and the *transitive* property $[(x, y) \in r \land (y, z) \in r \implies (x, z) \in r]$ $\forall x, y, z \in T$. More is to be said about these matters in the next chapter.

DEFINITION: If r is a relation between A and B, i.e., $r \subseteq A \times B$, then the <u>domain of r</u> is the set $\{x \in A \mid \exists y \in B \ni [(x, y) \in r]\}$ and is denoted dom(r). The <u>range of r</u> is given by $\{y \in B \mid \exists x \in A \ni [(x, y) \in r]\}$ and denoted ran(r). If $r \subseteq A \times B$ and dom(r) = A then r is a relation <u>from A to B</u>. If $r \subseteq A \times A$ and dom(r) = A then r is a relation <u>on</u> A.

If $r \subseteq A \times B$ then equivalent descriptions for dom (r) and ran (r) are dom (r) = $\{x \in A \mid (x, y) \in r\}$ and ran (r) = $\{y \in B \mid (x, y) \in r\}$.

EXAMPLE 12.4: The relation, "is congruent to," in Example 12.3, is a relation *on T*. The relation $a = \{(x, y) \mid x \in T \land y \in E_1 \land x \text{ has area } y\}$ is a *relation from T to E_1*. Why is it *from T to E_1*?

EXAMPLE 12.5: If r is the relation $r = \{(a, a), (a, b), (a, c), (b, b), (b, c)\}$ in $A = \{a, b, c\}$ then dom(r) = $\{x \in A \mid (x, y) \in r\} = \{a, b\}$. Now $a \in$ dom(r), since $(a, b) \in r$; and $b \in$ dom(r), since $(b, c) \in r$, but $c \notin$ dom(r), since $\sim (\exists y \ni y \in A \text{ and } (c, y) \in r)$. Similarly, ran($r$) = $\{a, b, c\}$.

DEFINITION: If r is a relation between A and B, i.e., $r \subseteq A \times B$, then the <u>inverse</u> of r is given by $\{(x, y) \mid (y, x) \in r\}$. The inverse of r is denoted by r^{-1}. That is $r^{-1} = \{(x, y) \mid (y, x) \in r\}$.

PROPERTY: If r is a relation between A and B, then r^{-1} is a relation between B and A. Can the same thing be said if r is a relation from A to B? Also notice $(y, x) \in r^{-1}$ iff $(x, y) \in r$.

EXAMPLE 12.6: Here are examples of inverses of relations.

a. Suppose T is a certain set of triangles and we are focusing on the, "is the area of," relation. That is, r is given by

$$r = \{(x, y) \mid x \in E_1 \land y \in T \land x \text{ is the area of } y\}.$$

Then

$$r^{-1} = \{(y, x) \mid (x, y) \in r\}$$
$$= \{(y, x) \mid x \in E_1 \land y \in T \land x \text{ is the area of } y\}.$$

Figure 12.1   *Figure 12.2*

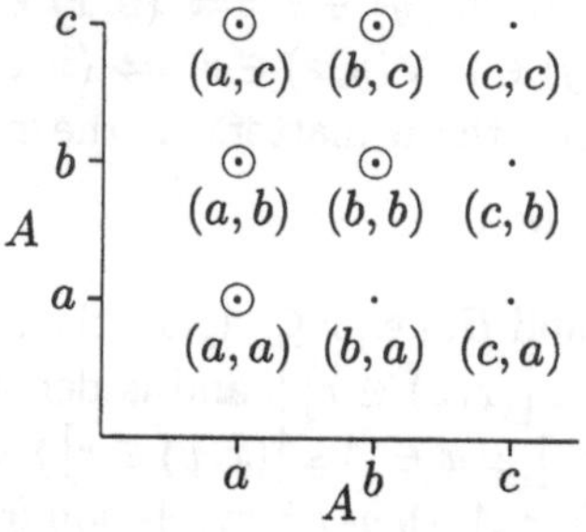

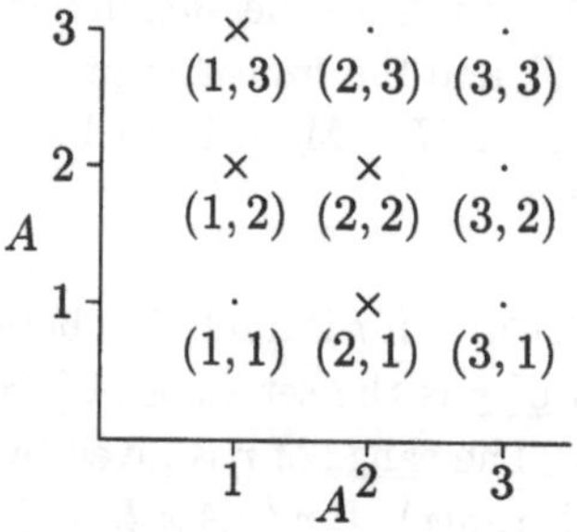

One way to read the inverse of this relation is, "y has area x."

b. Let $A = \{1, 2, 3\}$ and $r = \{(1,2), (2,1), (1,3), (2,2)\}$. Then

$$r^{-1} = \{(2,1), (1,2), (3,1), (2,2)\}.$$

c. Let $r = \{(x,y) \mid x, y \in E_1 \land x < y\}$ then

$$r^{-1} = \{(y,x) \mid (x,y) \in r\} = \{(y,x) \mid (x,y \in E_1) \land (x < y)\}$$

$$\overset{\text{why}}{=} \{(x,y) \mid (x,y \in E_1) \land (y < x)\}.$$

The relation r is the "is less than" relation while r^{-1} is the "is greater than" relation.

It is sometimes possible to denote a relation by a *lattice diagram* similar to *Figure 11.1*. Since a relation between A and B is a subset of $A \times B$, a typical way to graph the relation is to sketch the grid $A \times B$ and identify those ordered pairs (a, b) in r by circling or ×-ing the pairs in the given relation.

EXAMPLE 12.7: Sketch a lattice diagram of r in $A \times A$ in each of the following

   a. Sketch $r = \{(a,a), (a,b), (a,c), (b,b), (b,c)\}$ from Example 12.5. Circle those points of $A \times A$ that are in r. The graph is in *Figure 12.1*.

   b. Sketch a lattice diagram of $r = \{(1,2), (2,1), (1,3), (2,2)\}$ between A and A. See Example 12.6b. Put an × on each member of $A \times A$ that is in r. See the graph in *Figure 12.2*.

DEFINITION: Let A be a set and $I_A = \{(x,x) \mid x \in A\}$ then I_A is a relation on A called the <u>identity</u> <u>relation</u> on A.

Figure 12.3

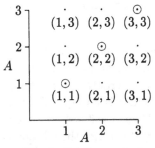

Identity Relation on $A \times A$

EXAMPLE 12.8: If $A = \{1,2,3\}$ then $I_A = \{(1,1),(2,2),(3,3)\}$. A lattice diagram of I_A is given in *Figure 12.3* with the circles identifying the identity relation.

From *Figure 12.3* it is clear why the identity relation is sometimes called the *diagonal* of $A \times A$.

DEFINITION: If $r \subseteq A \times A$, i.e., r is a relation in A, and $A \neq \emptyset$ then

i. r is <u>reflexive</u> on A if and only if $\forall x[x \in A \Longrightarrow (x,x) \in r]$.

ii. r is <u>symmetric</u> on A if and only if

$$\forall x, y\big[(x,y \in A \text{ and } (x,y) \in r) \Longrightarrow (y,x) \in r\big].$$

iii. r is <u>transitive</u> on A if and only if

$$\forall x, y, z\big[(x,y,z \in A \text{ and } (x,y) \in r \wedge (y,z) \in r) \Longrightarrow (x,z) \in r\big].$$

iv. r is an <u>equivalence relation</u> on A if and only if r is reflexive, symmetric, and transitive on A.

Notice we used the notation $\forall x, y$ for $\forall x \ \forall y$ in these definitions. Continuing with this notation and using another r notation there is a way to rewrite the first three conditions. In particular, in place of $(x,y) \in r$ if we write $x \, r \, y$, then we would get the following alternate descriptions.

i'. r is <u>reflexive</u> on A if and only if $\forall x[x \in A \Longrightarrow x \, r \, x]$.

ii'. r is <u>symmetric</u> on A if and only if

$$\forall x, y\big[(x,y \in A \wedge x \, r \, y) \Longrightarrow y \, r \, x\big].$$

iii'. r is <u>transitive</u> on A if and only if

$$\forall x,y,z\big[(x,y,z \in A \text{ and } x\,r\,y \wedge y\,r\,z) \Longrightarrow x\,r\,z\big].$$

There are several observations that should be made about the reflexive, symmetric, and transitive properties. The reflexive property is the <u>only</u> one of the three that asserts that something must belong in the relation. Suppose $A \neq \emptyset$ and r is an equivalence relation, i.e., satisfying conditions i, ii, and iii. Then $A \neq \emptyset \Longrightarrow \exists\, x \ni x \in A$. By Part i and the Law of Detachment, $(x,x) \in r$, hence the relation is <u>on</u> A. The symmetric and transitive properties are strictly conditional. For example, a symmetric relation r satisfies the conditional, "if $(x,y) \in r$ then $(y,x) \in r$." This does NOT say that $(x,y) \in r$ or any other pair is in r. For example, the identity relation $I_A = \{(a,a),(b,b),(c,c)\}$ on $A = \{a,b,c\}$ is symmetric and transitive as well as reflexive, even though there is NO pair (x,y) with $x \neq y$ in it.

EXAMPLE 12.9:   The relation $r = \big\{(x,y) \mid x,y \in E_1 \wedge x \leq y\big\}$ is reflexive and transitive, but NOT symmetric. The relation $\cong$ on the set T of triangles in a given plane is reflexive, symmetric, and transitive, hence an equivalence relation on T. The relation, "is the same age as," is an equivalence relation on a given set P of persons.

EXAMPLE 12.10:   Determine which of the reflexive, symmetric, and transitive properties is possessed by the relation r on the set $A = \{a,b,c\}$ given that $r = \{(a,a),(b,b),(c,c),(a,b),(b,a)\}$.

SOLUTION: Since $(a,a),(b,b),(c,c) \in r$ then $(x,x) \in r$ for all $x \in A$. Hence r is reflexive. Also, $(a,a),(b,b),(c,c),(a,b),(b,a) \in r$ implies $(a,a),(b,b),(c,c),(b,a),(a,b) \in r$, so r is symmetric. Now $(a,a) \in r \wedge (a,b) \in r \Longrightarrow (a,b) \in r$ and $(b,b) \in r \wedge (b,a) \in r \Longrightarrow (b,a) \in r$ and $(a,b) \in r \wedge (b,a) \in r \Longrightarrow (a,a) \in r$, and finally $(b,a) \in r \wedge (a,b) \in r \Longrightarrow (b,b) \in r$. Since all of the nontrivial ways to put these together according to the hypothesis of transitivity have been considered, and since the result satisfies transitivity in each case, then r is an equivalence relation on A.

EXAMPLE 12.11:   Is $r = \big\{(a,a),(a,b),(b,a),(b,b)\big\}$ an equivalence relation on $A = \{a,b,c\}$?

SOLUTION: The answer is NO. Since $c \in A$ and $(c,c) \notin r$, it is false that $\forall x[x \in A \Longrightarrow (x,x) \in r]$. Since r is not reflexive on A, it is not an equivalence relation on A. It is not necessary to check for symmetry and transitivity on A. It failed to be reflexive, hence it is not an equivalence relation.

EXAMPLE 12.12: If $r = \{(a, b), (b, a), (a, a), (b, b), (c, c), (a, c), (c, a)\}$, is r an equivalence relation on $A = \{a, b, c\}$?

SOLUTION: The answer is NO! The pairs (b, a) and (a, c) are in r, but $(b, c) \notin r$. Thus $\sim(\ \forall x, y, z \in A)[(x, y) \in r \wedge (y, z) \in r \implies (x, z) \in r]$. Hence, r is not transitive and so it is not an equivalence relation on A.

DEFINITION (*Partial Order Relation*): Let A be a set. A relation r on A is called a <u>partial</u> <u>order</u> <u>relation</u> <u>on</u> A if and only if r is reflexive, transitive, and satisfies the <u>antisymmetric</u> property, namely $a\,r\,b \wedge b\,r\,a \implies a = b$ $\forall a, b \in A$. Sometimes partial order relation is abbreviated as p.o.r.

EXAMPLE 12.13: Let N be the set of natural numbers. Let $a, b \in N$. Define the relation *divides* by "a <u>divides</u> b if and only if $\exists\, k \in N \ni ak = b$." To make it easier to write, we introduce a symbol to replace the words "divides." It is $|$, so the rewritten definition is, "$a \mid b$ if and only if $\exists\, k \in N \ni a \cdot k = b$." For example, $3 \mid 12$, since $\exists\, 4 \in N \ni 3 \cdot 4 = 12$. The relation $|$ is an example of a partial order relation on N. The "divides" relation is *reflexive* and *transitive*, but is not symmetric. IN FACT, it has the property of *antisymmetry* on N, for if $a, b \in N$ then $a \mid b \wedge b \mid a \implies a = b$. Note that the result of applying the relation $|$ to a pair, is a statement or predicate. It is NOT a number. "Divides" does not represent a quotient; it represents a state of affairs.

Exercises:

12.1 Redescribe the concept of a partial order relation (p.o.r.) using the ordered pair notation for each of its three requirements.

12.2 a. If $A = \{a, b, c\}$, show I_A is an equivalence relation on A.

b. Suppose A is a nonempty set then show I_A is an equivalence relation on A. (Caution, the student is not free to select the specific nature of A as for example, $A = \{1, 2, 3, 4\}$. He or she must leave it as A: nothing assumed except $A \neq \emptyset$.)

12.3 Suppose A and r are, in turn, as indicated in Parts a–j. Find $\mathrm{dom}(r)$, $\mathrm{ran}(r)$, and r^{-1} for each of the following.

a. $A = \{a, b, c\}$, $r = \{(a, b), (b, a), (c, b), (c, a)\}$.

b. $A = \{1, 2, 3, 4\}$, $r = I_A \cup \{(1, 2), (1, 3), (2, 4)\}$.

c. $A = Z, r = \{(x,y) \mid x,y \in Z \land x < y\}$.

d. $A = Z, r = \{(x,y) \mid x,y \in Z \land x^2 < y^2\}$.

e. $A = E_1, r = \{(x,y) \mid x,y \in E_1 \land 2x + 3y = 1\}$.

f. $A = E_1, r = \{(x,y) \mid x,y \in E_1 \land y = \ln x\}$.

g. $A = Z, r = \{(x,y) \mid x,y \in Z \land y = |x|\}$.

h. $A = Z, r = \{(x,y) \mid x,y \in Z \land x^2 + y^2 = 25\}$.

i. $A = N, r = \{(x,y) \mid x,y \in N \land x^2 + y^2 = 25\}$.

j. $A = E_1, r = \{(x,y) \mid x,y \in E_1 \land x^2 + y^2 = 25\}$.

12.4 Suppose A and r are as indicated in parts a–k below. Answer the following questions. Is r reflexive on A? Is r symmetric on A? Is r transitive on A?

a. $A = \{a,b,c\}, r = I_A$.

b. $A = \{a,b,c\}, r = A \times A$.

c. $A = \{a,b,c\}, r = \{(a,a),(b,b),(c,c),(a,b),(b,a)\}$.

d. $A = \{a,b,c,d\}, r = I_A \cup \{(a,b),(b,a),(b,c),(c,b)\}$.

e. $A =$ set of all persons in the U.S.,
   $r = \{(x,y) \mid x,y \in A \land x$ is at least as old as $y\}$.

f. Let A be the set of all persons in the world and let $P(x,y)$ be the predicate, "x is a citizen of the same country as y," and assume every person in the world is a citizen of exactly one country. Finally, let r be $\{(x,y) \mid x,y \in A \land P(x,y)\} = \{(x,y) \mid x,y \in A \land x$ is a citizen of the same country as $y\}$.

g. $A =$ set of all points in the plane,
   $r = \{(x,y) \mid x,y \in A \land x$ is one centimeter from $y\}$.

h. $A = Z, r = \{(x,y) \mid x,y \in Z \land P(x,y)\}$, where $P(x,y)$ is given by "(both x & y are even) $\lor$ (both x & y are odd)."

i. $A = Z$ and $r = \{(x,y) \mid x,y \in Z \land x - y = 3 \cdot k$ for some $k\}$.

j. $A = Z$ and
   $r = \{(x,y) \mid x,y \in Z \land x$ has the same units digit as $y\}$.

k. $A = \{a,b,c,d,e\}, r = I_A \cup \{(a,c),(c,a),(b,d),(d,b)\}$.

12.5 Suppose A and B are sets. Show the empty set $\emptyset$ is a relation between A and B.

12.6 Show $A \times B$ is a relation from A to B.

12.7 Prove if A is a nonempty set then the identity relation I_A is an equivalence relation on A.

12.8 Suppose $r = \{(x, y) \mid x, y \in E_1 \wedge y = e^x\}$. Find the domain and range of r.

12.9 Suppose $r = \{(x, y) \mid x, y \in E_1 \wedge x = e^y\}$. Find the domain of r, i.e., where is r "from?" What is the range of r?

12.10 Assume the usual arithmetic properties of the integers. Let $S = Z - \{0\}$. Define $(\forall a, b, c, d \in S) \left[(a, b) \sim (c, d) \text{ iff } ad = bc\right]$. Prove $\sim$ is an equivalence relation on S. Hint

$$\sim = \{((a, b), (c, d)) \mid a, b, c, d \in S \wedge ad = bc\}.$$

12.11 Prove: $A \times A$ is an equivalence relation on A.

12.12 Prove: If A is any nonempty set and r any equivalence relation on A, then $I_A \subseteq r \subseteq A \times A$. See Exercises 12.7 and 12.11 and note that this asserts that every equivalence relation on A is a set "between" I_A and the entire Cartesian product $A \times A$.

12.13 Prove: $I_A^{-1} = I_A$.

12.14 Prove: $A \times B = (B \times A)^{-1}$.

12.15 Write out example relations that satisfy exactly the following properties: a) reflexive only, b) symmetric only, c) transitive only, d) reflexive and symmetric only, e) reflexive and transitive only, f) symmetric and transitive only, g) all three, h) none of the three.

12.16 Prove: If r is a relation between A and B then

a. $\text{dom}(r) = \text{ran}(r^{-1})$.

b. $\text{dom}(r^{-1}) = \text{ran}(r)$.

c. $(r^{-1})^{-1} = r$.

12.17 Prove: If A is a set and r is an equivalence relation on A and $a, b \in A$ and $a\,r\,b$, $\left[\text{i.e., } (a, b) \in r\right]$ then $\{x \mid x \in A \wedge x\,r\,a\} = \{x \mid x \in A \wedge x\,r\,b\}$.

12.18 Using the definition of partial order relation, show that the relation $\subseteq$ is a partial order relation on a nonempty set S of sets.

12.19 Show that $\leq$ is a partial order relation on a nonempty set S of real numbers. You may assume order properties of the reals from Chapter 5.

12.20 Prove: If A is a set and r is a relation on A then $r = I_A$ if and only if r is an equivalence relation on A AND r is a partial order relation on A.

12.21 Is the relation $\neq$ for sets a transitive relation? Symmetric relation? Reflexive relation?

12.22 Suppose A and B are sets. Suppose also that A and B are <u>disjoint</u> iff $A \cap B = \emptyset$. Is the "disjoint relation" on a set of sets transitive? Is it reflexive? Is it symmetric?

12.23 Prove that the relation $\mid$ of Example 12.13 is a partial order relation.

12.24 For each of the following relations f, sketch a graph of f and its inverse relation f^{-1} on the same coordinate system. Is there any geometric relationship between f and f^{-1}?

    a. $f = \{(1,2),(2,4),(3,6)\}$.

    b. $f = \{(x,y) \mid y = 2x - 1\}$.

    c. $f = \{(x,y) \mid y = x^2 \wedge 0 \le x \le 2\}$.

    d. $f = \{(x,y) \mid y = x^2 \wedge -2 \le x \le 2\}$.

12.25 Prove: The empty set is a symmetric and transitive relation, but it is not reflexive. See Exercise 12.5.

12.26 For a set A and relation r on A either $a\,r\,b$ or $\sim(a\,r\,b)$. If $\sim(a\,r\,b)$ is it true that $a\,r^{-1}\,b$? Explain.

12.27   a. What is wrong with the proof of the argument, "If R is a symmetric relation then R^{-1} is symmetric," that begins as follows: Suppose R is symmetric. Let $(x,y) \in R$.

    b. Suppose at some point in the "proof" of the theorem in Part a, the theorem prover says, "Since R is reflexive, ..." What is wrong with that?

    c. What is wrong with the assertion: If r is symmetric then $r = \{(x,y) \mid \forall x, y (x,y) \in R \wedge (y,x) \in r\}$.

    d. Now suppose the "proof" begins with, "Suppose R is symmetric then $(a,b) \in R$ and $(b,a) \in R$." What is wrong with this?

12.28 Let $A = \{a,b,c,d,e\}$. Find the smallest subset S of $A \times A$ such that S is an equivalence relation on A and S contains the set $T = \{(a,c),(b,c)\}$.

CHAPTER 13

PARTITIONS

In the last chapter we dealt with relations and in particular the important category of relations called equivalence relations. A concept closely related to this is the concept of partitions. A partitioning of a set A is accomplished by subdividing A into categories so that each of its members is in one, and only one, category. A partitioning of the students at a university occurs, for example, by dividing them into academic classes: freshmen, sophomores, juniors, seniors, and graduate students. Neglecting exceptional cases, every student is in one of these categories and no student is in more than one. From this, it is easy to formalize the concept into a definition.

DEFINITION: Let A be a nonempty set. A <u>partition</u> of A is a set $\mathcal{P}$, of subsets S of A, such that

1. $\bigcup_{S \in \mathcal{P}} S = A$,
2. $S_1, S_2 \in \mathcal{P} \Longrightarrow (S_1 = S_2) \vee (S_1 \cap S_2 = \emptyset)$, and
3. $S \in \mathcal{P} \Longrightarrow S \neq \emptyset$.

The elements of $\mathcal{P}$ are called <u>cells</u>. The effect of Part 1 is that every element of A is in at least one cell S of the partition, and conversely every element of each cell is in A. Part 2 says that every element of A is in at most one cell of the partition. Part 3 says no cell is empty, because it would not be productive to allow a cell to be empty. Any element of a cell S is called a <u>representative</u> of that cell.

EXAMPLE 13.1: Let $A = \{1,2,3,4\}$ and $\mathcal{P} = \{\{1\},\{2,4\},\{3\}\}$. Then $\mathcal{P}$ is a partition of A and $\{1\}$, $\{2,4\}$, $\{3\}$ are the cells of that partition. The number 2 is a representative of the cell $\{2,4\}$.

EXAMPLE 13.2: Let $E = \{2n \mid n \in Z\}$ and $D = \{2n+1 \mid n \in Z\}$, then $\mathcal{P} = \{E, D\}$ is a partition of Z. D is a cell and -17 is a representative of that cell.

EXAMPLE 13.3: Let R be a set of persons. Subdivide R into subsets $S(p)$ where $S(p) = \{q \in R \mid p \text{ and } q \text{ have the same age }\}$. Then the set $\mathcal{P}$ given by

$$\mathcal{P} = \Big\{ S(p) \,\Big|\, p \in R \Big\}$$

151

is a partition of R and the cells of the partition are sets of persons all of whom are of the same age. That is, the persons p and q are in the same cell iff they have the same age.

EXAMPLE 13.4:   Let Z be the set of all integers and let n be a fixed positive integer. Then define the set S_j by

$$S_j = \{i \mid i \in Z \land (i - j = k \cdot n) \text{ for some } k \in Z\}.$$

Then the set $\mathcal{P} = \{S_j \mid j \in Z \land 0 \leq j < n\}$ is a partition of Z. In particular, if $n = 4$ then the cells S_j's are as follows

$$
\begin{aligned}
S_0 &= \{i \mid i \in Z \land (i - 0 = k \cdot 4 \text{ for some } k \in Z)\} \\
&= \{\ldots, -8, -4, 0, 4, 8, \ldots\}. \\
S_1 &= \{i \mid i \in Z \land (i - 1 = k \cdot 4 \text{ for some } k \in Z)\} \\
&= \{\ldots, -7, -3, 1, 5, 9, \ldots\}. \\
S_2 &= \{i \mid i \in Z \land (i - 2 = k \cdot 4 \text{ for some } k \in Z)\} \\
&= \{\ldots, -6, -2, 2, 6, 10, \ldots\}. \\
S_3 &= \{i \mid i \in Z \land (i - 3 = k \cdot 4 \text{ for some } k \in Z)\} \\
&= \{\ldots, -5, -1, 3, 7, 11, \ldots\}. \\
S_4 &= \{i \mid i \in Z \land (i - 4 = k \cdot 4 \text{ for some } k \in Z)\} \\
&= \{\ldots, -8, -4, 0, 4, 8, 12, \ldots\} \\
&= S_0. \\
S_5 &= S_1, \quad S_6 = S_2, \quad \ldots, \text{ and in general } S_{j+4} = S_j.
\end{aligned}
$$

Then $\{S_0, S_1, S_2, S_3\}$ is a partition of Z for the case $n = 4$. The partition in the general case n is $\{S_0, S_1, S_2, \ldots, S_{n-1}\}$. Notice that if $n = 2$, the partition $\{S_0, S_1\}$ is simply the partition of Example 13.2, namely $\{E, D\}$.

As mentioned in the introduction to this section, there is a connection between equivalence relations and partitions. We pursue a portion of this relationship by considering the relation

$$r = \{(a, a), (b, b), (c, c), (d, d), (e, e), (a, c), (c, a), (b, d), (d, b)\}$$

on the set $A = \{a, b, c, d, e\}$. A quick check reveals that r is indeed an equivalence relation. Now combine together into a single subset of A those elements which are related to one another by the relation r. In our example, the set of elements related to a is $\{x \in A \mid x\,r\,a\} = \{a,c\}$ since $a\,r\,a$ and $c\,r\,a$ AND no other elements of A are related to the element a by the relation r. The set of elements related to b is $\{x \in A \mid x\,r\,b\} = \{b,d\}$, since $b\,r\,b$ and $d\,r\,b$, exhausting all elements related to b. Finally, $\{x \in A \mid x\,r\,e\} = \{e\}$, since the only element related to e is e itself. Notice that the resulting sets are disjoint, nonempty, and their union is A. Do we know what this means? It means there is a partition $\mathcal{P}$ formed by using the equivalence relation as a way to collect together into subsets all elements of A which are related to one another. That partition $\mathcal{P}$ is $\mathcal{P} = \{\{a,c\},\{b,d\},\{e\}\}$ and is called the partition $\mathcal{P}$ induced by the equivalence relation r. The foregoing discussion is generalized in the succeeding definition and theorem.

DEFINITION: If A is a nonempty set and r and equivalence relation on A and $y \in A$ then the equivalence class of y with respect to r is the set $\{x \mid x \in A \wedge x\,r\,y\}$, i.e., $\{x \in A \mid x\,r\,y\}$ and is denoted by $[y]$.

It should be clear that the cells of the partition in the paragraph just discussed are precisely the equivalence classes $[a], [b]$, and $[e]$. That is, $[a] = \{a,c\}$, $[b] = \{b,d\}$ and $[e] = \{e\}$ are the cells of $\mathcal{P}$.

THEOREM 13.1: If A is a nonempty set and r an equivalence relation on A, then there is a partition $\mathcal{P}$ induced by r on A and that partition is $\mathcal{P} = \{[y] \mid y \in A\}$. (Here $[y] = \{x \in A \mid x\,r\,y\}$ is the cell containing y and is the equivalence class of y with respect to r.)

PROOF: Suppose $A \neq \emptyset$ and r is an equivalence relation on A. Let $y \in A$ then $[y] = \{x \in A \mid x\,r\,y\}$. Note that $[y] \subseteq A$ since $x \in [y] \implies x \in A \wedge x\,r\,y$, whence $x \in A$. The objective is to show that if $\mathcal{P} = \{[y] \mid y \in A\}$ then $\mathcal{P}$ is a partition of A. According to its definition, there are three things to be satisfied by a set $\mathcal{P}$ of subsets $[y]$ of A, in order for $\mathcal{P}$ to be a partition of A. They are

1. $\bigcup_{y \in A}[y] = A$,
2. If $[y_1], [y_2] \in \mathcal{P}$ then $[y_1] = [y_2]$ or $[y_1] \cap [y_2] = \emptyset$, and
3. If $[y] \in \mathcal{P}$ then $[y] \neq \emptyset$.

PART 1: Let $x \in A$. (We can let $x \in A$ since $A \neq \emptyset$.) Since r is reflexive then $x\,r\,x$, so $x \in [x]$. Thus $\exists\,[z] \in \mathcal{P} \ni x \in [z]$. Since $[z] \subseteq \bigcup_{y \in A}[y]$, (by Exercise 10.6) and since $x \in [z]$, $x \in \bigcup_{y \in A}[y]$. Thus $A \subseteq \bigcup_{y \in A}[y]$. Thus we have containment one way. On the other hand, $[y] \subseteq A$ for each $y \in A$, so by Exercise 10.4, $\bigcup_{y \in A}[y] \subseteq A$. Consequently, $\bigcup_{y \in A}[y] = A$.

PART 2: Suppose $[y_1], [y_2] \in \mathcal{P}$ then $[y_1] = [y_2]$ or $[y_1] \neq [y_2]$. (Notice how we will follow the suggested approach in Chapter 5 to prove an assertion of the form $Q \Longrightarrow P_1 \vee P_2$. See the discussion noted around Theorems 5.4 and 5.5) If $[y_1] = [y_2]$ the assertion is valid. If $[y_1] \neq [y_2]$ then the objective is to show $[y_1] \cap [y_2] = \emptyset$. Suppose on the contrary that $[y_1] \cap [y_2] \neq \emptyset$. Then $\exists x \ni (x \in [y_1] \wedge x \in [y_2])$. Thus $x \, r \, y_1 \wedge x \, r \, y_2$. Therefore, $y_1 \, r \, x \wedge x \, r \, y_2$ and consequently $y_1 \, r \, y_2$. In Exercise 13.7 you are asked to show that $[y_1] = [y_2]$ follows from $y_1 \, r \, y_2$. Thus $[y_1] \neq [y_2]$ and $[y_1] = [y_2]$ which is a contradiction. Hence, the assumption $[y_1] \cap [y_2] \neq \emptyset$ is false. So the intersection $[y_1] \cap [y_2]$ is empty. Therefore in any case $[y_1] = [y_2]$ or $[y_1] \cap [y_2] = \emptyset$.

PART 3:   Let $[y] \in \mathcal{P}$. Then since $y \in [y]$, (see Part 1), $[y] \neq \emptyset$. (Notice that we started with an element of $\mathcal{P}$ and showed that element was nonempty exactly as required by the definition of partition.)

So it has been established that $\mathcal{P}$ is a partition of A.

EXAMPLE 13.5: Let A be a nonempty set of students at XYZ university. Let r be the known equivalence relation defined by "$x \, r \, y$ if and only if x and y are in the same class" (i.e., Freshman, F; Sophomore, S; Junior, J; Senior, N; or Graduate Student, G). It is further assumed that every student is in one and only one class and there is at least one student in each class. Suppose *Jane Doe* is a freshman, *Johnny Smith* is a sophomore, *Jerry Side* a junior, *Mary Jones* a senior, and *Grant Lawson* a graduate student. The equivalence class of *Jane Doe* is $[Jane \, Doe]$, i.e.,

$\{x \mid x$ is a student at XYZ and x is in the same class as *Jane Doe*$\}$

which is the equivalence class $\{x \mid x$ is a freshman at XYZ university $\}$. The other equivalence classes are obtained in a similar manner. The class of sophomores is $[Johnny \, Smith]$; the class of juniors, $[Jerry \, Side]$; the class of seniors, $[Mary \, Jones]$; and the class of graduate students, $[Grant \, Lawson]$. Thus the partition induced by r is $\mathcal{P} = \{F, S, J, N, G\}$.

EXAMPLE 13.6:   Let A be the set Z of all integers and let r be the equivalence relation given by

$$\forall i, j \in Z \big(i \, r \, j \text{ iff } \exists k \in Z \ni i - j = k \cdot 4 \big).$$

The equivalence classes induced by this equivalence relation are the sets S_0, S_1, S_2, S_3 of Example 13.4. In other words, $[0] = S_0$, $[1] = S_1$, $[2] = S_2$, and $[3] = S_3$. So the induced partition is $\{[0], [1], [2], [3]\}$.

Theorem 13.1 does not give the entire picture of the relationship between partitions and equivalence relations. It turns out that one can begin with a partition $\mathcal{P}$ and get an equivalence relation r from it. And furthermore the equivalence relation r so obtained, has the property that the partition it induces, via Theorem 13.1, is exactly the original partition $\mathcal{P}$. Here is this theorem.

THEOREM 13.2: Suppose $A \neq \emptyset$ and $\mathcal{P}$ is a partition of A. If the relation r on A is given by

$$r = \left\{ (x,y) \in A \times A \,\middle|\, \exists\, P \ni (P \in \mathcal{P} \wedge x \in P \wedge y \in P) \right\} \tag{1}$$

then

    a. r is an equivalence relation, and

    b. The partition induced by r is $\mathcal{P}$.

PROOF: Let $A \neq \emptyset$ and $\mathcal{P}$ be a partition of A and r as given in Equation 1.

PART a: To do this part, r must be shown to be *reflexive, symmetric,* and *transitive* on A.

Since $A \neq \emptyset$ let $x \in A$. Since $\mathcal{P}$ is a partition of A, $A = \bigcup \mathcal{P}$ so $x \in \bigcup \mathcal{P}$. Thus $\exists\, P \ni (P \in \mathcal{P} \wedge x \in P)$ (Why?) and therefore, $\exists\, P \ni (P \in \mathcal{P} \wedge x \in P \wedge x \in P)$, (Why?). Consequently, $(x,x) \in r$, by definition of r. Thus r is reflexive. The symmetry part is an exercise.

To show transitivity, suppose $(x,y) \in r$ and $(y,z) \in r$. The objective is to show $(x,z) \in r$. From the supposition $\exists\, P \ni (P \in \mathcal{P} \wedge x \in P \wedge y \in P)$ and $\exists\, P \ni (P \in \mathcal{P} \wedge y \in P \wedge z \in P)$. By Rule E.S., applied twice (and therefore yielding <u>two</u> <u>different</u> P's. Why?) we get $P_1 \in \mathcal{P} \wedge x \in P_1 \wedge y \in P_1$ as well as $P_2 \in \mathcal{P} \wedge y \in P_2 \wedge z \in P_2$. Now $y \in P_1 \cap P_2$. (Why?) Since $\mathcal{P}$ is a partition, $P_1 = P_2$. (Why?) (Now and only now do we know they are the same cells. This is NOT because of Rule E.S. applied once.) Thus $x \in P_1 \wedge z \in P_1$. This implies $\exists\, P \ni (P \in \mathcal{P} \wedge x \in P \wedge z \in P)$ from which $(x,z) \in r$ arises. Thus r is transitive.

Consequently, the three things needed in Part a for r to be an equivalence relation have been established.

PART b: The second conclusion requires more effort. From Equation (1), i.e., $r = \left\{ (x,y) \,\middle|\, \exists\, P \ni (P \in \mathcal{P} \wedge x \in P \wedge y \in P) \right\}$, and Part a, r is an equivalence relation and from Theorem 13.1, this r induces a partition, say $\mathcal{P}'$ on A where $\mathcal{P}' = \{ [y] \mid y \in A \}$. The task is to show $\mathcal{P} = \mathcal{P}'$. This may be accomplished by showing $\mathcal{P} \subseteq \mathcal{P}'$ and $\mathcal{P}' \subseteq \mathcal{P}$. We show $\mathcal{P}' \subseteq \mathcal{P}$, deferring the other containment to the exercises (see Exercise 13.8).

SHOW: $\mathcal{P}' \subseteq \mathcal{P}$. To do this let $[y] \in \mathcal{P}'$. Then $y \in A$ because $\mathcal{P}'$ is the partition induced by r on A. Since $\mathcal{P}$ is also a partition of A, $\exists\, P$ such that $(P \in \mathcal{P} \wedge y \in P)$. (Why?) By Rule E.S. $P_0 \in \mathcal{P} \wedge y \in P_0$. To get $[y] \in \mathcal{P}$, we merely need to show that the cell P_0, which is in $\mathcal{P}$, and the equivalence class $[y]$ are the same, i.e., $[y] = P_0$. To that end, let $x \in [y]$ then $x \in A \wedge (x, y) \in r$, from the definition of $[y]$. Therefore, by the definition of r, $\exists\, P \ni (P \in \mathcal{P} \wedge x \in P \wedge y \in P)$. By Rule E.S. $P_1 \in \mathcal{P} \wedge x \in P_1 \wedge y \in P_1$. But since P_1 and P_0 are in $\mathcal{P}$ and since $y \in P_1$ and $y \in P_0$, P_1 must equal P_0. Thus $x \in P_0$, so $x \in [y] \implies x \in P_0$. Hence $[y] \subseteq P_0$. Now part of the equality, $[y] = P_0$ is established.

To complete the proof that $[y] = P_0$, suppose $[y] \neq P_0$. Then since $[y] \subseteq P_0$ from the preceeding paragraph, $\exists\, x \in P_0 \ni x \notin [y]$, say $x_0 \in P_0 \wedge x_0 \notin [y]$ (by Rule E.S.). Thus $x_0 \in P_0 \wedge (x_0, y) \notin r$. Since $\mathcal{P}$ is a partition of A and $P_0 \in \mathcal{P} \wedge x_0 \in P_0$ then we claim that $y \notin P_0$. For if y were in P_0 then $(x_0, y) \in r$, which it is not. Now since $[y] \subseteq P_0$, $y \in P_0$. But then $y \in P_0 \wedge y \notin P_0$. This contradiction leads to the denial of the assumption at the beginning of the paragraph, namely $[y] \neq P_0$. Thus $[y] = P_0$. The implication

$$\Big([y] \in \mathcal{P}' \implies [y] = P_0\Big) \wedge \Big(P_0 \in \mathcal{P}\Big),$$

which has just been done, establishes $\mathcal{P}' \subseteq \mathcal{P}$.

After showing $\mathcal{P} \subseteq \mathcal{P}'$, then Part b will be done (see Exercise 13.8b). Consequently, the theorem is established.

The equivalence relation r obtained on A from the partition $\mathcal{P}$ on A is called the <u>equivalence</u> <u>relation</u> <u>induced</u> <u>by</u> $\mathcal{P}$ <u>on</u> $\underline{A}$. The means of getting this equivalence relation is spelled out in the theorem. That is, one employs the equality

$$r = \Big\{(x, y) \,\Big|\, \exists\, P \ni (P \in \mathcal{P} \wedge x \in P \wedge y \in P)\Big\}$$

from Equation (1), by pairing elements x and y exactly when they occur in the same cell of that partition. Here is an example of how this is done.

EXAMPLE 13.7:   Let $A = \{a, b, c\}$ and $\mathcal{P} = \{\{a, c\}, \{b\}\}$. Then clearly $\mathcal{P}$ is a partition of A. To find the induced equivalence relation, simply

DO WHAT EQUATION (1) instructs. Let r be as in Equation (1). Then it is true that $(a, a) \in r$, for we know there is a cell, namely $P_1 = \{a, c\}$ of $\mathcal{P}$, such that $a \in P_1$ and $a \in P_1$. Likewise $(b, b) \in r$ since there is a cell $P_2 = \{b\}$ of $\mathcal{P}$ such that $b \in P_2$ and $b \in P_2$. And $(c, c) \in r$ since there is a cell $P_3 = \{a, c\}$ such that $c \in P_3$ and $c \in P_3$. Also, $(a, c) \in r$ and $(c, a) \in r$ because there is a cell $P_4 = \{a, c\}$ in $\mathcal{P}$ such that $a \in P_4$ and $c \in P_4$. No other pair $(x, y) \in r$, since there is no other way that x and y can share the same cell. So $r = \{(a, a), (b, b), (c, c), (a, c), (c, a)\}$. What cells are P_1, P_2, P_3, and P_4?

Exercises:

13.1 In each part, determine that the given r is an equivalence relation on the given set A then find the partition $\mathcal{P}$ induced by that r.

    a. $r = \{(a, a), (b, b), (c, c), (a, b), (b, a)\}$ on $A = \{a, b, c\}$.

    b. $r = I_A$ on $A = \{1, 2, 3, 4, 5\}$.

    c. $r = A \times A$ on $A = \{1, 2, 3, 4, 5\}$.

    d. $r = A \times A$ on A, for arbitrary A.

    e. Suppose $Q(x, y)$ is the predicate given by, "x and y have the same number of elements." And let $r = \{(x, y) \mid x, y \in A \land Q(x, y)\}$. And suppose A is the power set of $\{a, b, c\}$, i.e., $A = \{\emptyset, \{a\}, \{b\}, \{c\}, \{a, b\}, \{a, c\}, \{b, c\}, \{a, b, c\}\}$.

13.2 Below you will find partitions of the indicated sets. First verify they are indeed partitions. Then in each case find the equivalence relation r induced by the given partition.

    a. $\mathcal{P} = \{\{a, b\}, \{c\}, \{d\}\}$ on $A = \{a, b, c, d\}$.

    b. $\mathcal{P} = \{E, D\}$ where E and D are the even and odd integers, respectively (on the set Z of integers).

    c. $\mathcal{P} = \{\{a, b, c\}\}$ for $A = \{a, b, c\}$.

13.3 Let $A = \{1, 2, 3\}$ and $r = \{(1, 1), (2, 2), (3, 3), (1, 2), (2, 1)\}$. Verify that r is an equivalence relation on A. Next, find the equivalence classes $[1], [2], [3]$ and the partition $\mathcal{P}$ induced by r. Using the $\mathcal{P}$, just found, find the equivalence relation induced by $\mathcal{P}$ on A, via Theorem 13.2.

13.4 Let $A = \{1, 2, 3, 4\}$. Find $[1], [2], [3], [4]$ and the partition $\mathcal{P}$ induced by r on A in the case that

    a. $r = I_A$.

    b. $r = A \times A$.

13.5 Let $A = Z$ and n be fixed in Z^+. Define "x is congruent to y modulo n," denoted by "$x \equiv y \bmod n$," as follows

$$(\forall x, y \in Z)\left[(x \equiv y \bmod n) \text{ iff } (\exists k \in Z \ni x - y = k \cdot n)\right].$$

a. Prove that *"congruence modulo n"* is an equivalence relation on the set Z of integers.

b. If $n = 6$ find the equivalence classes.

c. If $n = 6$ find the partition induced by the equivalence relation *"congruence modulo 6."*

13.6 For each set A and partition $\mathcal{P}$ find the equivalence relation induced on A by $\mathcal{P}$.

a. $A = \{1, 2, 3\}$ and $\mathcal{P} = \{\{1, 3\}, \{2\}\}$.

b. $A = \{a, b, c, d, e\}$ and $\mathcal{P} = \{\{a, b, c\}, \{d, e\}\}$.

c. A is any set and $\mathcal{P} = \{\{a\} \mid a \in A\}$.

d. A is any set and $\mathcal{P} = \{A\}$.

13.7 Prove that if r is an equivalence relation on A and $[y_1]$ and $[y_2]$ are two equivalence classes with respect to r and if $x \in [y_1] \cap [y_2]$ then $[y_1] = [y_2]$. This is part of the proof of Theorem 13.1.

13.8 Prove

a. The symmetry part of Theorem 13.2.

b. The remainder of the proof that $\mathcal{P}' = \mathcal{P}$ in Theorem 13.2.

13.9 What happens if you try to induce an equivalence relation on a set A by some set of subsets of A that is not a partition of A? To help answer this, let $A = \{a, b, c\}$ and $Q = \{\{a, b\}, \{b, c\}\}$ then Q is certainly not a partition of A. Use Equation (1) to induce a relation r on A. Is that relation r an equivalence relation on A? If not, which of *reflexive*, *symmetric*, and *transitive* is true? Which are not?

13.10 What happens if you try to induce a partition on a set A by using a relation that is not an equivalence relation? See Exercise 9 for ideas on what you might try.

13.11 Let $C = \{f \mid f \text{ is differentiable on the real line}\}$ define $\sim$ by: If $f_1, f_2 \in C$ then $f_1 \sim f_2$ iff f_1 and f_2 differ by a constant, i.e., $f_1(x) = f_2(x) + c$ for some constant c.

a. Show: $\sim$ is an equivalence relation on C.

b. If $f_1 \sim f_2$ what can you say about f_1' and f_2'?

c. Find the equivalence class: $[x^2] = \{f \in C \mid f \sim x^2\}$ of x^2.

d. Find an alternate way to characterize $\sim$ using the derivative. That is $f_1 \sim f_2$ iff _____.

13.12 Let $C[0,1] = \{f \mid f : [0,1] \longrightarrow E_1$ and f is continuous$\}$. Define $\sim$ on $C[0,1]$ by: If $f_1, f_2 \in C[0,1]$ then

$$f_1 \sim f_2 \ \text{ iff } \ \int_0^1 f_1(x)\, dx = \int_0^1 f_2(x)\, dx.$$

a. Show $\sim$ is an equivalence relation.

b. Describe in words the equivalence class $[f]$ of f when f is the function defined by $f(x) = 2x$.

c. Consider the function f given by $f(x) = 2x$ on $[0,1]$. Find a function g on $[0,1]$ such that $g \sim f$ and g is:

   i. A constant function,

   ii. An exponential function of the form $y = ae^x$. Find a.

   iii. A sine function of the form $y = a\sin x$. Find a.

   iv. A quadratic function of the form $y = ax^2$. Find a.

   Each g found must be equivalent with respect to $\sim$ to $f(x) = 2x$ on $[0,1]$.

13.13 Let $\mathcal{P}$ be a partition of a set A. Define r by $x\, r\, y$ iff $\exists\, B \in \mathcal{P} \ni x \in B \wedge y \in B$. What is wrong with the "proof" that begins this way, "If $x\, r\, y \wedge y\, r\, z$ then $\exists\, B \in \mathcal{P} \ni x \in B \wedge y \in B \wedge z \in B$, etc."

CHAPTER 14

FUNCTIONS

One of the most important concepts used in mathematics is the notion of function. It is difficult to go very far in any area of mathematics (not to mention other disciplines) without using functions in some way. From the study of high school algebra through calculus, to the heights of graduate level mathematics, functions play a role that few, if any, mathematical notions exceed.

No doubt the student has already studied functions and has the kind of knowledge needed in courses like calculus. But in this text, a study of functions for their own sake is initiated. A foundation for their subsequent use in this text is established and a framework for their application to other mathematics is constructed. Since some of these uses of functions are deeper and more abstract than required for calculus, an appropriate definition in the framework of relations is needed.

DEFINITION: Let A and B be sets and $f \subseteq A \times B$ (i.e., f is a relation between A and B) then:

1. f is a <u>function</u> (or a <u>function between A and B</u>) if and only if $\forall a, b_1, b_2 \left[(a, b_1) \in f \text{ and } (a, b_2) \in f \implies b_1 = b_2 \right]$.
2. f is a <u>function from A to B</u> (denoted by $f : A \longrightarrow B$) if and only if f is a function in the sense of Part 1 and $\mathrm{dom}(f) = A$.

The first part says that a function between A and B cannot have any element of A paired with two different elements of B. See *Figure 14.1b*; no element $a \in A$ may be paired with two different elements, b_1 and b_2 of B. When Part 1 holds, f is said to be <u>single</u> <u>valued</u> or <u>well</u> <u>defined</u>.

Since $f \subseteq A \times B$, f is a relation between A and B, so the notion of the domain of f, as a relation, has already been explored. Thus the notion of a function *from A to B* should also be understood. Likewise, the *range* of a function is, as defined for relations; $\mathrm{ran}(f) = \{b \in B \mid (a, b) \in f\}$. The entire set B is called the <u>codomain</u> of f in this text, and not the "range" of f which has various different meanings in mathematical literature. The expression, "$f : A \longrightarrow B$," is read as, "f is a function from A to B."

EXAMPLE 14.1: Let $A = \{a, b, c\}$ and $B = \{1, 2, 3, 4\}$ then the set $\{(a, 1), (c, 4)\}$ is a function *between* A and B, but not *from A*. However, the function $f = \{(a, 2), (b, 1), (c, 2)\}$ between A and B is *from A to B*,

Figure 14.1 a   *Figure 14.1 b*

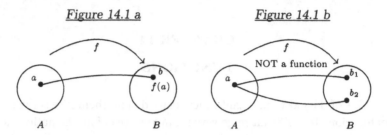

since its domain is A. Likewise $\{(a,a),(b,c),(c,b)\}$ is a function *from A to A*, since it is well defined and its domain and codomain are A.

EXAMPLE 14.2:   Let Z be the set of integers and $g = \{(x,2x) \mid x \in Z\}$. Is g a function from Z to Z? Now $g \subseteq Z \times Z$, and $\mathrm{dom}(g) = Z$ and the implication $\big[(x,y_1) \in g \wedge (x,y_2) \in g \implies y_1 = y_2\big]$ holds because $(x,y_1) \in g$ and $(x,y_2) \in g$ imply $y_1 = 2x$ and $y_2 = 2x$, hence $y_1 = 2x = y_2$, so the answer to the question is *yes*, g is a function from Z to Z.

EXAMPLE 14.3:   Let T be a set of triangles and E_1 the set of reals, then the relation $\{(x,y) \mid x \in T \wedge y \in E_1 \wedge x$ has area $y\}$ is a relation from T to E_1. It is a function because there is no triangle having two different areas. The domain is T because every triangle has an area, so it is a function from T to E_1.

Diagrams are frequently helpful in explaining certain features about functions. One typical type of diagram for a general function $f : A \longrightarrow B$ consists of two ovals, or circles, to denote A and B and an arc, denoted by f, perhaps with an arrowhead to indicate it is from A to B. And perhaps some labeled points are included. In some diagrams there may be an arc joining $a \in A$ with some $b \in B$ (see *Figure 14.1a*).

DEFINITION: Let $f : A \longrightarrow B$ and $(a,b) \in f$. Then b is called the image of a under f or the value of f at a and b is written in f *notation* as $b = f(a)$. In addition, a is called a preimage of b under f.

In *Figure 14.1a* the arc joining a and $b = f(a)$ signifies that $(a,b) \in f$. Here b is the image of a under f, while a is a preimage of b under f.

The terminology is, "b is the image of a under f" because, for a function, there is only one b for each $a \in A$. And in addition, the terminology is, "a is a preimage of b under f," because any given $b \in B$ may have more than one preimage a as illustrated in Examples 14.1 and 14.4 and *Figure 14.2*.

Figure 14.2

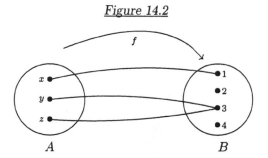

EXAMPLE 14.4:   A diagram of this example is given in *Figure 14.2*.
Suppose $A = \{x, y, z\}$, $B = \{1, 2, 3, 4\}$, and $f = \{(x, 1), (y, 3), (z, 3)\}$,
then 1 is the image of x under f, 3 is the image of y under f, and 3 is
also the image of z under f. So $f(x) = 1$, $f(y) = 3$, and $f(z) = 3$. Also,
we say x is a preimage of 1 under f, and y is a preimage of 3 under f
and z is also a preimage of 3 under f. It should now be clear why the
terminology is the image of a under f and a preimage of b under f.

One way to signify pairs in this f is by the arcs joining the elements
of A with elements of B as determined by the pairs in the given f. The
arcs in *Figure 14.2* illustrate that 1 is the image of x under f while 3 is
the image of both y and z under f. The diagram also illustrates that x
is a preimage of 1 under f and both y and z are preimages of 3 under f.
The elements 2 and 4 of B are not images of any element of A and do
not have preimages under f.

In some areas of mathematics, the terms *independent* and *dependent*
variables are used. If f is a function, then for the pairs $(x, y) \in f$, x
is called the independent variable and y the dependent variable. For
example, in each of the functions $\{(x, y) \mid x, y \in E_1 \text{ and } y = 2x\}$ and
$\{(x, y) \mid x, y \in E_1 \text{ and } y = x^2 - 2x - 1\}$, x is called the independent
variable and y is the dependent variable.

Another way of schematically illustrating certain functions is commonly used in mathematics, especially calculus. Functions f which are
subsets of $E_1 \times E_1$ can be "graphed" by plotting those points (x, y) on
$E_1 \times E_1$ that are in f. This method should seem quite familiar.

EXAMPLE 14.5:   The relation $f = \{(x, y) \mid x, y \in E_1 \wedge y = (x - 1)^2 - 2\}$
has the graph in *Figure 14.3*. From its definition f can be seen to be a
function *from* E_1, because for each x selected in E_1 there is exactly one
image y determined by the equation $y = (x - 1)^2 - 2$. The fact that there
are two different x's paired with certain y's, such as $(0, -1)$ and $(2, -1)$,
does not keep it from being a function, rather it keeps it from being $1 - 1$.

Figure 14.3

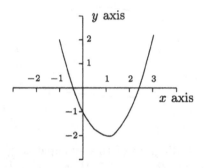

$$f=\{(x,y) \mid x,y \in E_1 \text{ and } y=(x-1)^2-2\}$$

See the following definition.

DEFINITION: If $f : A \longrightarrow B$ (i.e., f is a function from A to B) then f is said to be <u>one to one</u> (abbreviated $1-1$) if and only if the following statement is true

$$\forall a_1, a_2 \in A, \ \forall b \in B\Big[(a_1,b) \in f \ \wedge \ (a_2,b) \in f \Longrightarrow a_1 = a_2\Big].$$

A one–to–one function may be denoted by $f : A \xrightarrow{1-1} B$, which is read as, "$f$ is a $1-1$ function from the set A to the set B."

EXAMPLE 14.6:   The function $f = \{(x,1),(y,3),(z,3)\}$ of Example 14.4 is not $1-1$, since $(y,3) \in f$ and $(z,3) \in f$, but $y \neq z$.

A diagram of a $1-1$ function does not have two different arcs terminating at the same element of B. So the function in *Figure 14.2* is not $1-1$. As already observed, the function f in Example 14.5 and *Figure 14.3* is not $1-1$.

EXAMPLE 14.7:   The function $g = \{(x,2x) \mid x \in Z\}$ from Z to Z is $1-1$, for if $(x_1,y) \in g$ and $(x_2,y) \in g$ then $y = 2x_1$ and $y = 2x_2$, so $2x_1 = 2x_2$. By properties of the integers $x_1 = x_2$, so g is $1-1$. The graph of g, as a subset of $Z \times Z$, is given in *Figure 14.4a*.

Example 14.7 is a good place to illustrate another means of identifying that some functions are one–to–one. A function is a relation which is a subset of a Cartesian product. If $f \subseteq E_1 \times E_1$ then the $1-1$ property is equivalent to the so-called horizontal line rule.

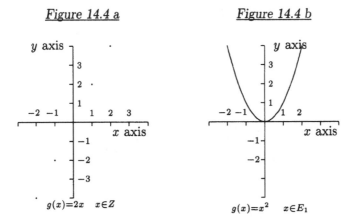

Figure 14.4 a — $g(x)=2x \quad x \in Z$

Figure 14.4 b — $g(x)=x^2 \quad x \in E_1$

HORIZONTAL LINE RULE: A function f satisfies the <u>horizontal line rule</u> if and only if no horizontal line intersects the graph of the function at two or more points.

The function given in *Figure 14.4a* is $1 - 1$ by the horizontal line rule, while the function $g = \{(x, x^2) \mid x \in E_2\}$ in *Figure 14.4b* does not appear to satisfy the horizontal line rule. It is clear that a function f is $1 - 1$ iff it satisfies the horizontal line rule, however that does not mean this is necessarily obvious from the graph (see Exercise 14.19).

DEFINITION: If $f : A \longrightarrow B$ then f is said to be <u>onto B</u> (or more completely $\underline{f}$ <u>is a</u> function <u>from</u> A **onto** B) if and only if

$$\forall b \Big[b \in B \implies \exists a \ni \big(a \in A \land f(a) = b\big)\Big].$$

This situation may be denoted by $f : A \overset{\text{onto}}{\longrightarrow} B$. When the context is clear, or when it is clear what A and B are, we may say "f is onto," with B understood as the range of the function.

EXAMPLE 14.8: The function $f = \{(x, 2x) \mid x \in E_1\}$ from E_1 to E_1 is "onto" E_1. For suppose $y \in E_1$ then $\frac{1}{2}y \in E_1$. Let $x = \frac{1}{2}y$ then $x \in E_1 \land f(x) = 2x = 2(\frac{1}{2}y) = y$. So it has been shown that

$$\forall b \Big[b \in E_1 \implies \exists x \big(x \in E_1 \land f(x) = b\big)\Big].$$

In other words, f is onto E_1 (see *Figure 14.5*). On the other hand, IS the function $g = \{(x, 2x) \mid x \in Z\}$ onto Z (see *Figure 14.4a*)?

Figure 14.5

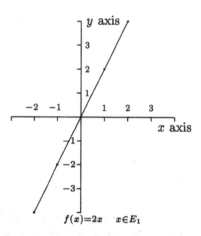

$f(x)=2x \quad x \in E_1$

There is a "graphical" way to interpret whether a function from E_1 to E_1 is "onto" a certain subset B of E_1. If every horizontal line through an element of B on the y axis, intersects the graph of f in at least one place then f is onto B. Although this is important to know, it may not be easy to determine that f is "onto" by this means alone (see Exercise 14.19).

EXAMPLE 14.9:
   a. The function $f = \{(x, 2x) \mid x \in E_1\}$ from Example 14.8 is $1-1$ as well as onto E_1. Its graph is given in *Figure 14.5*. Intuitively, each horizontal line through a point $b \in E_1$ (on the y axis) intersects the graph of f once and only once.

   b. The function $g = \{(x, 2x) \mid x \in Z\}$, graphed in *Figure 14.4a*, is not onto Z because $3 \in Z$, but $(x, 3) \notin g$ for any $x \in Z$. That is, the horizontal line through $y = 3$ does not intersect the graph of g.

EXAMPLE 14.10:
   a. The function $f = \{(x, 1), (y, 3), (z, 3)\}$ from $A = \{x, y, z\}$ to $B = \{1, 2, 3\}$, graphed in *Figure 14.2*, is not "onto B" since $2 \in B$, but $2 \neq f(x)$ for each $x \in A$.

   b. The function $g(x) = x^2$, $x \in E_1$, graphed in *Figure 14.4b*, is not "onto E_1" since $\sim \exists x(x \in E_1 \wedge g(x) = -1)$. This function is however, onto the subset $B = \{x \in E_1 \mid x \geq 0\}$ of E_1. This is because if $b \in B$ then $b \geq 0$, so $\sqrt{b} \in E_1$. If $x = \sqrt{b}$ then $x \in E_1 \wedge g(x) = x^2 = (\sqrt{b})^2 = b$. That is, $\forall b[b \in B \Longrightarrow \exists x(x \in E_1 \wedge g(x) = b)]$, which satisfies the definition of "onto

B." In other words, every horizontal line through an arbitrary point $b \in B$ (i.e., $b \geq 0$) on the y axis intersects the graph of g.

Sometimes, as in Example 14.9a and *Figure 14.5*, a function f from A to B is both $1-1$ and "onto B." In this case the following definition assigns a name to such a function.

DEFINITION: If A and B are sets then A and B are <u>in one–to–one correspondence</u> if and only if

$$\exists f \ni f : A \xrightarrow{\;1-1\text{ onto}\;} B.$$

Such a function f, is said to be a <u>one–to–one</u> <u>correspondence</u>.

EXAMPLE 14.11: The function $h = \{(a,1),(b,5),(c,3)\}$ is a $1-1$ function from $A = \{a,b,c\}$ onto $B = \{1,3,5\}$. So h is a $1-1$ correspondence between A and B. Also we say A and B are <u>in</u> $1-1$ correspondence.

EXAMPLE 14.12: We will now establish that $f = \{(x,y) \mid x,y \in E_1 \wedge y = e^x\}$ is a $1-1$ function from E_1 onto the set E_1^+ of positive real numbers. Thus E_1 and E_1^+ are <u>in</u> one to one correspondence.

PROOF: There are several things to ensure. f must be a function between E_1 and E_1. It must be from E_1. It has to be $1-1$. Finally, it has to be onto E_1^+.

We first show that f is a function. Let (x,y_1) and (x,y_2) be elements of f. Then $y_1 = e^x$ and $y_2 = e^x$, hence $y_1 = y_2$, so f is a function, i.e., single valued. Since $\operatorname{dom}(f) = \{x \mid x,y \in E_1 \wedge y = e^x\} = E_1$, f is a function *from* E_1 to E_1.

To establish that f is $1-1$, suppose $(s,y) \in f$ and $(t,y) \in f$. Then $y = e^s$ and $y = e^t$, so $e^s = e^t$. Does that mean that $s = t$? To answer that, consider the equation below. Since the natural logarithm is a function and $e^s = e^t$, then

$$s = \ln(e^s) = \ln(e^t) = t.$$

So f is $1-1$.

Finally, we show f is onto E_1^+. To do that, suppose $y \in E_1^+$, then $\ln y \in E_1$. (Why?) Let $x = \ln y$. Then $x \in E_1$ and $f(x) = e^x = e^{\ln y} = y$. Hence, $\forall y \in E_1^+ \; \exists x \in E_1 \ni f(x) = y$. (What is the x asserted to exist in the previous statement?) Thus f is onto E_1^+.

Consequently f is a $1-1$ correspondence between E_1 and E_1^+. That is, E_1 and E_1^+ are <u>in</u> $1-1$ correspondence, as we wanted. (Among other things, this establishes the remarkable fact that a set may be in $1-1$ correspondence with a proper subset of itself.)

Sometimes a function in $E_1 \times E_1$ is informally given by a "rule." For example, $y = f(x) = x^2 + 2x - 1$ is such a rule. A rule is typically a formula which says how the dependent variable and independent variable are related. It is easy to decide that the given $y = x^2 + 2x - 1$ is a function from E_1 to E_1 because for each $x \in E_1$, one and only one y can be obtained by going through the computation dictated by the "rule."

Since its graph is a parabola opening up in $E_1 \times E_1$, we know from calculus that it is not one–to–one and not onto E_1. It is, however, onto the set $\{x \mid x \in E_1 \land x \geq -2\}$. (Explain.)

The apparent absence of a domain and range of a function given by a *rule*, is not really an absence. The domain of such a function is understood (unless otherwise stated) to be the maximum possible set of values that the independent variable may assume and this domain is called the <u>natural</u> <u>domain</u>. The codomain may be any set containing the range of the function.

The function given by the rule $y = g(x) = \sqrt{\frac{x}{x-1}}$ has domain $A = \{x \mid x \in E_1 \land x \notin (0,1]\}$ and range $B = \{y \mid y \in E_1 \land y \geq 0\}$. And in fact, g is a one–to–one function from A onto B. Sketch the graph of g as an exercise. (See Exercise 14.18.)

The notions of $1-1$ and "onto" can be described in the f notation mentioned above. That is, if $y = f(x)$ then

$$f \text{ is } 1-1 \text{ iff } \left[f(x_1) = f(x_2) \implies x_1 = x_2\right].$$

This condition prevents two different x coordinates from being paired with the same y coordinate. Again, if $y = f(x)$ is a function from A to B then that

$$f \text{ is } onto \ B \text{ iff } \forall y \in B \ \exists \, x \in A \ni y = f(x).$$

As we draw near the end of this section, we consider a special mathematical function. It is called the characteristic function and it is used in a number of places in mathematics as well as later in this text.

Figure 14.6

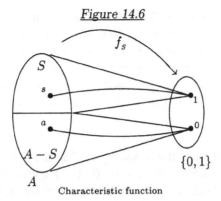

Characteristic function

EXAMPLE 14.13: Let A be a set and $S \subseteq A$. Define the <u>characteristic function</u> f_S, of S in A, to be $f_S : A \longrightarrow \{0,1\}$ where f_S is given by

$$f_S(x) = \begin{cases} 0, & \text{if } x \in A - S; \\ 1, & \text{if } x \in S. \end{cases}$$

Since the implication $x \in A \Longrightarrow x \in S \vee x \in A - S$ is true, it is clear that $\operatorname{dom}(f_S) = A$. It is also clear that a codomain is $\{0,1\}$. If $\emptyset \neq S$ and $S \neq A$ then $A - S \neq \emptyset$ so $\exists x_0 \in A - S$. Thus $f_S(x_0) = 0$. But since $S \neq \emptyset \; \exists x_1 \in S$. Thus $f_S(x_1) = 1$. So as long as $\emptyset \subset S \subset A$, f_S has range $\{0,1\}$. If $S = \emptyset$, then $f_S(x) = 0 \; \forall x \in A$. And if $S = A$ then $f_S(x) = 1$ $\forall x \in A$.

A schematic diagram of the characteristic function is given in *Figure 14.6*. As you can see, elements such as $s \in S$ are paired with 1 while elements such as $a \in A - S$ are paired with 0.

Finally, we know that functions are relations, i.e., sets of ordered pairs so if $f : A \longrightarrow B$ and $g : C \longrightarrow D$ are two functions then the question that might be asked is, "When are the functions f and g equal?" Now $f = g$ is their equality as sets of ordered pairs. In subsequent sections we will need another way to determine equality of functions. The following theorem provides an answer to this question. Its proof is left to be done in Exercise 14.10.

THEOREM 14.1: If $f : A \longrightarrow B$ and $g : A \longrightarrow B$ are functions then $f = g$ if and only if $\forall x \in A \big[f(x) = g(x) \big]$.

See Exercise 14.11 for a slightly more general result.

Let us summarize what we did in this section. The important concept of function was introduced and the means of diagramming and graphing them were considered as well. The notions of a function being $1-1$ and onto a set were also considered. Dependent and independent variables were defined and the concept of $1-1$ correspondence was initiated. A special function, the characteristic function, was defined and finally the question of determining whether two functions are equal was raised. Each of these concepts is important in the rest of this work.

Exercises:

14.1 Identify whether the following are functions, functions from A to B, $1-1$ functions from A to B, $1-1$ correspondences, or none of these. Explain your answers. In other words, if it is a function, show it. If it is $1-1$, show it, etc.

a. $A = B = \{1, 2, 3, 4\}$ and $f = \{(1, 2), (3, 1), (2, 4), (4, 3)\}$.

b. $A = B = Z$ and $f = \{(x, 2x) \mid x \in Z\}$.

c. $A = B = E_1$ and $f = \{(x, 2x) \mid x \in E_1\}$.

d. $A = E_1^+$ and $B = E_1$ and $f = \{(x, y) \mid x \in E_1^+ \wedge y = \ln x\}$.

e. $A = B = E_1^+$ and $f = \{(x, y) \mid x, y \in E_1 \wedge x^2 + y^2 = 1\}$.

14.2 Suppose $f = \{(1, 1), (2, 3), (3, 1), (4, 3), (5, 2)\}$. Find $\operatorname{dom}(f)$, $\operatorname{ran}(f)$, $f(1)$, $f(2)$, $f(3)$, $f(4)$, and $f(5)$. Is f 1–1? Is f onto $\{1, 2, 3, 4, 5\}$?

14.3 If $g = \{(1, 2), (2, 3), (3, 4), (4, 1), (5, 5)\}$, then is g $1-1$? Is it onto $\{1, 2, 3, 4, 5\}$? Explain. What is $\operatorname{dom}(g)$? $\operatorname{ran}(g)$? What is g^{-1}? Is g^{-1} a function?

14.4 If $f : A \longrightarrow A$ is one-to-one, is f necessarily onto A? If f is onto A, is it necessarily $1-1$? Explain your answers.

14.5 If $f(x) = x^3$ in $E_1 \times E_1$ then what are its domain and range? Is f a $1-1$ function? What set is f onto?

14.6 Repeat Exercise 14.5 for $f(x) = x$.

14.7 Repeat Exercise 14.5 for $f(x) = \sin x$.

14.8 Repeat Exercise 14.5 for $f(x) = \tan x$.

14.9 Prove that if A is a set, then $I_A : A \longrightarrow A$ is a $1-1$ function from A onto A.

14.10 (*Alternate Equality of Functions*) Prove that if $f : A \longrightarrow B$ and $g : A \longrightarrow B$ are functions then

$$f = g \text{ if and only if } \forall x \in A\big[f(x) = g(x)\big].$$

14.11 (*Alternate Equality of Functions*) Prove that if $f : A \longrightarrow B$ and $g : C \longrightarrow D$ are functions then

$$f = g \text{ if and only if } A = C \text{ and } f(x) = g(x) \ \forall x \in A.$$

14.12 a. If f is a function from A to B and f is not $1-1$, does f^{-1} necessarily exist? (Be careful. Remember a function is a relation.)

b. If f^{-1} does exist, is f^{-1} a function? Explain.

c. If f is a function from A onto B, what does that say about the relation f^{-1}?

14.13 Prove there exists a function f such that $f : A \times B \longrightarrow B \times A$ and such that f is $1-1$ and onto.

14.14 Prove there exists a function g such that $g : A \times (B \times C) \longrightarrow (A \times B) \times C$ is a $1-1$ function onto.

14.15 Prove there exists a function h such that $h : A \times \{0\} \longrightarrow A$ is a function and such that h is $1-1$ and onto A.

14.16 Let $A = \{a, b, c\}$ and S as indicated. Find $f_S(x)$ for each x in A. Then write f_S as a set of ordered pairs.

a. $S = A$ $\qquad\qquad f_s = $ _____

b. $S = \{a, b\}$ $\qquad f_s = $ _____

c. $S = \{b\}$ $\qquad\quad f_s = $ _____

d. $S = \emptyset$ $\qquad\qquad f_s = $ _____

14.17 a. Let $f_1 : A_1 \longrightarrow B$ and $f_2 : A_2 \longrightarrow B$. Prove: If A_1 and A_2 are disjoint then $f_1 \bigcup f_2 : A_1 \bigcup A_2 \longrightarrow B$.

b. Construct an example, with f_1 and f_2 having only finitely many pairs, illustrating that the disjointness condition is essential.

c. Suppose $f : A_1 \overset{1-1}{\longrightarrow} B_1$, $f : A_2 \overset{1-1}{\longrightarrow} B_2$, $A_1 \cap A_2 = \emptyset$, and $B_1 \cap B_2 = \emptyset$. Prove $f_1 \cup f_2$ is a $1-1$ function from $A_1 \cup A_2$ to $B_1 \cup B_2$.

14.18 The two functions $f(x) = 2x, x \in E_1$ and $f(x) = 2x, x \in Z$ are given by rules. Are they the same function? (Remember, a function is a certain set of ordered pairs.)

14.19 Define $f(x)$ by

$$f(x) = \begin{cases} x^3, & \text{if } x \text{ is rational and positive;} \\ -x^3, & \text{if } x \text{ is irrational and positive;} \\ x^3, & \text{if } x \text{ is irrational and negative;} \\ -x^3, & \text{if } x \text{ is rational and negative.} \end{cases}$$

Is f a function? If it is, is it $1 - 1$ and is it onto E_1?

14.20 Let A be a set and $\emptyset$ the empty set. Is $\emptyset : \emptyset \longrightarrow A$ a function (see Exercise 12.5)?

14.21 Prove that if $f : A \xrightarrow{\ 1 - 1 \text{ onto}\ } B$ then $f^{-1} : B \longrightarrow A$ is a function AND is $1 - 1$ AND is onto. (Onto what?) Since a function is a relation, see the definition of *inverse* of a relation.

14.22 If a and b are real and $a \neq 0$ and $f(x) = ax + b \ \forall x \in E_1$ then prove f is a function from E_1 to E_1 and is $1 - 1$ and onto E_1. Find f^{-1} (see Exercise 14.21).

14.23 Sketch a graph of the function $y = \sqrt{\frac{x}{x-1}}$. Prove it is a function.

Find its domain. Use the graph to see that it is $1 - 1$. Verify this analytically. Then, without verification, find an appropriate set B to which it is "onto."

14.24 Write a paragraph describing how to use "horizontal lines" to determine whether a function $f \subseteq E_1 \times E_1$ is $1 - 1$, and whether it is onto some particular subset B of E_1. Following this line of thinking, find a similar "geometric" way to determine whether such an f is "from" a subset A of E_1.

CHAPTER 15

COMPOSITION OF FUNCTIONS

The development of functions thus far has been limited to looking at individual functions and investigating properties they possess. The next matter to be pursued is a means of putting functions together or modifying them in various ways to get new functions. One of these ways of assembling functions to form a new function is by composition.

DEFINITION: Let A, B, and C be sets and let $f \subseteq A \times B$ and $g \subseteq B \times C$ be relations. Then the <u>composition</u> of f and g, denoted $g \circ f$, is given by

$$g \circ f = \{(x, z) \mid (x, z) \in A \times C \wedge \exists\, y \in B \ni (x, y) \in f \wedge (y, z) \in g\}.$$

Composition is really not new, since it has been explored in calculus in regard to the chain rule.

Clearly $g \circ f \subseteq A \times C$ so $g \circ f$ is a relation between A and C. If g and f are really functions then it natural to ask whether $g \circ f$ is a function. The answer to this question is in Theorem 15.1. But first let us look at a diagram of what is involved.

In *Figure 15.1* the function f pairs elements of A with elements of B while g pairs elements of B with elements of C. Their composition pairs elements of A directly with elements of C. The next theorem says the composition of functions is a function and it says exactly how the elements of A are paired with elements of C.

THEOREM 15.1: If $f : A \longrightarrow B$ and $g : B \longrightarrow C$ then $g \circ f : A \longrightarrow C$. Also $(g \circ f)(x) = g\big(f(x)\big)$ for all $x \in A$.

PROOF: As observed above, $g \circ f \subseteq A \times C$. Although it seems clear that $\text{dom}(g \circ f) \subseteq A$, it is an exercise to prove it. Conversely, if $a \in A$ then $\exists\, b \in D \ni (a, b) \in f$ because $f : A \longrightarrow B$. Since $g : B \longrightarrow C$ and since $b \in B$, $\exists\, c \in C \ni (b, c) \in g$. Putting this together, $(a, c) \in A \times C \wedge \exists\, b \in B \ni (a, b) \in f \wedge (b, c) \in g$. Therefore, by the definition of composition, $(a, c) \in g \circ f$, so $a \in \text{dom}(g \circ f)$ and consequently $A = \text{dom}(g \circ f)$.

If $(a, c_1) \in g \circ f$ and $(a, c_2) \in g \circ f$ then it is an exercise to show $c_1 = c_2$. Thus $g \circ f : A \longrightarrow C$.

If $f, g, A, B,$ and C are as above and $(a, b) \in f$ and $(b, c) \in g$, we can write $b = f(a)$ and $c = g(b)$ using f notation. So $c = g\big(f(a)\big)$ by

173

Figure 15.1

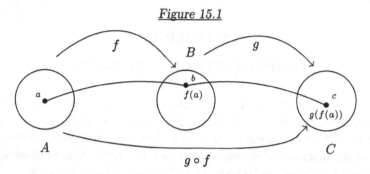

substitution. Since $(a, c) \in g \circ f$, we also know $c = (g \circ f)(a)$. Putting the two together, we get $(g \circ f)(a) = g(f(a))$. Consequently, by Rule U.G., $(g \circ f)(x) = g(f(x))$ for all $x \in A$.

EXAMPLE 15.1: Consider the functions f and g given by the rules $f(x) = \cos x$ and $g(x) = e^x$ then $(g \circ f)(x) = g(f(x)) = g(\cos x) = e^{\cos x}$ while $(f \circ g)(x) = f(g(x)) = \cos(g(x)) = \cos(e^x)$. Clearly, $g \circ f$ and $f \circ g$ are not equal functions since $(g \circ f)(x) > 0$ for all $x \in E_1$. On the other hand, $f(g(x)) = \cos(e^x)$ is negative for some x, in particular for $x = 1$. So composition of functions is not commutative.

The lack of commutativity of composition might lead one to question whether composition is associative. That is, is $h \circ (g \circ f) = (h \circ g) \circ f$? The affirmative answer is established in the following theorem. Refer to Exercise 14.10 to see how this is to be approached.

THEOREM 15.2: If $f : A \longrightarrow B$ and $g : B \longrightarrow C$ and $h : C \longrightarrow D$ are functions, then $h \circ (g \circ f) = (h \circ g) \circ f$.

Before doing the proof, let us look at a diagram of what is involved. Now f, g, and h are each functions from one set to another as indicated in the hypothesis. $g \circ f$ takes an element a in A to $g(f(a))$ in C. Then h takes that element to $h(g(f(a)))$ in D. This may be done in a different order as indicated in the diagram.

PROOF: Using Exercise 15.8, $\text{dom}(h \circ (g \circ f)) = \text{dom}(f) = \text{dom}((h \circ g) \circ f)$. Thus by Exercise 14.10, there is only one thing left to do. That is, show $\forall x \in A$, $(h \circ (g \circ f))(x) = ((h \circ g) \circ f)(x)$. Let $x \in A$.

$$((h \circ g) \circ f)(x) = (h \circ g)(f(x)) = h(g(f(x)))$$
$$= h((g \circ f)(x)) = (h \circ (g \circ f))(x)$$

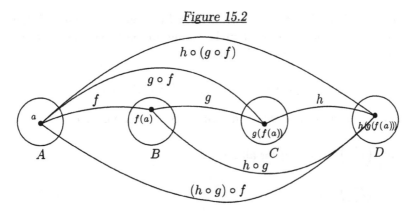

Figure 15.2

Thus by Rule U.G., composition is associative (see *Figure 15.2*).

EXAMPLE 15.2: If $f = \{(1,1),(2,3),(3,1),(4,3),(5,2)\}$ and $g = \{(1,2),(2,1),(3,3),(4,4),(5,1)\}$ then it is clear that f is a function from $A = \{1,2,3,4,5\}$ to A and likewise $g : A \longrightarrow A$. What is $g \circ f$? From its definition

$$g \circ f = \{(a,c) \mid (a,c) \in A \times A \land \exists b[b \in A \land (a,b) \in f \land (b,c) \in g]\}.$$

Now $(1,2) \in g \circ f$ since it is true that $(1,2) \in A \times A$ and there exists an element b (namely $b = 1$) such that $b \in A$ and $(1,b) \in f$ and $(b,2) \in g$. Likewise the points $(2,3),(3,2),(4,3),(5,1) \in g \circ f$. Note, for example, that $(3,4) \notin g \circ f$ since there is no element $b \in B$ such that $(3,b) \in f$ and $(b,4) \in g$.

Restriction of Functions:

It is common to take a function defined on a certain set and consider its "restriction" to a subset of the domain of the function. This is essentially what is being done in calculus, for example, when a function such as $y = \sin x$ for $x \in E_1$ is *restricted to* $y = \sin x$ for $0 \leq x \leq 2\pi$. The definition of restriction simply generalizes this concept.

DEFINITION: If $f : A \longrightarrow B$ and $S \subseteq A$ then the <u>restriction of f to S</u>, denoted by $f\big|_S$, is given by

$$f\big|_S = \{(x,y) \mid (x,y) \in f \land x \in S\}.$$

It is obvious that the restriction of a function is a subset of that function. The following theorem lists that property as well as other properties of the restriction of a function.

THEOREM 15.3:   Suppose $f : A \longrightarrow B$ and $S \subseteq A$ then

    a. $f\big|_S$ is a function whose domain is S and codomain B.

    b. $\left(f\big|_S\right)(x) = f(x)$ for all $x \in S$.

    c. $\left(f\big|_S\right) \subseteq f$.

    d. If f is $1 - 1$ then $f\big|_S$ is $1 - 1$.

    e. If $f\big|_S$ is onto B then f is onto.

PROOF: The proof is left to the reader.

EXAMPLE 15.3:   If $f = \{(1,1),(2,3),(3,1),(4,3),(5,2)\}$ and $S = \{1,2,3\}$ then S is a subset of $\{1,2,3,4,5\}$ which is the domain of f and $f\big|_S = \{(x,y) \mid (x,y) \in f \wedge x \in S\} = \{(1,1),(2,3),(3,1)\} \subseteq f$. Also, $(f\big|_S)(1) = 1 = f(1)$, $(f\big|_S)(2) = 3 = f(2)$ and $(f\big|_S)(3) = 1 = f(3)$. This illustrates Parts a, b, and c.

The function $g(x) = e^x$ on E_1 is a $1 - 1$ function, so $g\big|_{[-1,1]}$ is also a $1 - 1$ function. This seems obvious, if you draw the two graphs. The function $f = \{(a,1),(b,2),(c,1),(d,3)\}$, when restricted to $\{a,b,d\}$ is onto $\{1,2,3\}$, so f is also onto $\{1,2,3\}$.

In Chapter 12 the notion of the inverse r^{-1} of a relation r was defined and explored. Since a function f is a special relation, the concept of f^{-1} is already known: $f^{-1} = \{(y,x) \mid (x,y) \in f\}$.

The following theorem spells out some properties of functions and their inverses.

THEOREM 15.4:   Suppose $f : A \longrightarrow B$ then

    a. f^{-1} is a relation between B and A.

    b. If f is onto B then f^{-1} is a relation from B to A.

    c. If f is $1 - 1$ then f^{-1} is a function between B and A.

    d. If $f : A \xrightarrow{\; 1-1 \text{ onto} \;} B$, then $f^{-1} : B \xrightarrow{\; 1-1 \text{ onto} \;} A$.

PROOF: Suppose A and B are sets and $f : A \longrightarrow B$.

a. Since f is a function from A to B, f is a subset of $A \times B$. So f^{-1} is a subset of $B \times A$, hence a relation between B and A.

b. If f is onto B then every $b \in B$ is in the range of f. Thus in f^{-1}, every $b \in B$ is an element in the domain of f^{-1}, so f^{-1} is from B to A.

c. If f is $1-1$ then (a_1, b), (a_2, b) in f imply $a_1 = a_2$. So, turning the pairs around, (b, a_1) and $(b, a_2) \in f^{-1}$ imply $(a_1, b), (a_2, b) \in f$, but since f is $1-1$, $a_1 = a_2$. Thus f^{-1} is a function between B and A.

d. If f is $1-1$ and onto B then from Part c, f^{-1} is a function between B and A and from Part b, f is from B, hence f is a function from B to A. It is an exercise to show that f^{-1} is a $1-1$ function from B onto A.

EXAMPLE 15.4: Suppose $A = \{1, 2, 3, 4\}$, $B = \{a, b, c\}$, and $f = \{(1, a), (2, b), (3, c), (4, c)\}$. Then f is onto B and f^{-1} is found to be $\{(a, 1), (b, 2), (c, 3), (c, 4)\}$; a relation from B to A. Since f isn't $1-1$, it is clear that f^{-1} is only a relation from B to A. On the other hand, if $g = \{(1, a), (2, b), (3, c), (4, d)\}$ and $B' = \{a, b, c, d\}$ then g is a $1-1$ function from A onto B'. Now $g^{-1} = \{(a, 1), (b, 2), (c, 3), (d, 4)\}$ which is a $1-1$ function from B' onto A, just as asserted in Theorem 15.4d.

Exercises:

15.1 If f and g are given by: $f = \{(1, 1), (2, 3), (3, 2), (4, 5), (5, 4)\}$ and $g = \{(1, 2), (2, 3), (3, 4), (4, 1), (5, 5)\}$. Answer the following questions

a. Is f $1-1$? Is f onto $\{1, 2, 3, 4, 5\}$?

b. Is g $1-1$? Is f onto $\{1, 2, 3, 4, 5\}$?

c. $\text{dom}(f) = $ _____ $\text{ran}(f) = $ _____

d. $\text{dom}(g) = $ _____ $\text{ran}(y) = $ _____

e. $f^{-1} = $ _____

f. $g^{-1} = $ _____

g. $g \circ f = $ _____

h. $(g \circ f)^{-1} = $ _____

i. $f^{-1} \circ g^{-1} = $ _____

j. Is $f^{-1} \circ g^{-1} = (g \circ f)^{-1}$? _____

15.2 Find $g \circ f$ given that:

$$f = \{(a,1), (b,3), (c,2), (d,3)\}, \text{ and}$$
$$g = \{(1,x), (2,y), (3,z), (4,y), (5,x)\}.$$

15.3 If $A = \{1,2,3,4,5\}$, $B = N$, and $f = \{(x,x^2) \mid x \in A\}$, and $S = \{1,3,5\}$ then find $f|_S$.

15.4 If $f : A \xrightarrow{\text{1 - 1 onto}} B$ and $g : B \xrightarrow{\text{1 - 1 onto}} C$ then prove:

   a. $g \circ f : A \xrightarrow{\text{1 - 1 onto}} C$.

   b. $(g \circ f)^{-1} : C \xrightarrow{\text{1 - 1 onto}} A$.

   c. $f^{-1} \circ g^{-1} : C \xrightarrow{\text{1 - 1 onto}} A$.

   d. $f^{-1} \circ g^{-1} = (g \circ f)^{-1}$.

15.5 If $f : A \longrightarrow B$ is a function and $I_B = \{(x,x) \mid x \in B\}$ is the identity relation on B then prove:

   a. I_B is a function. What is its domain?

   b. $I_B \circ f = f$.

   c. $f \circ I_A = f$.

   d. If $f : A \xrightarrow{\text{1 - 1 onto}} B$ then $f \circ f^{-1} = f^{-1} \circ f = I_A$.

15.6 Do the following parts.

   a. Prove Theorem 15.3a.

   b. Prove Theorem 15.3b.

   c. Prove Theorem 15.3c.

   d. Prove Theorem 15.3d.

   e. Prove Theorem 15.3e.

15.7 Prove that if $f : A \longrightarrow B$ and $S_1 \subseteq S_2 \subseteq A$ then $f|_{S_1} \subseteq f|_{S_2}$.

15.8 Prove the rest of Theorem 15.1 by showing:

   a. $\text{dom}(g \circ f) \subseteq \text{dom}(f)$.

   b. If $\text{ran}(f) \subseteq \text{dom}(g)$ then $\text{dom}(g \circ f) = \text{dom}(f)$.

   c. $(a, c_1) \in g \circ f \land (a, c_2) \in g \circ f \Longrightarrow c_1 = c_2$.

CHAPTER 16

IMAGE AND PREIMAGE FUNCTIONS

Now attention is turned to the formation of two new functions from a given function, called the *image* and *preimage* functions. First we define the image of a set under a function.

DEFINITION: If $f : A \longrightarrow B$ and $S \subseteq A$, i.e., $S \in \mathcal{P}(A)$, then the <u>image</u> <u>of the set S under f</u>, denoted $f[S]$ is given by

$$f[S] = \{f(s) \mid s \in S\}.$$

A particular example of the image of a set in the case where $f \subseteq N \times N$ is treated in Example 16.1.

EXAMPLE 16.1: If $A = \{1, 2, 3, 4, 5\}$, $B = N$, $f = \{(x, x^2) \mid x \in A\}$, and $S = \{1, 2, 4\}$ then $f[S] = \{f(s) \mid s \in S\} = \{f(1), f(2), f(4)\} = \{1, 4, 16\}$. If $S = \{3\}$ then $f[S] = \{f(s) \mid s \in S\} = \{f(3)\} = \{9\}$. Also, $f[\emptyset] = \{f(s) \mid s \in \emptyset\} = \emptyset$. And $f[A] = \{x^2 \mid x \in A\} = \{1, 4, 9, 16, 25\}$.

Notice that $f[S]$ can be computed for each $S \in \mathcal{P}(A)$; we just evaluated $f[S]$ for several different S's. Also notice that for each S there is exactly one $f[S]$ and that $f[S] \subseteq B$ or, in other words, $f[S] \in \mathcal{P}(B)$. These observations are formalized in the following theorem. It is beginning to look like we have some kind of new function here, which takes elements from $\mathcal{P}(A)$ and yields elements in $\mathcal{P}(B)$.

THEOREM 16.1: If $f : A \longrightarrow B$ then
  a. For each $S \in \mathcal{P}(A)$, $f[S] \in \mathcal{P}(B)$.
  b. If $S_1 = S_2$ then $f[S_1] = f[S_2]$.
  c. $\{(S, f[S]) \mid S \in \mathcal{P}(A)\}$ is a function from $\mathcal{P}(A)$ to $\mathcal{P}(B)$.

PROOF: Suppose $f : A \longrightarrow B$.
  a. If $S \in \mathcal{P}(A)$ then $S \subseteq A$. Now $f : A \longrightarrow B$ so $f(s) \in B$ $\forall s \in S$. Thus $f[S] = \{f(s) \mid s \in S\} \subseteq B$. So $f[S] \in \mathcal{P}(B)$.
  b. If $S_1 = S_2$ then $f[S_1] = \{f(s) \mid s \in S_1\} = \{f(s) \mid s \in S_2\} = f[S_2]$. Thus Part b is proven.
  c. Notice that $\{(S, f[S]) \mid S \in \mathcal{P}(A)\}$ is a subset of $\mathcal{P}(A) \times \mathcal{P}(B)$ from Part a. Then Parts a and b guarantee that our set of ordered pairs has domain $\mathcal{P}(A)$ and is a function.

Figure 16.1

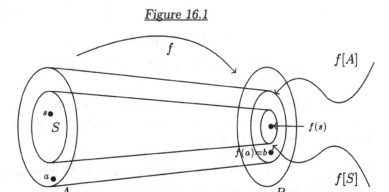

In the general case, if $f : A \longrightarrow B$ and $S \subseteq A$ then the image $f[S]$, of the set S under f, can be illustrated schematically as in *Figure 16.1*. Here is how. Suppose $f : A \longrightarrow B$ and $S \subseteq A$ then $f[S]$ consists of all of those elements of B that are images $f(s)$, of elements $s \in S$. $f[S]$ is denoted in the diagram by the smallest oval on the right. Similarly, $f[A]$ is the set of all images of elements of A under f. $f[A]$ is the middle oval and it contains $f[S]$. The lines going from the set A over to B are meant to suggest how f treats elements in A; things in S go to $f[S]$ and things in A go over to $f[A]$. For example, if $a \in A$ then $b = f(a)$ is in $f[A]$.

DEFINITION: If $f : A \longrightarrow B$ and $f[\ ] = \{(S, f[S]) \mid S \in \mathcal{P}(A)\}$ then $f[\ ]$ is called the <u>image function</u> from $\mathcal{P}(A)$ to $\mathcal{P}(B)$. In other words, if $f : A \longrightarrow B$ then $f[\ ] : \mathcal{P}(A) \longrightarrow \mathcal{P}(B)$ is the image function.

Although it is a little awkward, the notation is somewhat like the normal function notation in that, if $f : A \longrightarrow B$ and $x \in A$ then $f(x) \in B$. In the image function notation, if $f[\ ] : \mathcal{P}(A) \longrightarrow \mathcal{P}(B)$ and if $S \in \mathcal{P}(A)$ then $f[S] \in \mathcal{P}(B)$. The difference is that S is put inside the brackets, instead of after the symbol $f[\ ]$ and enclosed in parentheses, [e.g., $f[\ ](S)$] as is done in normal function notation.

EXAMPLE 16.2:   Let A be the set $\{1, 2, 3\}$, B the set N, and let $f = \{(1, 1), (2, 4), (3, 2)\}$ then

$$f[\ ] = \{(S, f[S]) \mid S \subseteq A\} = \{(\emptyset, \emptyset), (\{1\}, f[\{1\}]), ([\{2\}], f[\{2\}]),$$
$$([\{3\}], f[\{3\}]), \ldots, (\{2, 3\}, f[\{2, 3\}]), (A, f[A])\}.$$

Figure 16.2 a   *Figure 16.2 b*

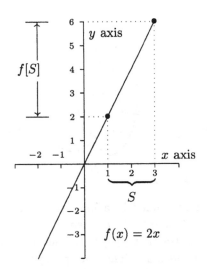

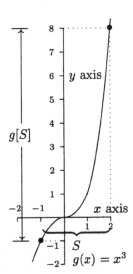

If f and g are two functions from E_1 to E_1, and S is an interval on the x axis, then $f[S]$ and $g[S]$ are as indicated in *Figure 16.2*, for the two functions $f(x) = 2x$ and $g(x) = x^3$ with S equal to the interval $[1,3]$, for f in Part a; and $S = [-1,2]$ for g in Part b.

In *Figure 16.2a* the diagram should suggest that the function $f(x) = 2x$ takes each of the points in the interval $[1,3]$ on the x axis and yields points in the interval $[2,6]$ on the y axis. In *Figure 16.2b*, g takes points in the interval $[-1,2]$ and yields the points in the interval $[-1,8]$.

Another function that goes hand–in–hand with the image function is the preimage function. Just as the image function takes a subset of A and yields a subset of B, the preimage function will take a subset of B and yield a subset of A. As we did with the image function, we first define the preimage of a set. This function is very important in the study of continuity of functions in the reals as well as more general spaces.

DEFINITION: If $f : A \longrightarrow B$ and $T \subseteq B$, i.e., $T \in \mathcal{P}(B)$, then the preimage of the set T under f, denoted $f^{-1}[T]$ is given by

$$f^{-1}[T] = \{x \mid x \in A \land f(x) \in T\}.$$

This definition says that $f^{-1}[T]$ consists of all elements of A whose images under f lie in the set T. In other words, it is the set of all preimages of elements of T.

Figure 16.3

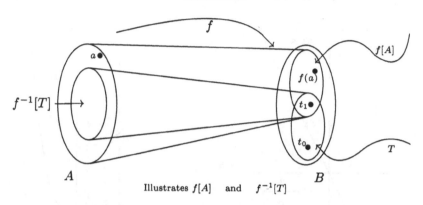

Illustrates $f[A]$ and $f^{-1}[T]$

The preimage of a set can be diagrammed in a manner similar to the image of a set. Let $f : A \longrightarrow B$ and $T \subseteq B$ then $f^{-1}[T]$ is as in *Figure 16.3*. The figure also illustrates that $f[A]$ is the subset of B indicated by the upper inner oval on the right. The lines going from A over to that oval are meant to suggest this. If $T \subseteq B$ (indicated by the lower inner oval on the right), *Figure 16.3* also illustrates that $f^{-1}[T]$ is the inner oval on the left. This is because that inner oval is the set of all elements $a \in A$ whose image lies in T. Notice that some elements of T, such as t_0, may not have a preimage in A. In fact, only those elements in T, such as t_1, which are also in $f[A]$, have preimages in A.

EXAMPLE 16.3:   If $f : E_1 \longrightarrow E_1$, say $f(x) = x^2$ and $T = [1,4]$ (on the y axis) then we claim that $f^{-1}[T] = [-2,-1] \cup [1,2]$ as in *Figure 16.4*. Let us see how this works. We are seeking the preimages, under f, of numbers $y \in [1,4]$ on the y axis. We get those numbers x, on the x axis, whose image $f(x) = x^2$ is in $[1,4]$. The resulting x's are in $[-2,-1] \cup [1,2]$. For example, the set of preimages of the number 4 on the y axis is the subset $\{-2,2\}$ on the x axis and the set of preimages of 1 is the subset $\{-1,1\}$ on the x axis. One might draw a horizontal line through various points in the interval $[1,4]$ on the y axis and look to see what coordinates they identify on the x axis.

Now a theorem analogous to Theorem 16.1 is considered.

THEOREM 16.2:   If $f : A \longrightarrow B$ then
    a. For each $T \in \mathcal{P}(B)$, $f^{-1}[T] \in \mathcal{P}(A)$.
    b. If $T_1 = T_2$ then $f^{-1}[T_1] = f^{-1}[T_2]$.
    c. $\{(T, f^{-1}[T]) \mid T \in \mathcal{P}(B)\}$ is a function from $\mathcal{P}(B)$ to $\mathcal{P}(A)$.

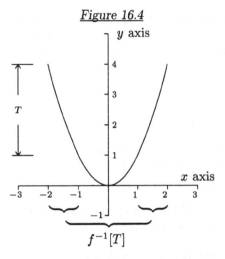

Figure 16.4

$T=[1,4]$ on y axis   and $f^{-1}[T]=[-2,-1]\cup[1,2]$ on x axis

The proof of Theorem 16.2 is left as an exercise. See the proof of Theorem 16.1 for ideas of how to proceed.

DEFINITION: The function in Part c of Theorem 16.2, is called the preimage function and is denoted by $f^{-1}[\ ]$. In other words, if $f : A \longrightarrow B$ then the function $f^{-1}[\ ]$ is $f^{-1}[\ ] : \mathcal{P}(B) \longrightarrow \mathcal{P}(A)$ and is called the preimage function.

Both $f[\ ]$ and $f^{-1}[\ ]$ are called set functions since their domains and ranges are sets of sets and their independent and dependent variables are also sets.

EXAMPLE 16.4:   Let $A = \{1,2,3,4,5\}$, $B = N$, $T = \{1,3,5,7\}$, and $f = \{(1,1),(2,4),(3,1),(4,3),(5,2)\}$. Then $f : A \longrightarrow B$. The preimage of T under f is $f^{-1}[T] = \{x \mid x \in A \wedge f(x) \in T\} = \{1,3,4\}$. For example, $3 \in f^{-1}[T]$ because $f(3) = 1$ and $1 \in T$. And $4 \in f^{-1}[T]$ because $f(4) = 3$ and $3 \in T$. The elements 5 and 7 of T do not have preimages in A, but this is of no consequence in computing $f^{-1}[T]$. $f^{-1}[T]$ simply consists of those elements x in A such that $f(x) \in T$. These x's are 1 and 3. Notice also that if T had been any other element S, of $\mathcal{P}(B)$, say $\emptyset$ or B, then $f^{-1}[S] \subseteq A$ would still hold. For example $f^{-1}[\emptyset] = \emptyset$ and $f^{-1}[B] = A$.

The following theorem illustrates some important properties relating set functions and union and intersection. It also shows how easy it is to incorrectly conjecture relationships based on superficial understanding. By looking at Parts a, c, and d of Theorem 16.3, one might make an incorrect conjecture for Part b. See Exercise 16.5.

THEOREM 16.3:   If $f : A \longrightarrow B$, $S_1, S_2 \in \mathcal{P}(A)$ and $T_1, T_2 \in \mathcal{P}(B)$, then

    a. $f[S_1 \cup S_2] = f[S_1] \cup f[S_2]$.

    b. $f[S_1 \cap S_2] \subseteq f[S_1] \cap f[S_2]$.

    c. $f^{-1}[T_1 \cup T_2] = f^{-1}[T_1] \cup f^{-1}[T_2]$.

    d. $f^{-1}[T_1 \cap T_2] = f^{-1}[T_1] \cap f^{-1}[T_2]$.

PROOF: The proofs of Parts a, b, and d are left to the student. Part c is sketched here with the most essential ideas.

$$y \in f^{-1}[T_1 \cup T_2] \equiv f(y) \in T_1 \cup T_2 \qquad \text{Why?}$$
$$\equiv f(y) \in T_1 \ \lor \ f(y) \in T_2 \qquad \text{Why?}$$
$$\equiv y \in f^{-1}[T_1] \ \lor \ y \in f^{-1}[T_2] \qquad \text{Why?}$$
$$\equiv y \in f^{-1}[T_1] \cup f^{-1}[T_2]. \qquad \text{Why?}$$

Thus Part c is true.

Exercises:

16.1 Let $A = \{1, 2, 3, 4, 5\}$, $B = \{a, b, c, d\}$, $S = \{1, 2, 3, 4\}$, $T = \{b, c, d\}$, and $f = \{(1, a), (2, b), (3, c), (4, b), (5, b)\}$, then find $f[S]$, $f^{-1}[T]$, $f^{-1}[\{b\}]$, and $f^{-1}[\{d\}]$.

16.2 Suppose $A = B = E_1$, $f = \frac{1}{2}x + 1$, $S = [2, 4]$, and $T = [\frac{1}{2}, 2]$, then

    a. Compute $f[S]$ and graph it with f and S on $E_1 \times E_1$ in a manner similar to *Figure 16.2*.

    b. Compute $f^{-1}[T]$ and graph it on $E_1 \times E_1$ with T and f in a manner similar to *Figure 16.4*.

16.3 Carefully prove that if $f : A \longrightarrow B$ and $S_1 \subseteq S_2 \subseteq A$ then $f[S_1] \subseteq f[S_2]$.

16.4 Prove Theorem 16.2. Hint: see the proof of Theorem 16.1.

16.5 Look carefully at Parts a, c, and d of Theorem 16.3. Suppose the student had been asked to conjecture a relationship for Part b based on what was in Parts a, c, and d, without being told what the relationship actually was. What would the student conjecture? Why is it important to pay close attention to details and follow the definitions and previously established results?

16.6 a. Prove Theorem 16.3a.

b. Prove Part b of Theorem 16.3. Explain why it is containment only, rather than equality? In other words, why doesn't containment go the other way?

c. Prove Theorem 16.3d.

16.7 Prove: If $f : A \longrightarrow B$ and $\forall \lambda \in \Lambda,\ T_\lambda \in \mathcal{P}(B)$ then

    a. $f^{-1}\left[\bigcup_{\lambda \in \Lambda} T_\lambda\right] = \bigcup_{\lambda \in \Lambda} f^{-1}[T_\lambda]$.

    b. $f^{-1}\left[\bigcap_{\lambda \in \Lambda} T_\lambda\right] = \bigcap_{\lambda \in \Lambda} f^{-1}[T_\lambda]$.

16.8 Prove: If $f : A \longrightarrow B$ and $\forall \lambda \in \Lambda,\ S_\lambda \in \mathcal{P}(A)$ then

    a. $f\left[\bigcup_{\lambda \in \Lambda} S_\lambda\right] = \bigcup_{\lambda \in \Lambda} f[S_\lambda]$.

    b. $f\left[\bigcap_{\lambda \in \Lambda} S_\lambda\right] \subseteq \bigcap_{\lambda \in \Lambda} f[S_\lambda]$.

c. Considering Theorem 16.3a and b, and Parts a and b in this exercise, are the connectives $=$ and $\subseteq$ to be expected?

16.9 Prove: If f, A, B, S_1, S_2 are as in Theorem 16.3 and f is $1 - 1$ then $f[S_1 \cap S_2] = f[S_1] \cap f[S_2]$. This says the containment symbol, $\subseteq$ of Theorem 16.3b, becomes $=$ when f is $1 - 1$.

16.10 If $f : A \longrightarrow B$, $S_1 \subseteq A$, $S_2 \subseteq A$ then is the equation $f[S_1 - S_2] = f[S_1] - f[S_2]$ true or false? Why? Explain.

SECTION IV

ALGEBRAIC AND ORDER PROPERTIES
OF
NUMBER SYSTEMS

CHAPTER 17

BINARY OPERATIONS

Before starting the development of the natural number system a few concepts, beginning with binary operations, need to be explored.

The operations of addition and multiplication are examples of binary operations on certain sets of numbers. Union and intersection of sets are also binary operations on the power set of a set. In general, a binary operation may be applied to a pair of elements from a set. The result of that application is a unique element in that set. If this sounds like a function is involved, it is.

DEFINITION: Let A be a nonempty set. A binary operation on A is a function, denoted (for now) by $\circ$ from $A \times A$ to A. That is $\circ : A \times A \longrightarrow A$.

Since a binary operation $\circ$ is a function from $A \times A$ to A, when $\circ$ is applied to a pair $(x, y) \in A \times A$, an element of A results. Also since $\circ$ is a function, the element of A which results from the application of $\circ$ is unique. In other words, there is exactly one output, which may be denoted by $x \circ y$ for each input $(x, y) \in A \times A$. Because $\circ$ is a function, the image of (x, y) under $\circ$, in function notation, is $\circ((x, y))$, so that $\circ((x, y)) = x \circ y$. If $\circ$ is a binary operation on any subset S of A then S is said to be closed with respect to $\circ$.

EXAMPLE 17.1:   Let $A = N$ and let $\circ$ be ordinary addition, i.e., $+$. For purposes of illustration, suppose we know all about the arithmetic of N with respect to addition. Then $+((2, 3)) = 2 + 3 = 5$. In other words, $+$ takes the pair $(2, 3)$ as an input and yields the unique natural number $2 + 3$ or 5 as an output. More generally, if $a, b \in N$ then $+((a, b)) = a + b$ which is a unique element of N. Hence $+$ is a binary operation on N.

EXAMPLE 17.2:   Let T be a nonempty set and $A = \mathcal{P}(T)$ (the power set of T.) Define $\circ$ on A by $\circ((S_1, S_2)) = S_1 \cap S_2$ for each $(S_1, S_2) \in A \times A$. Then $\circ$ assigns to each ordered pair (S_1, S_2) in $A \times A$, the unique set $S_1 \cap S_2$ in A. Thus $\circ$, or more to the point $\cap$, is a binary operation on A. Note that A is closed with respect to $\cap$. In a similar way $\cup$ is a binary operation on A, so $A = \mathcal{P}(T)$ is closed with respect $\cup$.

EXAMPLE 17.3:   Let D be the set of odd natural numbers and let $\circ$ be ordinary addition. Is D closed with respect to $+$? Consider the

189

example, $+\big((3,5)\big) = 3 + 5 = 8 \notin D$. Thus $+$ is NOT a binary operation on the subset D of N. A conclusion to be drawn from this and the preceding examples, is that $\circ$ may be a binary operation on a set A, but not necessarily on every subset of A.

The binary operations in the previous examples are familiar operations which are commutative and associative. But commutativity and associativity are not necessarily properties possessed by a given binary operation. For example, if A is the set Z of integers then subtraction is a binary operation that is not commutative (see Example 17.4). Since binary operations are not necessarily commutative and not necessarily associative, these properties may be of some interest. So a definition is in order.

DEFINITION: Let A be a nonempty set and $\circ : A \times A \longrightarrow A$ a binary operation on A, then

    a. $\circ$ is <u>commutative</u> on A iff $x \circ y = y \circ x$ for all $x, y \in A$.

    b. $\circ$ is <u>associative</u> on A iff $x \circ (y \circ z) = (x \circ y) \circ z$ for all $x, y, z \in A$.

EXAMPLE 17.4: Let A be the set of rational numbers and let $\circ$ be given by $x \circ y = x - y$. Then $\circ$ is a binary operation on Q, but $\circ$ is neither commutative nor associative on A. It is an exercise to verify these assertions (see Exercise 17.3).

DEFINITION: Let A be a nonempty set and $\circ$ a binary operation on A. The set A <u>has</u> <u>an</u> <u>identity</u> with respect to $\circ$ on A if and only if $(\exists i \in A) \ni (\forall x \in A)\big[x \circ i = i \circ x = x\big]$. If the element i exists it is called <u>an</u> <u>identity</u> <u>element</u> with respect to $\circ$ on A.

EXAMPLE 17.5: If $A = N$ and $\circ$ is ordinary multiplication of natural numbers then $\circ$ is commutative and associative. Also 1 is an identity with respect to $\circ$ because $1 \circ x = x \circ 1 = x$ $\forall x \in A$. We say 1 is a *multiplicative identity* in N.

EXAMPLE 17.6: Let $T \neq \emptyset$ and $A = \mathcal{P}(T)$ and define $S_1 \circ S_2 = S_1 \cap S_2$. Then $\circ$ is commutative, associative, and the set T in A is an identity with respect to $\circ$ because $S \cap T = T \cap S = S$ $\forall S \in A$.

An important fact, of which to be aware, is that for A to have an identity with respect to $\circ$, there must exist <u>an</u> element $i \in A$ such that $\forall a \in A, \ i \circ a = a \circ i = a$. In other words, the i does not change as the element a changes and that same i works for all $a \in A$. See the discussion of the effect of placing $\exists$ before $\forall$ in a predicate involving two variables. Specifically, refer to Examples 3.1–3.3 and Exercise 3.12.

Since the article "an" precedes "identity," the door is left open for there to be more than one identity on a set A. Is this a real possibility? The answer is found in Theorem 17.1.

DEFINITION: Let A be a nonempty set and $\circ$ a binary operation on A and let i be an identity with respect to $\circ$ on A. If $x \in A$ then x is said to have an inverse with respect to $\circ$ on A if and only if there exists an element $\overline{x} \in A$ such that $\overline{x} \circ x = x \circ \overline{x} = i$. The element $\overline{x}$ is called an inverse of x with respect to $\circ$ on A.

When the context is clear one can simply say that "$\overline{x}$ is an inverse of x" with the understanding that it is with respect to $\circ$ on A. Contrary to the situation previously mentioned, in which there was a single element i which worked for all $x \in A$, here there are several differences for inverses to consider.

- In order for a discussion to even occur, regarding whether or not an element x has an inverse, there must be an identity in the set A. Without an identity, there can be NO inverses.

- For those x's which do have inverses, the inverses may change as x changes. The reason the inverse may change as x changes but the identity does not change as x changes, is found in the definitions. The quantifiers spell out this distinction. Refer to Chapter 3 on quantifiers. See how the universal and existential quantifiers interact with each other.

- Not all x's necessarily have inverses with respect to $\circ$ on A.

THEOREM 17.1: If A is a nonempty set and $\circ$ a binary operation on A and if i is an identity with respect to $\circ$ on A, then

a. The element i is unique. That is, if $j \in A$ and $j \circ x = x \circ j = x$ $\forall x \in A$ then $j = i$.

b. If $\circ$ is associative, $x \in A$ and x has an inverse $\overline{x}$, then $\overline{x}$ is unique. That is, if $\circ$ is associative and if $y \in A$ and $x \circ y = y \circ x = i$ then $y = \overline{x}$.

The proof is an exercise. Because of uniqueness, we now can say the identity and the inverse of x rather than an identity or an inverse of x.

EXAMPLE 17.7:   Let $A = Z$ and $\circ$ be ordinary $+$. Temporarily, we assume that the usual properties of addition of integers are known, so we may illustrate identities and inverses. Observe that 0 is the identity with respect to $+$, since $x + 0 = 0 + x = x$ $\forall x \in Z$. We say 0 is an *additive identity.* Also for each x, the inverse of x with respect to addition is $-x$, because $x + (-x) = (-x) + x = 0$. In this case, the inverse of x is called the *negative* of x. Notice the additive identity, i.e., 0, works for all x's, but $-x$ changes as x varies through Z. The reason centers on the interplay between the existential and universal quantifiers.

On the other hand, if $\circ$ is ordinary multiplication on Z, then 1 is the identity of Z with respect to $\circ$. But not every element of Z has an inverse in Z with respect to this operation. In particular, 2 does not have such an inverse in Z, because for NO integer x is it true that $2 \cdot x = 1 = x \cdot 2$. In fact, the only elements of Z which have inverses with respect to multiplication are 1 and -1. As a consequence, one of the elements having an inverse is the identity element itself. This results from the more general property that if i is the identity with respect to $\circ$ on A then $i \circ i = i \circ i = i$. So i is its own inverse.

Finally we consider a set on which there are simultaneously two binary operations. This is not really unusual, since on $\mathcal{P}(B)$ we have both union and intersection and on the set N of natural numbers we have both addition and multiplication. The next definition names a common situation that occurs when two binary operations on a set interrelate with one another.

DEFINITION: If A is a nonempty set and $\circ$ and $\star$ are binary operations on A then $\star$ <u>distributes</u> over $\circ$ if and only if $x \star (y \circ z) = (x \star y) \circ (x \star z)$ $\forall x, y, z \in A$.

EXAMPLE 17.8:   Let $A = \mathcal{P}(B)$ be the set of all subsets of a nonempty set B. Then $\cap$ distributes over $\cup$ because

$$C \cap (D \cup E) = (C \cap D) \cup (C \cap E) \quad \forall C, D, E \in \mathcal{P}(B).$$

Similarly $\cup$ distributes over $\cap$.

EXAMPLE 17.9:   In the system of integers, multiplication distributes over addition in Z because $a \cdot (b + c) = a \cdot b + a \cdot c$. However, addition does not distribute over multiplication because $a + b \cdot c$ is not equal to $(a + b) \cdot (a + c)$.

Exercises:

17.1 Let $A = N$ and $\circ$ be usual $+$. Is $\circ$ a $1 - 1$ function from $N \times N$ to N? Is $\circ$ onto N? Explain.

17.2 If $B \neq \emptyset$ and $A = \mathcal{P}(B)$ and $\circ$ is the usual binary operation of $\cup$ then verify that $\circ$ is commutative and associative and that there is an element of A which serves as an identity with respect to $\circ$. Does any element of A have an inverse with respect to $\circ$? Use previous results to answer these questions, but just find and reference them.

17.3 a. If $A = Z$ and $x \circ y = x - y$ $\forall x, y \in Z$, is $\circ$ a binary operation on Z? If it is, is it also commutative? Associative?

b. If $A = N$ and $x \circ y = x - y$ $\forall x, y \in N$, is $\circ$ a binary operation on N? Is it commutative? Associative?

17.4 Let $A = N$. For the purposes of this exercise, assume the usual properties of $\leq$ and $>$. Then define $\forall x, y \in A$

$$x \circ y = \begin{cases} x, & \text{if } x \leq y; \\ y, & \text{if } y < x. \end{cases}$$

a. Prove $\circ$ is a binary operation on A.

b. Prove $\circ$ is commutative.

c. Is there an identity with respect to $\circ$ on A? Explain.

17.5 Let $A = N$ and define $\forall x, y \in A$

$$x \circ y = \begin{cases} x, & \text{if } x \geq y; \\ y, & \text{if } y > x. \end{cases}$$

a. Prove $\circ$ is a binary operation on A.

b. Prove $\circ$ is commutative.

c. Is there an identity with respect to $\circ$ on A? Explain.

17.6 Let $A \neq \emptyset$ and define $\circ$ by $\forall x, y \in A$, $x \circ y = x$. Is $\circ$ a binary operation on A? Is $\circ$ commutative? Associative? Do any elements of A have inverses with respect to $\circ$?

17.7 Complete the proof of Part a of Theorem 17.1. That is, prove if i is an identity in A with respect to $\circ$ then i is unique.

17.8 Complete the proof of Part b of Theorem 17.1. That is, if $\circ$ is associative, $x \in A$ and A has an identity i and if x has an inverse $\bar{x} \in A$ then $\bar{x}$ is unique.

17.9 Justify the assertions and answer the question posed in Example 17.8. Specifically, find the exact reason for the equality in Example 17.8. Then make the analogous statement for ∪ distributing over ∩ and cite a specific reason.

CHAPTER 18

THE SYSTEMS OF WHOLE AND NATURAL NUMBERS

In this section a brief, non–rigorous treatment of the system of whole and natural numbers is developed. Rather than follow one of the common approaches of obtaining these systems through Peano's Postulates (Giuseppe Peano, 1858–1932), an alternative development of cardinal numbers is pursued. The rationale for selecting this means to develop the whole and natural numbers is because it ties together, in an elementary manner, so many wonderful ideas that have been developed in the earlier chapters. In particular, it makes good use of properties of sets, functions, and one–to–one correspondences to arrive at the desired properties in very short order. In addition, the same ideas are employed later to develop transfinite cardinal numbers.

To begin, it is assumed that every set S has a cardinal number, which is denoted by $\#(S)$. And furthermore it is assumed that each of the cardinal numbers $0, 1, 2, 3, \ldots$, is defined as indicated below.

$$0 = \#(\emptyset)$$
$$1 = \#(\{0\})$$
$$2 = \#(\{0, 1\})$$
$$3 = \#(\{0, 1, 2\})$$
$$4 = \#(\{0, 1, 2, 3\})$$
$$\vdots \qquad \vdots$$
$$m + 1 = \#(\{0, 1, 2, \ldots, m\})$$
$$\vdots \qquad \vdots$$

Notice that each cardinal number is defined in terms of the set of all of the preceding cardinal numbers. Because of the way we defined the finite cardinal numbers, it will be easy to develop addition for them.

Overlooking possible subtle complications, it is sufficient for our work to take $W = \{0, 1, 2, 3, \ldots, m, m+1, \ldots\}$ to be the set of "finite" cardinal numbers defined above. W is called the set of whole numbers and $N = W - \{0\}$ is the set of natural numbers.

The notion of cardinal numbers has now been established for the following very specific sets.

$$\emptyset, \{0\}, \{0, 1\}, \{0, 1, 2\}, \{0, 1, 2, 3\}, \ldots, \{0, 1, 2, 3, \ldots, m\}, \ldots \qquad (1)$$

But what is the cardinal number of a set like $\{6,7,8\}$ or $\{a,b,c,d\}$? The definitions of the cardinal numbers $0, 1, 2, 3, \ldots$ given above, do not address this problem. So we choose to establish, in a more careful manner, the cardinal number of sets other than those in Equation (1). This is accomplished as follows.

DEFINITION: Let A and B be sets. Then A and B are said to <u>have the same cardinal number</u> if and only if $\exists\, f : A \longrightarrow B \ni f$ is $1-1$ and onto the set B.

In other words, sets A and B have the same cardinal number iff there is a $1-1$ correspondence between A and B. Sets A and B, which have the same cardinal number, are said to be <u>cardinally equivalent</u> and, in this case, we write $\#(A) = \#(B)$.

EXAMPLE 18.1: If A is a set of persons in an auditorium and B the set of seats in the auditorium then, neglecting exceptional cases, such as two persons sitting in the same chair or one person in two chairs, the number of persons is the same as the number of chairs iff every seat is occupied and no one is standing. Here, a suitable $1-1$ correspondence is the function which pairs each person with the seat he or she occupies. There is an important point here, worthy of belaboring. Neither the number of seats nor the number of chairs need to be determined to ascertain cardinal equivalence. In fact, it would be foolish to do so, if all that was needed was whether the number of persons was the same as the number of chairs. Remember two sets such as A and B have the same cardinal number if and only if they are in $1-1$ correspondence.

In the event that the number of elements in a set A needs to be determined, it also is done by establishing a $1-1$ correspondence. The mere process of counting the elements of A is the establishment of a $1-1$ correspondence between the set A, to be counted, and a set of numbers, say $\{1, 2, 3, \ldots, k\}$. The latter set is a set each person carries around at all times for such purposes. Such is the essence of counting.

Arithmetic is not limited to counting; there are binary operations such as addition and multiplication yet to be developed. In Chapter 17, the notion of the binary operation of addition was superficially exploited to provide an example for binary operations. The way addition on N or W comes into being, needs to be defined and more carefully developed.

In order to impose an addition on W, a little care must be taken. For example, since $2 = \#(\{0,1\})$ and $3 = \#(\{0,1,2\})$ then if $2+3$ were defined as $\#(\{0,1\}\cup\{0,1,2\})$, the sum would be $2+3=3$ which, to say the least, is inconsistent with tradition. The reason for the faulty result is clear–the sets $\{0,1\}$ and $\{0,1,2\}$ are not disjoint. So no distinction was made of the 0 and 1 in the second set from the 0 and 1 of the first set. However, if we were to find a set B such that B is in $1-1$ correspondence with $\{0,1,2\}$ and $B\cap\{0,1\}=\emptyset$ then $\#(B)=3$ and then $2+3=\#(\{0,1\}\cup B)$. In particular, one could choose $B=\{2,3,4\}$, hence $2+3=\#(\{0,1\}\cup\{2,3,4\})=\#(\{0,1,2,3,4\})=5$ by definition. Obviously, the set in parenthesis is exactly the set needed for the number 5. It was no accident that the set B was chosen to consist of numbers that began, where the set $\{0,1\}$ left off. That too, is exactly what a person does when counting all of the members of several sets–the counting of the second set resumes where the counting left off in the first set. The definition reflects a generalization of the process outlined previously.

DEFINITION (*Addition in W*): If $m,n\in W$, and A and B are sets such that $m=\#(A)$ and $n=\#(B)$ and $A\cap B=\emptyset$ then $m+n=\#(A\cup B)$.

Now we are prepared to list and prove some algebraic properties of addition of whole numbers.

THEOREM 18.1: For all $m,n,p\in W$:
   a. $m+n\in W$
   b. $m+n=n+m$
   c. $m+(n+p)=(m+n)+p$
   d. $m=m+0$

PROOF: Parts b and d are done below while a and c are exercises.
   b. Let A and B be disjoint sets such that $m=\#(A)$ and $n=\#(B)$ then, since $A\cup B=B\cup A$, and since B and A are also disjoint, we have $m+n=\#(A\cup B)=\#(B\cup A)=n+m$.
   d. Since $A=A\cup\emptyset$ and $A\cap\emptyset=\emptyset$ we have $m=\#(A)=\#(A\cup\emptyset)=m+0$. This part says 0 is an *additive identity*.

DEFINITION (*Multiplication in W*): If $m,n\in W$, and A and B are sets such that $m=\#(A)$ and $n=\#(B)$ then $mn=\#(A\times B)$. Sometimes $m\cdot n$ is used instead of mn.

EXAMPLE 18.2: If $m=2$ and $n=3$ then $m=\#(\{0,1\})$ and $n=\#(\{0,1,2\})$ and

$$mn=\#\big(\{(0,0),(0,1),(0,2),(1,0),(1,1),(1,2)\}\big)=\#(\{0,1,\ldots,5\})=6.$$

THEOREM 18.2:   For all $m, n, p \in W$:

  a. $mn \in W$

  b. $mn = nm$

  c. $m(np) = (mn)p$

  d. $m0 = 0$

  e. $m \cdot 1 = m \cdot \#(\{0\}) = m$

  f. $m(n + p) = mn + mp$

PROOF: Let A, B, and C be sets such that $m = \#(A)$, $n = \#(B)$ and $p = \#(C)$. We do Parts b, c, and e.

  b. It is easy to prove that there exists a $1 - 1$ function from $A \times B$ onto $B \times A$ (see Exercise 14.13). Consequently,

$$mn = \#(A \times B) \overset{\text{Why?}}{=} \#(B \times A) = nm.$$

  c. By Exercise 14.14 $\#(A \times (B \times C)) = \#((A \times B) \times C)$. Then

$$m(np) \overset{\text{Why?}}{=} m \cdot \#(B \times C) \overset{\text{Why?}}{=} \#(A \times (B \times C))$$

$$\overset{\text{Why?}}{=} \#((A \times B) \times C) \overset{\text{Why?}}{=} \#(A \times B) \cdot p \overset{\text{Why?}}{=} (mn)p.$$

  e. Using Exercise 14.15, $m = \#(A) = \#(A \times \{0\}) = m \cdot \#(\{0\})$ which is $m \cdot 1$ since $1 = \#(\{0\})$.

The arithmetic of W has been limited to addition and multiplication. Subtraction is not a binary operation on W because the equation $a + x = b$ cannot always be solved for x in W. As a particular example, it is impossible in W to solve the equation $3 + x = 2$. This shortcoming will be alleviated soon when the system Z of integers is developed. But first the nature of *order*, i.e., $<$ or $>$, in W is investigated. After the definition of order, some of its properties are enumerated and proven.

The definition of *is less than* is motivated from the notion that if some cardinal number is added to another non-zero cardinal number, the sum so determined is "greater than" the original number.

DEFINITION (*"Is Less Than" in W*): Let $m, n \in W$ then $m < n$ if and only if $\exists p \in N \ni m + p = n$.

EXAMPLE 18.3:   Let $m = 3$ and $n = 5$ then $\exists p \in W$ (namely $p = 2$) such that $m + p = n$. Thus $3 < 5$. In other words, $3 < 5$ because $2 \in N$ and $3 + 2 = 5$.

THEOREM 18.3: Let $k, m, n \in W$, then

a. $m, n \in N \implies [m + n \in N \wedge mn \in N]$

b. $m < n \implies m + k < n + k$

c. $m < n \wedge k \in N \implies mk < nk$

d. $k < m \wedge m < n \implies k < n$

e. $k, n \in N \implies [k < n \text{ or } n < k \text{ or } n = k]$ Only one part holds.

f. $m + k = n + k \implies m = n$

Some of these properties have names. The first part says N is closed with respect to (w.r.t.) addition and multiplication. Part d says $<$ is transitive. Part e says the trichotomy law holds, i.e., <u>exactly</u> one of the three conditions $k < n$, $k = n$ and $n < k$ must hold.

PROOF: a. Let A, B be sets such that $m = \#(A)$ and $n = \#(B)$ and $A \cap B = \emptyset$. Then since $m, n \in N$, $A \neq \emptyset$ and $B \neq \emptyset$. Thus $A \cup B \neq \emptyset$. This means that $m + n = \#(A \cup B)$ and $\#(A \cup B) \in N$. Since $A \neq \emptyset$ and $B \neq \emptyset$ then we also know $A \times B \neq \emptyset$. (Why?) Thus $mn = \#(A \times B)$ and $\#(A \times B) \in N$.

b. If $m < n$ then $\exists p \in N \ni m + p = n$. Therefore

$$(m + k) + p \overset{\text{Why?}}{=} m + (k + p) \overset{\text{Why?}}{=} m + (p + k)$$

$$\overset{\text{Why?}}{=} (m + p) + k \overset{\text{Why?}}{=} n + k.$$

Thus $m + k < n + k$. (Why?)

c. This is an exercise.

d. $k < m$ and $m < n$ imply $\exists k_1 \in N \ni k + k_1 = m \wedge \exists k_2 \in N \ni m + k_2 = n$. Thus $k + (k_1 + k_2) = (k + k_1) + k_2 = m + k_2 = n$. Since $k_1 + k_2 \in N$ (Why?) and since $k + (k_1 + k_2) = n$, we have $k < n$.

e. This part requires additional lemmas to establish, but we choose not to do them in this text. Those who are interested can find a nice development of the trichotomy law in the *Grundlagen der Analysis* by Edmund Landau. In that work, however, the development of the natural numbers uses Peano's postulates, a different and more detailed way to develop the natural numbers.

f. This is an indirect proof. If $m \neq n$, then by the trichotomy law, $m < n$ or $n < m$. If $m < n$ then $m + k < n + k$, by Part b, whence $m + k \neq n + k$. And if $n < m$ then $n + k < m + k$, hence $n + k \neq m + k$. In either case, a contradiction is obtained, so $m = n$.

It is also worth mentioning that $<$ can be reversed in Parts b, c, and d and the corresponding results for $>$ will be true. Also $<$ can be replaced by $\leq$ in Parts b, c, and d.

Exercises:

18.1 In a fashion similar to the example computations of this section, show that

    a. $5 + 4 = 9$

    b. $0 + m = m \;\; \forall m \in N$

    c. $0 < n \;\; \forall n \in N$

    d. $k < n \Longrightarrow k + 1 \leq n$

18.2 Using the definitions of this section

    a. Show that $3 < 5$.

    b. Show that $2 \cdot 3 = 6$.

18.3 Prove Theorem 18.1, Parts a and c.

18.4 Prove Theorem 18.2, Parts a, d, and f.

18.5 Prove Theorem 18.3, Part c.

18.6 Suppose $a, b \in W$. The difference between b and a, if it exists, is given by $b - a$ where $b - a = k$ if and only if $b = k + a$. Use this definition to find:

    a. $3 - 1$

    b. $6 - 5$

    c. $4 - 4$

    d. $5 - 7$

CHAPTER 19

THE SYSTEM Z OF INTEGERS

The systems of natural numbers and whole numbers are closed with respect to addition and multiplication and they have the basic properties mentioned in the previous chapter. However, these two systems are inadequate to solve every linear equation $a + x = b$, for x if given a and b in N. The difficulty is that solving $a + x = b$ is essentially a problem of subtracting. In Exercise 18.6, one can see why this cannot always be done. The inability to subtract a from b for arbitrary a and b is overcome by forming a new system from N in which subtraction is always possible. The system so formed is the system of integers.

To begin the development of the system of integers, we take the system N of natural numbers and its properties for our starting point. We form the Cartesian product $N \times N$ and define a relation $\sim$, on $N \times N$, which relates two natural numbers if they have a common difference. The relation $\sim$ turns out to be an equivalence relation.

DEFINITION ($\sim$ for pairs in N): Let $D = N \times N$ and define $\sim$ by

$$\forall (m,n) \in D, \left[(m,n) \sim (p,q) \text{ if and only if } m + q = n + p \right].$$

Notice that $(3,1) \sim (4,2)$ and $(4,2) \sim (7,5)$. (Why?) Is $(3,1) \sim (7,5)$? Since $3 + 5 = 1 + 7$, it is. It appears that $\sim$ is transitive? What other properties of relations might it satisfy? See the following theorem.

THEOREM 19.1: $\sim$ is an equivalence relation on $D = N \times N$.

PROOF: Let N be the natural numbers and $D = N \times N$. Show $\sim$ is reflexive, symmetric, and transitive.

    i. If $(m,n) \in D$ then $m, n \in N$ and $m + n = n + m$ from Theorem 18.1b. Thus $(m,n) \sim (m,n)$. This says $\sim$ is reflexive.

    ii. If $(m,n) \sim (p,q)$ then $m,n,p,q \in N$ and $m + q = n + p$. Thus $p + n = q + m$. (Why?) Consequently, $(p,q) \sim (m,n)$. (Why?) Thus $\sim$ is symmetric.

    iii. The proof that $\sim$ is transitive, is left to the exercises.

Now that we know $\sim$ is an equivalence relation, we can use it to form equivalence classes. Let $(m,n) \in D$ and let $[(m,n)]$ be the equivalence class of (m,n), i.e., $[(m,n)] = \{(x,y) \mid (x,y) \in D \wedge (x,y) \sim (m,n)\}$.

For simplicity, let $[m, n] = [(m, n)]$. Then by properties of partitions and equivalence relations $[m, n] = [p, q]$ if and only if $(m, n) \sim (p, q)$ if and only if $m + q = n + p$. In other words, $[m, n] = [p, q]$ iff $m + q = n + p$. The elements $[m, n]$ for each $(m, n) \in D$ are called <u>integers</u> and $Z = \{[m, n] \mid (m, n) \in D\}$ is called the <u>set of all</u> <u>integers</u>. Now we know what the integers $[m, n]$ and $[p, q]$ are and when they are equal.

It is desirable to have the arithmetic operations of addition and multiplication on the set Z of integers consistent with those operations on the set W of whole numbers. In addition, it is desirable to have capability to solve equations such as $a + x = b$ where $a, b \in Z$. It was the inability to solve all such equations in W that was a motivating factor in developing the integers. To that end, the definitions of addition and multiplication in Z are given in terms of the operations of $+$ and $\cdot$ in N. Then the algebraic properties of the integers are established by using various properties in N.

An important requirement that must be met by the operations of $+$ and *times* in Z is that they be *well defined*. This means that in an expansion involving addition or multiplication, an element may be replaced by its equal. The following lemma will insure that these binary operations are well defined.

LEMMA: If $[m, n]$, $[p, q]$, $[r, s] \in Z$ and $[p, q] = [r, s]$ then

1. $[m + p, n + q] = [m + r, n + s]$
2. $[mp + nq, np + mq] = [mr + ns, nr + ms]$

PROOF: Since $[p, q] = [r, s]$ then $p + s = q + r$. Our objective is to show the equality in Part 1.

$$(m+p)+(n+s) \overset{\text{Why}}{=} m+n+p+s \overset{\text{Why}}{=} m+n+q+r \overset{\text{Why}}{=} (n+q)+(m+r).$$

Thus $[m+p, n+q] = [m+r, n+s]$. (This is true even though the equations $p = r$ and $q = s$ do NOT necessarily hold.) Similarly,

$$\begin{aligned}
(mp + nq) + (nr + ms) &= m(p + s) + n(q + r) \\
&= m(q + r) + n(p + s) \\
&= mq + mr + np + ns \\
&= (np + mq) + (mr + ns).
\end{aligned}$$

Thus Part 2 is established. Provide reasons for each equality above.

DEFINITION (*Addition & Multiplication in Z*): If $[m, n], [p, q] \in Z$ then:

$$[m, n] + [p, q] = [m + p, n + q] \qquad \text{Addition in } Z$$
$$[m, n] \cdot [p, q] = [mp + nq, np + mq] \qquad \text{Multiplication in } Z$$

The lemma above guarantees that $+$ and *times* are well defined in the following way. If $[p, q] = [r, s]$ then

$$[m, n] + [p, q] = [m + p, n + q] \overset{1}{=} [m + r, n + s] = [m, n] + [r, s]$$

where step 1 is by the lemma. Similarly $[m, n][p, q] = [m, n][r, s]$. In other words, we have the substitution principle.

SUBSTITUTION PRINCIPLE: In any expression involving addition or multiplication of integers, any integer may be replaced by its equal. That is, addition and multiplication of integers are *well defined*.

This property is essential in almost any work with sums or products of integers, and it is a property that is probably familiar to you from at least as far back as geometry.

EXAMPLE 19.1: Because the sums and products may still seem a bit obscure, let us see what effect the definitions have on addition and multiplication of particular pairs. Consider the pairs $[3, 1]$ and $[1, 4]$. From the definition of equality, $[3, 1] = [4, 2] = [5, 3] = [6, 4] = \cdots = [n + 2, n]$ for any $n \in N$. Any one of these denotes the same integer, $+2$. Note that the first coordinate exceeded the second by 2. Similarly, $[1, 4] = [2, 5] = [3, 6] = \cdots = [n, n + 3]$ for any $n \in N$. What integer does this denote? Here the second exceeded the first by 3, so it represents -3. Let us see how they add together. $[3, 1] + [1, 4] = [3 + 1, 1 + 4] = [4, 5]$, which when translated as above, would assert that $(+2) + (-3) = (-1)$. In a similar manner $[3, 1] + [6, 3] = [9, 4]$ says $(+2) + (+3) = (+5)$.

As an example of multiplication by the definition, consider the product $[3, 1][1, 4]$.

$$[3, 1][1, 4] = [3 \cdot 1 + 1 \cdot 4, 1 \cdot 1 + 3 \cdot 4] = [7, 13]$$

which translates to $(+2)(-3) = (-6)$.

One of the desirable features of this kind of development is that the algebraic and order structures of Z, which are more complex than those of W, are completely derivable from $\underline{W}$ through a few definitions. All we needed to get to the present point with Z was to carefully define what integers are in terms of equivalence classes. Then define the binary operations of addition and multiplication in Z using previously developed operations in W. Furthermore, quite similar methods when applied to Z will yield the next system, the system of rational numbers.

In the present situation with integers and in the next one with rational numbers, there is a connection between the problem to be solved and the development of the system that admits the solution to the problem. We observed that, in the whole number system, we were unable to solve equations of the form $a + x = b$ for $a, b \in W$. When we formed Z, we wanted to be able to subtract to get $x = b - a$. But $b - a$ isn't always in W. In our new system, we will demonstrate that we have made the integers to do exactly that. The positive integer 3 is the equivalence class $[4, 1]$ using only whole numbers 4 and 1 and whole number arithmetic. Likewise the integer -2 is the class $[3, 5]$ using only whole numbers. We did not artificially put negative signs in front of whole numbers; we carried out the process very carefully using only a few definitions, equivalence classes, partitions, and properties developed in W.

Turning again to the main thrust of this section, let us consider a theorem which spells out the most basic algebraic properties of the system of integers, namely closure, commutativity, associativity, etc. for addition and multiplication.

THEOREM 19.2:   Suppose $x, y, z \in Z$ then

a. $x + y \in Z$ and $xy \in Z$ $\qquad\qquad$ Z is closed w.r.t. $+$ and $\cdot$

b. $x + y = y + x$ and $xy = yx$ $\qquad\quad$ $+$ and $\cdot$ are commutative

c. $x + (y + z) = (x + y) + z$ $\qquad\qquad$ $+$ is associative
$\quad$ $x(yz) = (xy)z$ $\qquad\qquad\qquad\qquad$ $\cdot$ is associative

d. $(\exists w \in Z) \ni (\forall x \in Z)[x + w = x]$ $\qquad$ Rename w by 0.
$\quad$ $(\exists u \in Z) \ni [u \neq w \wedge xu = x \ \forall x \in Z]$ $\quad$ Rename u by 1.

e. $x(y + z) = xy + xz$ $\qquad\qquad\qquad$ Distributive law

f. $(\forall x \in Z)(\exists v \in Z) \ni [x + v = 0]$. $\qquad$ Rename v by $-x$.

These properties include those that hold in W, but there is a new property in Z that is not in W. In particular, Part f says every x has an additive inverse v, denoted by $-x$, which is traditionally called the negative of x. The other parts are the familiar commutative, associative, and distributive laws and the existence of additive and multiplicative identities, 0 and 1, (with $0 \neq 1$) to use the traditional symbols, 0 and 1.

PROOF: As previously mentioned, the proof of this theorem follows from the facts we established about W and N and the definitions pertaining to Z. Parts c and e are exercises, the others are done below.

    a. $x, y \in Z \Longrightarrow \exists p, q, r, s \in N \ni x = [p, q] \wedge y = [r, s]$. Thus $x + y = [p + r, q + s]$. But since $p + r \in N$ and $q + s \in N$ and $x + y$ is the equivalence class of the pair $(p + r, q + s)$, $x + y \in Z$. By doing the analogous things, $xy \in Z$ as well.

    b. Using the notation from Part a,

$$x + y = [p + r, q + s] = [r + p, s + q] = [r, s] + [p, q] = y + x.$$

    The commutative property for multiplication is similar and is an exercise.

    d. Let $x = [p, q]$ be an arbitrary element of Z and let $w = [1, 1]$ then $w \in Z$ and $x + w = [p, q] + [1, 1] = [p + 1, q + 1] = [p, q] = x$. The equality $[p + 1, q + 1] = [p, q]$ holds because $(p + 1) + q = (q + 1) + p$. Therefore, $x + w = x$ for all $x \in Z$. Now let $u = [2, 1]$ then $xu = [p, q][2, 1] = [2p + q, 2q + p] = [p, q] = x$, where $[2p + q, 2q + p] = [p, q]$ because $2p + q + q = 2q + p + p$. Therefore $xu = x \ \forall x \in Z$. Now $u \neq w$, i.e., $[2, 1] \neq [1, 1]$ since $2 + 1 \neq 1 + 1$ in N, so Part d of Theorem 19.1 is established.

    f. Let $x = [p, q]$ and $v = [q, p]$. Then it is an exercise to show $v \in Z$ and $x + v = [1, 1]$, which is the "zero" described in Part d. (Note the order of the quantifiers–"For all" precedes "there exists." The student may want to review what that means in Chapter 3, specifically in Examples 3.1–3.3.)

Order Properties For Z:

    The properties of $<$ for W can be extended to Z in the following manner. Let the integer $z = [q, r]$ be called <u>negative</u> if and only if $q < r$ (as natural numbers). And let z be called <u>positive</u> if and only if $r < q$ (as natural numbers). Also let $Z^+ = \{z \in Z \mid z \text{ is positive }\}$. The definition of "less than" in Z is now possible. Note the similarity to $<$ in N.

DEFINITION (*Order in Z*): Let $x, y \in Z$ then $x < y$ if and only if $\exists p \in Z^+ \ni x + p = y$. Also $y > x$ iff $x < y$ and $x \leq y$ iff $(x < y \vee x = y)$ and $(y \geq x \text{ iff } x \leq y.)$

LEMMA: Z^+ is closed with respect to addition and multiplication.

    The proof is an exercise. Notice that the lemma is quite similar to Part a of Theorem 18.3, concerning $<$ for N. The similarity does not stop here. The remaining properties of $<$ in the context of N also have counterparts in Z. The next theorem gives you some of these.

THEOREM 19.3: Suppose $x, y, z \in Z$. Then

    a. If $x < y$ then $x + z < y + z$.

    b. If $x < y$ and $y < z$ then $x < z$.   Transitivity of $<$.

    c. If $x < y$ and $p \in Z^+$ then $xp < yp$.

    d. If $x, y \in Z$ then $[x < y$ or $y < x$ or $x = y]$ AND only one holds.

The proof of Parts a, b, and c are quite similar to the proof of the corresponding properties in Theorem 18.3. Note, while going through the following argument, that although we are proving certain properties for $<$ in Z, we are using properties of $<$ and addition and multiplication in N. In some cases we are using the analogous property in N. In any case we are using previously established facts.

PROOF: We will establish a portion of Part d. The proof is based on the trichotomy law for N whose proof was not included in this text.

Let $x, y \in Z$ then $\exists m, n, p, q \in N \ni x = [m, n] \wedge y = [p, q]$. Now $z \in Z^+$ if and only if $\exists t \in N \ni t > 1 \wedge z = [t, 1]$. Thus

$$
\begin{aligned}
(x < y) &\equiv \exists z \in Z^+ \ni (x + z = y) \\
&\equiv \exists t \in N \ni (t > 1 \wedge [m, n] + [t, 1] = [p, q]) \\
&\equiv \exists t \in N \ni (t > 1 \wedge (m + t) + q = (n + 1) + p) \\
&\equiv \exists t \in N \ni (t > 1 \wedge (m + t) + q = n + (1 + p))
\end{aligned}
$$

Similarly

$$
\begin{aligned}
(x > y) &\equiv \exists z \in Z^+ \ni (x = z + y) \\
&\equiv \exists t \in N \ni (t > 1 \wedge [m, n] = [t, 1] + [p, q]) \\
&\equiv \exists t \in N \ni (t > 1 \wedge m + (1 + q) = n + (t + p))
\end{aligned}
$$

For the x and y we are given, we show $x < y$ or $x = y$ or $y < x$. If $x = y$ then the assertion of the theorem is valid. On the other hand, if $x \neq y$ then we must show either $x < y$ or $y < x$. To do this, there are four cases to consider.

             Case 1. $m < n$ and $p < q$
             Case 2. $m < n$ and $q < p$
             Case 3. $n < m$ and $q < p$
             Case 4. $n < m$ and $p < q$

CASE 1: If $m < n$ and $p < q$ then $\exists r, s \in N \ni m + r = n \wedge p + s = q$. Using the equivalences above and substituting for n and q in the preceding equivalence, we get

$$(x < y) \equiv \exists t \in N \ni (t > 1 \wedge m + t + p + s = m + r + 1 + p)$$
$$\equiv \exists t \in N \ni (t > 1 \wedge t + s = r + 1) \qquad \text{Why?}$$

and in a similar manner

$$(x > y) \equiv \exists t \in N \ni (t > 1 \wedge m + 1 + p + s = m + r + t + p)$$
$$\equiv \exists t \in N \ni (t > 1 \wedge 1 + s = r + t).$$

Since $r, s \in N$ then by the trichotomy law for N, $r < s$, $r = s$ or $r > s$. If $r < s$ then $\exists t \in N \ni (t > 1 \wedge 1 + s = r + t)$. Hence $x > y$. If $r > s$ then $\exists t \in N \ni (t > 1 \wedge t + s = r + 1)$. Then $x < y$. If $r = s$ then

$$m + q = m + p + s = m + p + r = m + r + p = n + p,$$

so $[m, n] = [p, q]$. Then $x = y$. Consequently, in Case 1, the trichotomy law holds in Z.

CASE 2: If $m < n$ and $q < p$ then $x < 0$ and $y > 0$. Hence, $x < y$ by transitivity of $<$.

CASE 3: If $n < m$ and $q < p$ then an analogous situation to Case 1 arises. The proof proceeds by reversing the roles of m and n as well as p and q.

CASE 4: If $n < m$ and $p < q$ then $x > 0$ and $y < 0$. Hence $y < x$ by transitivity.

To establish that, ONLY one of the outcomes in the trichotomy law holds, i.e., $x < y$ or $x = y$ or $y < x$. So suppose on the contrary that

i. $x < y$ and $x = y$ or,

ii. $x < y$ and $x > y$ or,

iii. $x > y$ and $x = y$. Then find a contradiction.

If Part i were true then $[m, n] < [p, q]$ and $[m, n] = [p, q]$. Then $\exists t \in N \ni t > 1 \wedge m + t + q = n + 1 + p$ and $m + q = n + p$. But then $t = 1$. Since $t > 1$ and the trichotomy law holds for N there is a contradiction. The eager student might want to complete this already somewhat lengthy proof by doing Parts ii and iii.

It is also worth mentioning that $<$ can be reversed in Parts a, b, and c of this theorem, and the corresponding results for $>$ will be true. Also $<$ can be replaced by $\leq$ or $\geq$ in these parts as well.

Recall that is not always possible to solve for $z \in W$ for equations of the form $x + z = y$ if given $x, y \in W$. In fact, this amounts to having the ability to <u>subtract</u>. The definition of subtraction in Z is analogous to the previous definition of subtraction in W: if $x, y, z \in Z$ then <u>$y - x = z$ if and only if $z + x = y$</u>. Here is the theorem.

THEOREM 19.4:   If $x, y \in Z$ then there is a unique $z \in Z$ such that $z + x = y$. (That is, if $x, y \in Z$ then $y - x \in Z$.) (Note that $\exists! z \in Z$ means that $\exists z \in Z$ and that z is the only z satisfying $z + x = y$.)

PROOF: If $x, y \in Z$ then $-x \in Z$ and $y \in Z$. By closure for addition, $y + (-x) \in Z$ so let $z = y + (-x)$. Then $z \in Z$ and

$$z + x = \big(y + (-x)\big) + x = y + \big((-x) + x\big) = y + 0 = y.$$

Thus the existence part has been established.

To verify the uniqueness of the z in the theorem, suppose $\exists z_1 \in Z \ni z_1 + x = y$. Then $z_1 + x = y = z + x$. Consequently,

$$z_1 = z_1 + 0 = z_1 + \big(x + (-x)\big) = (z_1 + x) + (-x)$$
$$= (z + x) + (-x) = z + \big(x + (-x)\big) = z + 0 = z.$$

Thus the theorem is established.

As a result of this theorem, Z is closed for subtraction. This feature was lacking in N and W, so Z has some significant improvements over N and W.

Also note that there are two distinct uses of the $-$ symbol in Theorem 19.4. One is a subtraction symbol, as in $y - x$, and the other is the additive inverse symbol, as in $-x$. The equation $y - x = y + (-x)$ spells out their relationship.

The next property is a very useful one. It says one can "cancel" a term x from both sides of an equation, if that x is a summand of each side of that equation.

THEOREM 19.5   (*Cancellation Law of Addition*): If $x, y, z \in Z$ and $x + y = x + z$ then $y = z$.

PROOF: The proof is an exercise.

Some similarities are shared by N and Z, as can be noted by returning to Theorems 18.1, 18.2, and 19.2 as well as Theorems 18.3 and 19.3. But, as mentioned before, there are some additional properties true in Z but not in N. In a sense Z is more versatile than N.

THEOREM 19.6: The following properties hold for all $x, y \in Z$.

a. $-(x+y) = (-x) + (-y)$ d. $(-x)(-y) = xy$

b. $x \cdot 0 = 0 \cdot x = 0$ e. $(xy = 0) \equiv (x = 0 \lor y = 0)$

c. $-(xy) = (-x)y = x(-y)$ f. $-(-x) = x$

The proof refers back to various parts of earlier theorems in this chapter. Those theorem numbers are cited to the right of the corresponding step. Parts d, e, and f are exercises.

PROOF: Let $x, y \in Z$.

PART a: $-(x+y) = -(x+y) + 0$ 19.2d

$= -(x+y) + [(x+y) + (-y) + (-x)]$ 19.2f

$= [-(x+y) + (x+y)] + (-y) + (-x)$ 19.2c

$= 0 + (-y) + (-x)$ 19.2f

$= (-x) + (-y)$ 19.2b, 19.2d

PART b: From Theorem 19.2d, $0 + 0 = 0$. Then

$x \cdot 0 + 0 = x \cdot 0$ 19.2d

$= x \cdot (0 + 0)$ previously mentioned

$= x \cdot 0 + x \cdot 0$ 19.2e

Thus $x \cdot 0 + 0 = x \cdot 0 + x \cdot 0$, so by Theorem 19.5 (cancellation), $0 = x \cdot 0$. The other part of b is an exercise.

PART c: $-(xy) = -(xy) + 0$ 19.2d

$= -(xy) + 0y$ 19.6b

$= -(xy) + [x + (-x)] \cdot y$ 19.2f

$= -(xy) + [xy + (-x)y]$ 19.2e

$= [-(xy) + xy] + (-x)y$ 19.2c

$= 0 + (-x)y$ 19.2f

$= (-x)y$ 19.2d

The other half of this part and Part d are exercises.

PART e: The left implication, i.e., $x = 0 \lor y = 0 \implies xy = 0$, was established in Part b. The other implication can be verified by assuming that $\sim [x = 0 \lor y = 0]$. Then $x \neq 0 \land y \neq 0$. Splitting this into cases $(x > 0$ or $x < 0$ and $y > 0$ or $y < 0)$ and using Theorem 19.3, yields the result $xy \neq 0$. This establishes the contrapositive.

Some Aspects of Divisibility (Optional):

This section is closed with a brief introduction to some concepts that are in the realm of number theory. Particular attention is paid to the notions of divisibility and primeness and use is made of the many properties we have developed thus far.

There is a property of W, or of Z^+, which will be considered in more detail in a later chapter. The property is that Z^+ is *well ordered*, meaning that every nonempty subset of Z^+ has a least element. This result, which is readily established using Peano's postulates, or finite induction, will simply be stated here for future reference.

WELL ORDERING PRINCIPLE OF Z^+:   To say Z^+ is <u>well ordered</u> means that every nonempty subset S of Z^+ has a unique least element, i.e., an element $m \in S$ such that $m \leq s$ for all $s \in S$.

For example, the set P of positive prime numbers is a nonempty subset of Z^+. The least element of P is 2. The set $\{10^n \mid n \in N\}$ is a nonempty subset of Z^+ so it has a least element, $10^1 = 10$. The set $\{n \in Z^+ \mid (1.01)^n > 5\}$ is a nonempty subset of Z^+, so it has a least element. It is 162. The well ordering of Z^+ is our guarantee that such a least element exists, whether we are able to find it or not.

DEFINITION: If $x, y \in Z$ and $x \neq 0$ then x <u>divides</u> y, denoted $x \mid y$, if and only if $\exists\, k \in Z \ni xk = y$. Also, $\sim (x \mid y) \equiv x \nmid y$.

THEOREM 19.7:   If $x \neq 0$ and $y \neq 0$ and $x \mid y$ and $y \mid z$ then $x \mid z$. i.e., the relation $\mid$ is transitive.

PROOF: If $x, y, z \in Z$ and $xy \neq 0$ and $x \mid y$ and $y \mid z$ then $\exists k_1 \in Z \ni xk_1 = y \land \exists k_2 \ni yk_2 = z$. Thus $x(k_1 k_2) = (xk_1)k_2 = yk_2 = z$. Since $k_1 k_2 \in Z$ by Theorem 19.2a, $x \mid z$.

THEOREM 19.8:   If $x, y, z \in Z$ and $x \neq 0$ and $x \mid y$ and $x \mid z$ then $x \mid (ym + zn) \ \forall m, n \in Z$.

PROOF: Suppose $x, y, z \in Z$ and $xy \neq 0$ and $x \mid y$ and $x \mid z$ then $\exists k_1 \in Z \ni xk_1 = y \land \exists k_2 \ni xk_2 = z$. If $m, n \in Z$ then $x(k_1 m + k_2 n) = xk_1 m + xk_2 n = ym + xn$. Since $k_1 m + k_2 n \in Z$ then $x \mid (ym + xn)$.

As an example of the use of this theorem, suppose that $x \mid 6 \wedge x \mid 15$ then $x \mid \left(6 \cdot (-2) + 15 \cdot 1 \right)$, i.e., $x \mid 3$.

However, not every integer is divisible by a given non–zero integer. For example, $2 \nmid 5$ since $2n \neq 5$ $\forall n \in Z$. The next best thing to this kind of divisibility is to "divide" and get a remainder. The justification for this familiar process is given in the so-called division algorithm.

THEOREM 19.9 (*Division Algorithm*): If $a \in Z$, $b \in Z$, and $b > 0$ then $\exists! q, r \in Z$ such that

$$a = bq + r \quad \text{and} \quad 0 \le r < b.$$

PROOF: This two–part proof is a classical example of a proof by contradiction, involving the well ordering principle of Z^+. The first part establishes that q and r exist; the next part shows they are unique.

EXISTENCE: Let $S = \{a - bx \mid x \in Z \wedge a - bx \ge 0\}$ then $S \subseteq Z$. To show $S \neq \emptyset$ two cases are considered: $a > 0$ and $a \le 0$. If $a > 0$ then, by letting $x = 0$, $a \in S$, hence $S \neq \emptyset$. If $a \le 0$ then $-a \ge 0$ and since $b \ge 1$, $(a - ab) = (-a)(b - 1) \ge 0$. So if $x = a$ then $a - bx = a - ba = a - ab \ge 0$. Therefore $a - ab \in S$. Thus $S \neq \emptyset$ in any case.

If $0 \in S$ then 0 is the least element of S. If $0 \notin S$ then $S \subseteq Z^+$ so by the well ordering principle of Z^+, the S defined above has a least element. Let the least element be r. Then $r \ge 0$ and $r = a - bq$ for some $q \in Z$. Thus $a = bq + r$ and $0 \le r$. There remains to show that $r < b$.

If $r \not< b$ then $r \ge b$, (Why?). So $r - b \ge 0$. Now $a - b(q + 1) = a - bq - b = r - b \ge 0$, so $a - b(q + 1) \in S$. However, since $q < q + 1$, (see Exercise 19.14c), $-(q + 1) < -q$. Thus $-b(q + 1) < -bq$, so $a - b(q + 1) < a - bq = r$. Since $a - b(q + 1) \in S$ and $a - b(q + 1) < r$, r is not the least element of S. This contradiction assures us that the assumption $r \not< b$ is false. Thus $r < b$. Consequently $\exists q, r \in Z$ such that

$$a = bq + r \quad \text{and} \quad 0 \le r < b.$$

UNIQUENESS: Suppose $\exists q', r' \in Z$ such that

$$a = bq' + r' \quad \text{and} \quad 0 \le r' < b.$$

Uniqueness is established by showing $q = q'$ and $r = r'$. Using the equations above yields $bq + r = a = bq' + r'$, whence $b(q - q') = r' - r$. But, by the definition of divides, $b \mid (r' - r)$. Since $0 \le r < b$ and $0 \le r' < b$

then $r' - r \le r' < b$ and $r - r' \le r < b$. Thus $|r' - r| < b$. But since $b \mid |r' - r|$ and $|r' - r| < b$, then $|r' - r| = 0$ (see Exercise 19.17). Thus $r' = r$ and so $q' = q$. Consequently, uniqueness is established.

Another theorem that uses the well ordering principle of Z^+ is now considered. It is called the *greatest common divisor theorem*.

THEOREM 19.10 (*Greatest Common Divisor Theorem*): If $a, b \in Z$ and $ab \ne 0$ then there exists a unique $d \in Z^+$ such that

  a. $d \mid a$ and $d \mid b$, and
  b. If $(c \in Z$ and $c \mid a$ and $c \mid b)$ then $c \mid d$.

PROOF: Let $S = \{ax + by \mid x, y \in Z\}$. Then it is an exercise (Exercise 19.15a) to show S has a positive element. This follows readily from the assumption that $ab \ne 0$, hence $a > 0$ or $a < 0$.

By the well ordering principle of Z^+, S has a least positive element, say d. Since $d \in S$ there exists $x', y' \in Z \ni d = ax' + by'$.

PART a: Show: $d \mid a \wedge d \mid b$.

By Exercise 19.15b, $a, b \in S$. To show $d \mid a$ and $d \mid b$, show that d divides each non–zero element of S. To do that, let z be a non–zero element of S. Then $z = ax + by \in S$ with $ax + by \ne 0$. Then by the division algorithm, since $d \in Z^+$, $\exists\, q, r \in Z \ni$

$$ax + by = dq + r \quad \wedge \quad 0 \le r < d.$$

But $d = ax' + by'$ from above, so we get

$$ax + by = (ax' + by')q + r.$$

After rewriting the expression above we get

$$a(x - x'q) + b(y - y'q) = r.$$

But then $r \in S$, since it is a sum of an integer multiple of a and an integer multiple of b, i.e., it has the general form shared by each element of S. Now since d is the least positive element of S and $0 \le r < d$, then $r = 0$. Since $ax + by = dq + r$ from the equation above, we get

$$ax + by = dq.$$

Consequently, d divides each element of the form $ax + by$ with $x, y \in Z$. But then $d \mid a$ and $d \mid b$. (Why?) Thus Part a has been established.

It is an exercise to verify Part b and show d is unique.

The integer d, which we now know exists and is unique, is called the greatest common divisor of a and b. Note $\exists\, x, y \in Z \ni d = ax + by$.

EXAMPLE 19.2: Let $a = 12$ and $b = 20$. The unique d mentioned in Theorem 19.10 is $d = 4$. This is because $4 \mid 12$ and $4 \mid 20$, and, if $c \mid 12$ and $c \mid 20$ then the choices for c have to be $c = \pm 1$ or $c = \pm 2$ or $c = \pm 4$. But 4 is the only value of c that each c divides. That is, $d = 4$ is the greatest common divisor of 12 and 20.

DEFINITION: An integer p is prime if and only if

$$p > 1 \text{ and } (x \in Z^+ \wedge x \mid p) \implies (x = 1 \vee x = p).$$

Thus an integer p (with $p > 1$) is a prime iff its only positive divisors are 1 and p. The first six primes are $2, 3, 5, 7, 11$, and 13.

THEOREM 19.11 (*Euclid's Lemma*): If p is a prime and $a, b \in Z$, and $p \mid ab$ then $p \mid a$ or $p \mid b$.

PROOF: Suppose p is a prime and $a, b \in Z$, and $p \mid ab$. Now either $p \mid a$ or $p \nmid a$. If $p \mid a$, the theorem is true.

SHOW: If $p \nmid a$ then necessarily $p \mid b$.

To do this, suppose $p \nmid a$ then the greatest common divisor of p and a is 1. Using the proof of Theorem 19.10 with a as a and p as b we get $1 = ax + py$. Multiplying by b yields $b = bax + bpy = abx + pby$. Now $p \mid ab$ and $p \mid p$ (Why?), so by Theorem 19.8, $p \mid (abx + pby)$. That is $p \mid b$. Thus we have shown $p \mid ab \implies p \mid a \vee p \mid b$.

EXAMPLE 19.3: If $3 \mid ab$ then $3 \mid a$ or $3 \mid b$. For example, $3 \mid 6 \cdot 4 \implies 3 \mid 6 \vee 3 \mid 4$ is valid. Notice that the "divisor" must be a prime for this theorem to apply. For example, $6 \mid 3 \cdot 4$ is true, but $6 \nmid 3$ and $6 \nmid 4$. The reason for this difficulty is that 6 is not prime.

Exercises:

19.1 Prove the transitivity part of Theorem 19.1.

19.2 a. Prove Theorem 19.2c.   b. Prove Theorem 19.2e.

19.3 Prove the addition part of the lemma preceding Theorem 19.3.

19.4 Referring to the lemma preceding Theorem 19.3, prove the multiplication part. (This is a trickier problem.)

19.5 Prove $(x, y, z \in Z \wedge x < y \wedge z < 0) \implies (yz < xz)$.

19.6 Prove the "other part" of Theorem 19.6b.

19.7 Prove the "other part" of Theorem 19.6c.

19.8 Prove Theorem 19.6d.

19.9 Fill in the details of the proof of Part e of Theorem 19.6.

19.10 Prove Theorem 19.6f.

19.11 Prove that $(x, y, z \in Z \wedge x \cdot y = x \cdot z \wedge x \neq 0) \implies y = z$. Be careful! This is called the cancellation law of multiplication.

19.12 Is there a least element in each of the following sets? If so, why? If not, why not? If possible, find the least element when it exists.

    a. $\{n \mid n \in Z^+ \wedge n^2 > 15n\}$.

    b. $\{n \mid n \in Z \wedge 2^n > 1000\}$.

    c. $\{n \mid n \in Z \wedge n^2 > n\}$.

    d. $\{n \mid n \in Z^+ \wedge \left(1 + \frac{1}{n}\right)^n > 2.7\}$. Hint: Use a well–known result from calculus concerning limits.

19.13 Complete the proof of Theorem 19.3d (the trichotomy law for Z) by doing the following. Some parts may be a bit tedious.

    a. Do Case 3.   b. Do Case ii.   c. Do Case iii.

    d. Answer the "why" in the proof of Case 1 of Theorem 19.3.

19.14 a. Prove: If $x \in Z$ then $x^2 \geq 0$.

    b. Prove: $0 < 1$.

    c. Prove: If $x \in Z$ then $x < 1 + x$.

19.15 a. Prove: The set S of Theorem 19.10 has a positive element.

    b. Show $a, b \in S$ as asserted in Part a of the proof of Theorem 19.10.

    c. Show $d \mid a$ and $d \mid b$ as asserted in the proof of Theorem 19.10.

    d. Show Part b of Theorem 19.10.

    e. Show d is unique.

19.16 Prove if a is any non–zero integer then $a \mid a$.

19.17 Prove if $a, b \in Z^+$, and $a \mid b$ then $a \leq b$.

19.18 Prove every integer greater than 1 is a prime or a product of primes. This result is called the *Fundamental Theorem of Arithmetic*. You may assume the following: If $n \in Z^+$ and $n > 1$ then n is *prime* iff there are no positive integers a and b such that $1 < a < n$ and $1 < b < n$ and $n = ab$.

CHAPTER 20

THE SYSTEM Q OF RATIONAL NUMBERS

We are now ready to develop the system Q of rational numbers using the system Z of integers as the point of departure. A principal motivation for this development centers upon the notion of divisibility. In Z the definition of the relation "divides" is, "If $a, b \in Z$, and $a \neq 0$ then a <u>divides</u> b, (denoted $a \mid b$) if and only if $\exists k \in Z$ such that $ak = b$." (See Example 12.13 and Exercise 12.23 as well as Theorems 19.7 and 19.8 for an earlier mention of this concept.)

In the system of integers one can always solve the equation $a + x = b$ if $a, b \in Z$. The solution is $x = b + (-a)$; however, the equation $ax = b$ with $a \neq 0$ is not always solvable in Z. For example, the equation $2x = 3$ has no solution in Z. Of course, this is because $2x$ is even while 3 is odd. More generally, Z is not closed with respect to division by its non–zero elements. Our objective in this section is to derive a number system in which division by non–zero elements is possible

Let $F = \{(x, y) \mid x, y \in Z \wedge y \neq 0\}$. Let $(x, y) \sim (p, q)$ if and only if $xq = yp$ for all $(x, y), (p, q) \in F$. As in the case of the development of the integers, we have defined an equivalence relation. This relation makes pairs equivalent if they have the same "quotient." The proof of the assertion that $\sim$ is an equivalence relation is left to the student.

THEOREM 20.1: $\sim$ is an equivalence relation.

At this point it is worthwhile to observe similarities between this development and the one for the integers. In the case of the formation of the integers from the natural numbers, the objective was to obtain the ability to solve all equations of the form $a + x = b$, for x. This was done by considering ordered pairs $(p, q) \in N \times N$ and defining an equivalence relation $\sim$ whose equivalence classes admit subtraction. In the present case it is desired to solve all equations of the form $ax = b$ with $a \neq 0$. Even though $ax = b$ with $a \neq 0$ is not always solvable in Z, we form equivalence classes in $Z \times Z$ with respect to $\sim$ above. These classes will form a new system in which all equations of the form $ax = b$ with $a \neq 0$ can be solved. In other words, since the ability to subtract was attained in Z using an equivalence relation and addition in N, we do the analogous thing, using addition and multiplication in Z in order to divide.

DEFINITION: Let $Q = \{[p,q] \mid (p,q) \in F\}$ and call Q the set of <u>rational</u> numbers. Here $[p,q] = \{(x,y) \mid (x,y) \sim (p,q) \wedge (x,y) \in F\}$.

Thus in a very similar manner to equality of integers, equality of rational numbers is determined by $[p,q] = [r,s]$ iff $ps = qr$. Also in a manner analogous to the development of $+$ and times for Z, we have the following lemma to ensure these operations are *well defined*

LEMMA: If $[a,b],[c,d],[e,f] \in Q$ and $[c,d] = [e,f]$ then

1. $[ad + bc, bd] = [af + be, bf]$
2. $[ac, bd] = [ae, bf]$

The lemma guarantees that $+$ and $\cdot$ are well defined, i.e., that a rational number may be replaced by its equal in a sum or product. The proof is quite similar to the proof of the corresponding lemma for Z. Examples 20.1a,b,e, and f illustrate that the lemma works.

DEFINITION (*Addition and Multiplication in Q*): If $[a,b],[c,d] \in Q$ then

$$[a,b] + [c,d] = [ad + bc, bd] \qquad \text{Addition in } Q$$
$$[a,b][c,d] = [ac, bd] \qquad \text{Multiplication in } Q$$

In Z the pairs $[p,q]$ were made to represent "differences" between natural numbers so we would be able to subtract, while in Q the pairs $[a,b]$ represent "fractions" so we will be able to "divide." For example, $\frac{1}{4}$ is represented by $[1,4]$ or equivalently, $[-3,-12]$. Actually the rational number $\frac{1}{4}$ represents many pairs, because

$$[1,4] = \{(1,4),(-1,-4),(2,8),(-2,-8),(3,12),(-3,-12),\ldots\}.$$

EXAMPLE 20.1: From the definition, notice that $[1,4] = [2,8]$ and $[3,7] = [-3,-7]$. Then from the definition of $+$ and *times* notice the following

a. $[1,4] + [3,7] = [1\cdot 7 + 4\cdot 3, 4\cdot 7] = [19,28]$.
b. $[2,8] + [-3,-7] = [(-14) + (-24),(-56)] = [-38,-56] = [19,28]$. Thus replacing "equals by equals" in a particular addition problem, does not change the sum. This is true because the lemma above guarantees that addition and multiplication are *well defined*.
c. $[1,4] + [0,1] = [1+0,4] = [1,4]$. So it appears that the "zero" of Q is $[0,1]$.
d. $[1,4] + [-1,4] = [4+(-4),16] = [0,16] = [0,1]$. (Why does the last equation hold?) It appears that the "inverse with respect to $+$" of $[1,4]$ is $[-1,4]$.

e. $[1,4][2,3] = [1 \cdot 2, 4 \cdot 3] = [2,12] = [1,6]$.

f. $[2,8][-6,-9] = [-12,-72] = [1,6]$. Since $[1,4] = [2,8]$ and $[2,3] = [-6,-9]$, the two previous examples illustrate that re-placing "equals by equals" in a multiplication problem does not change the product. The lemma above guarantees this can be done and this example illustrates the idea for particular rational numbers.

g. $[0,1][1,4] = [0,4] = [0,1]$. This provides a further hint that the zero of Q is $[0,1]$.

h. $[1,1][2,3] = [2,3]$. It appears that $[1,1]$ serves as the identity for multiplication.

i. $[1,4][4,1] = [4,4] = [1,1]$. It appears that $[4,1]$ serves as the "multiplicative inverse" of $[1,4]$.

These examples suggesting certain properties relating to addition and multiplication, are true. But in order to provide real validity to these observations as well as other fundamental properties of Q, the counterparts of Theorems 18.1, 18.2, and 19.2 are now established.

THEOREM 20.2: Let $r, s, t \in Q$, then the following are true

a. $r + s \in Q$ and $r \cdot s \in Q$ Closure

b. $r + s = s + r$ and $r \cdot s = s \cdot r$ Commutativity

c. $r + (s + t) = (r + s) + t$ and $r \cdot (s \cdot t) = (r \cdot s) \cdot t$ Associativity

d. $r(s + t) = rs + rt$ Distributivity

e. $\exists w \in Q \ni r + w = r \ \forall r \in Q$ w is the "zero"
   $\exists u \in Q \ni (u \neq 0 \wedge \forall r \in Q \, (ru = r))$ u is the "one"

f. $\forall r \in Q \, \exists y \in Q \ni r + y = 0$ Denote y by $-r$

g. $\forall r \in Q [r \neq 0 \implies \exists v \in Q \ni rv = 1]$ Denote v by r^{-1}

PROOF: Let $r = [a, b]$, $s = [c, d]$ and $t = [e, f]$. Then $a, b, c, d, e, f \in Z$ and $b \neq 0$, $d \neq 0$ and $f \neq 0$.

a. Since all forms such as $ad + bc$, ac, and bc are in Z and $bd \neq 0$ (why is this?) then $r + s = [ad + bc, bd]$ and $rs = [ac, bd]$ belong to Q. Thus Q is closed with respect to addition and multiplication.

b. $r + s = [a, b] + [c, d] = [ad + bc, bd] = [cb + da, db] = [c, d] + [a, b] = s + r$. The justification for the middle equality stems from the commutative laws of addition and multiplication in Z. Commutativity for multiplication in Q is left to the exercises.

c. The associative laws are left to the exercises.

d. The distributive law is left to the exercises.

e. The second part is an exercise. Let $w = [0, 1]$ and $r \in Q$ then $w \in Q$ and $r = [a, b]$ for some $a, b \in Z$. Then $r + w = [a, b] + [0, 1] = [a \cdot 1 + b \cdot 0, b \cdot 1] = [a, b] = r$. Thus $\exists\, w \in Q \ni r + w = r \;\; \forall r \in Q$, establishing the first assertion of Part e. Hereafter w is called an *additive identity* and u a *multiplicative identity*.

f. This is left to the exercises. The y or $-r$ is called the *additive inverse of* r.

g. Let $r \in Q$ and $r \neq 0$. Then $r = [a, b]$ where $a, b \in Z \wedge a \neq 0 \wedge b \neq 0$. Thus $[b, a] \in Q$ (Why is this true?) and if $[b, a] = v$ then $rv = [a, b][b, a] = [ab, ba] = [ab, ab] = [1, 1] = u$. Thus $\forall r \in Q\,[r \neq 0 \Longrightarrow \exists\, v \in Q \ni rv = 1]$.

Notice that the observations made in Example 20.1 turned out to be correct, in particular, the "zero" mentioned in Parts c and g is indeed the zero element. The additive inverse mentioned in Example 20.1d and the multiplicative inverse in Part i of the example were also the proper choices. The example illustrates how experimentation with a new concept can help in the formulation of general properties concerning that concept.

The algebraic properties featured in Theorem 20.2 include many properties corresponding to properties in Z, which again correspond to properties in N. With each number system in succession, however, a few additional properties are established. Although closure, commutativity, associativity, and distributivity hold in all three number systems: N, Z, and Q, the systems Z and Q have a zero and inverses with respect to addition, while N does not. Q has inverses with respect to multiplication, for non–zero elements, while N and Z do not.

One can now establish in Q, a theorem analogous to Theorem 19.6. The proof of Parts a through d of the following theorem will be the same as in 19.6, except for obvious symbol changing. However, Part e can be done more easily in Q than in Z because of the existence of x^{-1} for each non–zero element x in Q, so this part may be proven differently than in Z. These proofs will be left as exercises.

THEOREM 20.3: Let $x, y \in Q$. Then

a. $-(x + y) = (-x) + (-y)$ d. $(-x)(-y) = xy$

b. $x \cdot 0 = 0 \cdot x = 0$ e. $(xy = 0) \equiv (x = 0 \vee y = 0)$

c. $-(xy) = (-x)y = x(-y)$ f. $-(-x) = x$

Now one can establish the property which could be considered as a motivation for constructing the systems of rational numbers.

THEOREM 20.4: Let $x, y \in Q$ and suppose $x \neq 0$ then $\exists ! z \in Q \ni xz = y$.

PROOF: Let $x, y \in Q$ and suppose $x \neq 0$. Then $x^{-1} \in Q$ (see Theorem 20.2g). Also since $y \in Q$ then $x^{-1}y \in Q$. And if $z = x^{-1}y$ then $xz = x(x^{-1}y) = (xx^{-1})y = 1 \cdot y = y$. Thus an element z exists which meets the desired requirements. Now we settle the uniqueness part. If $z_1 \in Q \wedge xz_1 = y$ then $xz_1 = xz$, so $x^{-1}(xz_1) = x^{-1}(xz)$ hence

$$z_1 = 1 \cdot z_1 = (x^{-1}x)z_1 = x^{-1}(xz_1) \overset{\text{why?}}{=} x^{-1}(xz) = (x^{-1}x)z = 1 \cdot z = z.$$

So any element z_1 with the same property as z must be z. That is, z is unique. (Is the step marked, "why," begging the question? Why?)

This theorem says essentially that every rational number y is divisible by each non–zero rational number x. The rational number z in the theorem is usually called the quotient, "y divided by x." A common notation for z is $\frac{y}{x}$. In other words, $x^{-1}y = \frac{y}{x}$. Thus, for example, $x^{-1} = x^{-1} \cdot 1 = \frac{1}{x}$.

Order in Q:

Following the pattern established with N and Z, we now seek an order relation for Q and properties analogous to the order properties of the previous number systems. As with the integers, the beginning point is the definition of "positive." Then we follow a scheme for ordering Q that is similar to the one used for ordering Z.

DEFINITION: A rational number $[a, b]$ is positive if and only if ab is positive as an integer. Let Q^+ be the set of all positive rational numbers. (So if $r \in Q$ and $r = [a, b]$, i.e., $r = \frac{a}{b}$, where $a, b \in Z$, and $b \neq 0$ then $r \in Q^+$ iff ab is a positive integer.)

By the definition above $[-3, -12] \in Q^+$ since the integer $(-3)(-12) = 36$ is a positive integer. Likewise $[1, 4] \in Q^+$ because $1 \cdot 4$ is a positive integer. Since $[1, 4] = [-3, -12]$, $\frac{1}{4}$ and $\frac{-3}{-12}$ are the same positive rational.

DEFINITION $(<, >, \geq, \leq$ in $Q)$: Let $r, s \in Q$. Then $r > s$ iff $r + (-s) \in Q^+$. Also $r < s$ iff $s > r$. In addition $r \geq s$ iff $r > s$ or $r = s$. And $r \leq s$ iff $s \geq r$. Also $r < s < t$ iff $r < s$ and $s < t$, while $r \leq s \leq t$ iff $r \leq s$ and $s \leq t$. Finally, $r = s < t$ means that $r = s$ and $s < t$.

As a consequence of these definitions, Q has all of the order properties that Z has in Theorem 19.3, and some additional ones.

LEMMA: Q^+ is closed with respect to $+$ and multiplication.

The proof is an exercise.

THEOREM 20.5:   Let $r, s, t \in Q$ then

    a. If $r < s$ then $r + t < s + t$.

    b. If $r < s$ and $s < t$ then $r < t$.

    c. If $r < s$ and $t \in Q^+$ then $rt < st$.

    d. $r, s \in Q \implies [r < s \text{ or } s < r \text{ or } s = r]$ AND only one is true.

    e. If $r < s$ and $-t \in Q^+$ then $st < rt$.

    f. If $0 < r < s$ then $0 < s^{-1} < r^{-1}$.

    g. If $r, s \in Q^+$ then $r < s$ iff $r^2 < s^2$.

Part b is called transitivity of $<$ and Part d is the trichotomy law.

PROOF: The proofs of parts of this theorem are quite similar to parts of the proofs to Theorems 18.3 and 19.3. The remaining parts are left to the exercises.

It is also worth mentioning that $<$ can be reversed in Parts a, b, c, e, and g so the corresponding results for $>$ will be true. Also $<$ can be replaced by $\leq$ in Parts a, b, c, e, and g.

Now Z may be thought of as a kind of subset of Q. Though Z is not literally a subset of Q there is a subset of Q having an algebraic structure like Z. To identify that subset, observe the $1-1$ correspondence

$$n \longleftrightarrow \frac{n}{1} \longleftrightarrow [n, 1] \in Q.$$

The set $\hat{Z} = \{\frac{n}{1} \mid n \in Z\}$ is a subset of Q which for all practical purposes is the same as Z. The result of addition and multiplication in $\hat{Z}$ is the same as it is in Z. This is because of the $1-1$ correspondence defined above.

$$n + m \longleftrightarrow \frac{n+m}{1} = \frac{n}{1} + \frac{m}{1} \qquad \text{and} \qquad nm \longleftrightarrow \frac{nm}{1} = \frac{n}{1} \cdot \frac{m}{1}.$$

Such a correspondence between Z and $\hat{Z}$ which preserves $+$ and $\cdot$ is called an *isomorphism* and we say Z and $\hat{Z}$ are *isomorphic*. Also notice that

$$0 \longleftrightarrow \frac{0}{1} = 0 \qquad \text{and} \qquad 1 \longleftrightarrow \frac{1}{1} = 1.$$

More generally, any property true of Z will be true of $\hat{Z}$. For example, thinking of 0 and 1 as rational numbers, $0 < 1$, just as it was in Z.

Exercises:

20.1   Use the definitions of $+$, $\cdot$ and $>$ to work the following problems:

    a. $[2,3] + [-1,5] =$

    b. $[3,4] + [-2,3] =$

    c. $[7,2] + [0,1] =$

    d. $[-5,3] + [0,4] =$

    e. $[7,6] > [9,8]$ because _____

20.2   Prove Theorem 20.1.

20.3   Prove Theorem 20.2f. That is show $\forall x \in Q \; \exists \, y \in Q \ni x + y = 0$.

20.4   In Q prove that multiplication is commutative. In other words, complete the proof of Theorem 20.2b.

20.5   In Q, prove multiplication distributes over addition. That is, prove Theorem 20.2d.

20.6   Prove Theorem 20.3e, i.e., $xy = 0$ iff $x = 0$ or $y = 0$ for all $x, y \in Q$. See if it can be done in a non–analogous way to Theorem 19.6e.

20.7   Prove Theorem 20.3f, i.e., prove $-(-x) = x$ for all $x \in Q$. (Be careful; do not beg the question.)

20.8   For each $x, y \in Q$ prove $(-x)(-y) = xy$. (It is easy to beg the question, so be careful.)

20.9   Thinking of 0 and 1 as rational numbers, prove $0 < 1$. (Try not to use equivalence classes of ordered pairs. Instead use the results of Theorem 20.5.)

20.10   Prove Theorem 20.5e, i.e.,
$$\left[r, s \in Q \wedge r < s \wedge -t \in Q^+ \Longrightarrow st < rt \right].$$

20.11   Prove Theorem 20.5f, i.e., if $0 < x < y$ then $0 < y^{-1} < x^{-1}$. Use this result and properties of $<$ in Q to verify that $\frac{1}{3} < \frac{1}{2}$.

20.12   Prove Theorem 20.5g, i.e., prove that if $x, y \in Q^+$ then $x < y$ iff $x^2 < y^2$. Use this result to verify that $\frac{1}{4} < \frac{25}{64}$. You may assume $x^2 = x \cdot x$.

20.13 Prove the *cancellation laws of addition and multiplication* in Q. That is prove:

    a. If $x, y, z \in Q$ and $x + y = x + z$ then $y = z$.

    b. If $x, y, z \in Q$ and $x \neq 0$ and $xy = xz$ then $y = z$. Do it two different ways.

20.14 Prove Q^+ is closed with respect to addition and multiplication.

20.15 If $a, b, c, d \in Z^+$ then $\frac{a}{b} < \frac{c}{d}$ iff $ad < bc$.

CHAPTER 21

OTHER ASPECTS OF ORDER

The algebraic structure of Q consists of addition and multiplication together with those familiar properties pertaining to addition and multiplication in Q. The algebraic structure of Q is called a *field*. More generally, a field is any set having the properties listed in Theorem 20.2, so not only does Q form a field, but the system E_1 forms a field as well. The system of real numbers is discussed in the next section. If in addition to the algebraic properties of a field there is the presence of an order relation which has the properties listed in Theorem 20.5, the resulting structure is called an *ordered field*. Therefore, the system Q of rational numbers is an ordered field. Here is the working definition of an "ordered" field.

DEFINITIONS: Let F be a field. F is an <u>ordered</u> field iff $\exists F^+ \subseteq F \ni F^+$ is closed with respect to addition and multiplication and if $x \in F$ then $x = 0$ or $x \in F^+$ or $-x \in F^+$. Let F be an ordered field (for example, the rationals or the reals) with the order relation $<$. Let $\leq$ be defined by $(x \leq y) \equiv (x < y \vee x = y)$. Let $\emptyset \neq T \subseteq F$ and $u \in F$. Then u is an <u>upper</u> <u>bound</u> <u>of</u> T in F if and only if $t \leq u \ \forall t \in T$.

To illustrate how simple the concept of upper bound really is, several examples are considered.

EXAMPLE 21.1: Let $S = \{2, -3, \frac{17}{3}, -9.1, 7, 234\}$. Then $S \subseteq Q$. Note that 236 is an upper bound of S because $t \leq 236$ for all $t \in S$. Similarly 237, 5001, 235.1, and 234 are upper bounds of S. Clearly, there can be more than one upper bound for a set. Also an upper bound for a set S may belong to S, though it is possible that no upper bound for S actually belongs to S. This is the case in the next example.

EXAMPLE 21.2: Let $T = \{1 - \frac{1}{n} \mid n \in N\}$ then $T \subseteq Q$ and 1 is an upper bound of T, because $1 - \frac{1}{n} < 1$ for all $n \in N$. In addition, notice that $T = \{\frac{n-1}{n} \mid n \in N\} = \{\frac{n}{n+1} \mid n \in Z \wedge n \geq 0\}$. We now show that no rational number less than 1 is an upper bound of T. If $b < 1$ and $b \in Q^+$ then $b = \frac{p}{q}$ where $p, q \in Z^+$ and $p < q$. Thus $p + 1 = q$ or $p + 1 < q$. If $p + 1 = q$ then

$$\frac{p}{q} = \frac{p}{p+1} < \frac{p+1}{p+2} \in T.$$

The latter inequality holds since $p^2 + 2p < p^2 + 2p + 1 \ \forall p \in Z^+$. (Why?) Thus $\frac{p}{q}$ is not an upper bound of T when $p + 1 = q$.

If $p + 1 < q$ then $q > 2$, and $p < q - 1$ and $p, q - 1 \in Z^+$. Thus $b = \frac{p}{q} < \frac{q-1}{q} \in T$. Again, b is not an upper bound of T. In any case, if $b \in Q \ \wedge \ b < 1$ then b is not an upper bound of $T = \{1 - 1/n \, | \, n \in N\}$.

In a sense soon to be defined, 1 is the *least upper bound* for T even though $1 \notin T$. Similarly the rational number 234 is the "least" upper bound of the set of Example 21.1. Thus as mentioned above, an upper bound may or may not belong to the set.

To define a "least" upper bound k, one would surely require that k be an upper bound and that somehow, it be the least. This is accomplished in the two parts of the following definition.

DEFINITION: Let F be an ordered field with order $\leq$ and let $\emptyset \neq S \subseteq F$. If $k \in F$ then k is called the <u>least</u> <u>upper</u> <u>bound</u> of S in F if and only if

1. k is an upper bound of S, and

2.   If u is any upper bound of S then $k \leq u$.

By this definition, the least upper bounds in Q for the sets S and T in the examples above, are 234 and 1, respectively.

PROPERTY: If F is an ordered field and $\emptyset \neq S \subseteq F$ and S has a least upper bound k in F, then k is unique. (And therefore we can say "the" least upper bound. See Exercise 21.1.)

EXAMPLE 21.3:   Let $S = \{r \, | \, r \in Q \ \wedge \ r < 5\}$ then clearly 5 is an upper bound of S. If $b < 5$ and $b \in Q$ then $\frac{b+5}{2} \in Q$ and $b < \frac{b+5}{2} < 5$. (Why?) Thus $\frac{b+5}{2} \in S$ (Why?) Since $b < \frac{b+5}{2}$, b is NOT an upper bound for S. (Why?) Hence 5 is the least upper bound of S in Q. (Why?)

Now an example of somewhat more complexity is considered. In addition, this example points out a type of "incompleteness" that Q has, and as a result, it indicates a fundamental difference between Q and the next number system, the system of real numbers, treated in Chapter 22.

EXAMPLE 21.4:   Consider the set

$$T = \{x \, | \, x \in Q \ \wedge \ x > 0 \ \wedge \ x^2 < 2\}.$$

It is easy to determine that $1.4 \in T$, while $1.5 \notin T$. In fact, 1.5 is an upper bound for T, because if $t \in T$ then $t^2 < 2 < 2.25 = (1.5)^2$. Then, by Theorem 20.5g, $t < 1.5$, i.e., 1.5 is an upper bound for T. Now, since $(1.41)^2 = 1.9881 < 2$ and $1.41 > 0$ and $1.41 \in Q$, then $1.41 \in T$. However, since $1.4 < 1.41$, 1.4 is not an upper bound of T. Thus the least upper bound, if any, of T will be between 1.4 and 1.5. By an argument quite similar to the above, it is possible to show that 1.42 is an upper bound for T, $1.42 \notin T$ and $1.414 \in T$. Thus the least upper bound of T, if any, will be between 1.41 and 1.42.

In fact, by continuing this kind of argument, two sequences of rational numbers are formed.

$$1 , 1.4, 1.41, 1.414, 1.4142, 1.41421, 1.414214, \ldots$$
$$\text{and}$$
$$2 , 1.5, 1.42, 1.415, 1.4143, 1.41422, 1.414215, \ldots$$

Each element of the first sequence is in T since it is a positive rational number with square less than 2. But none of the elements in the first sequence is an upper bound of T because each is smaller than the element to its right which is also in T. Each element of the second sequence is an upper bound of T, but is not in T, since its square exceeds 2.

In calculus terminology, each of these sequences "converges" to the real number $\sqrt{2}$. Although we won't do it here, it can be argued that $\sqrt{2}$ is the least "number" that can serve as an upper bound for T. But $\sqrt{2} \notin T$ since $\sqrt{2} \notin Q$, in contrast to the elements of T. According to the definition, T does not have a least upper bound <u>in</u> Q. The reason is that the ordered field in question is Q and the least upper bound of T "should be" $\sqrt{2}$, but that number $\sqrt{2} \notin Q$. So Q is not "complete."

One technique for "enlarging" the system Q to E_1 of real numbers involves augmenting Q with each such "least upper bound" of nonempty subsets S of Q having upper bounds in Q. In other words, each nonempty subset S of Q having an upper bound in Q is considered. Then the infinite decimals serving as least upper bounds of each of these S's is included with Q to form the set E_1 of real numbers.

Another technique, created by Richard Dedekind (1831–1916), involves the formation of subsets of Q, called *Dedekind cuts*. Then two binary operations and an order are defined on these cuts resulting in a number system that is, for all practical purposes, the system of real numbers. The procedures outlined above, while not especially difficult, do not suit the purposes of this text and thereby are not pursued. Instead, the more direct approach of stating the necessary axioms for E_1 is followed.

The system E_1 of real numbers, whether developed by Dedekind cuts or created by axioms, satisfies the least upper bound property (LUB).

DEFINITION: (LUB): An ordered field F is said to satisfy the least upper bound property (LUB), if and only if it is true that, for each non-empty subset S of F, if S has an upper bound in F then S has a least upper bound in F. Furthermore, an ordered field F with LUB is said to be a complete ordered field.

In Example 21.4, the given subset T of Q illustrates that Q, while it is an ordered field, is not a *complete* ordered field. The irrational number $\sqrt{2}$ that serves as the least upper bound of the set T in that example, is not in Q, so Q is not "complete." The next chapter treats the least upper bound principle and completeness in some depth. More on these ideas follow.

Exercises:

21.1 Prove that if a subset S of E_1 has a least upper bound then that least upper bound is unique. Thus the least upper bound is appropriate terminology.

21.2 Rework Example 21.3, filling in the reasons marked by, "why?"

21.3 If $S = \{x \mid x \in Q \land 1 < x < 2\}$, find the least upper bound for the set S. Verify the answer.

21.4 Review the definitions of upper bound and least upper bound. Then define lower bound and greatest lower bound analogously.

21.5 Find a lower bound and "the" greatest lower bound of the set $\{\frac{1}{n} \mid n \in N\}$ in Q.

21.6 Let $S = \{x \in Q \mid x > 0 \land x^2 < 3\}$. Find three upper bounds b_1, b_2, b_3 for the set S in Q and three elements x_1, x_2, x_3 in S where $b_i - x_i = 10^{-i}$. Argue that the choices for the b_i's are upper bounds and the x_i's are in S.

CHAPTER 22

THE REAL NUMBER SYSTEM

As mentioned in the previous section, the approach which is followed in this text, is to obtain the reals through axiomatic means. Those axioms follow. It is interesting to note that in a certain sense there is only one complete ordered field. It is the field E_1 of real numbers no matter how that field is obtained, whether by axioms or by Dedekind cuts.

Real Number Axioms:

There is a set, denoted by E_1 of real numbers subject to the following.

A. There are two binary operations $+$ and $\cdot$ on E_1 such that:
   1. E_1 is closed with respect to $+$ and $\cdot$.
   2. The operations of $+$ and $\cdot$ are associative.
   3. The operations of $+$ and $\cdot$ are commutative.
   4. The distributive law holds (*times* over $+$).
   5. There exists $z \in E_1$ such that $a + z = a$ for all $a \in E_1$. (Denote z by 0.)
   6. For all $r \in E_1$ there exists $y \in E_1$ such that $r + y = 0$. (Denote y by $-r$.)
   7. There exists $u \in E_1$ such that $u \neq 0$ and for all $r \in E_1, r \cdot u = r$. (Denote u by 1.)
   8. For all $r \in E_1$ such that $r \neq 0$ there exists $s \in E_1$ such that $r \cdot s = s \cdot r = 1$. (Denote s by r^{-1}.)

B. There exists a relation, denoted $<$ on E_1 such that:
   1. If $x, y, z \in E_1$ and $x < y$ and $y < z$ then $x < z$.
   2. (*Trichotomy law*) If $x, y \in E_1$ then one and only one of the following are true:

   $$x < y \quad \text{or} \quad x = y \quad \text{or} \quad y < x.$$

   3. If $x, y, z \in E_1$ and $y < z$ then $x + y < x + z$.
   4. If $x, y \in E_1$ and $0 < x$ and $0 < y$ then $0 < xy$.
   5. (*LUB*) For all $S \subseteq E_1$, if $S \neq \emptyset$ and if S has an upper bound in E_1 with respect to $\leq$ (defined by $x \leq y$ if and only if $x < y$ or $x = y$), then S has a least upper bound in E_1 with respect to $\leq$.

From these axioms all algebraic and order properties can be established for real numbers. The properties considered in the next several pages, while not an exhaustive list, form the central core of theorems about real numbers. By using A5 and B3 the property, "If $0 < x$ and $0 < y$ then $0 < x+y$ is true." Then by letting $E_1^+ = \{x \in E_1 \mid x > 0\}$ and using B4 we have that E_1 is an ordered field in the sense of the definition in Chapter 21.

The first theorem to be considered is the cancellation law for addition and multiplication. Note the limitation on the cancellation law for multiplication.

THEOREM 22.1:   Let $x, y, z \in E_1$.

   a. If $x + y = x + z$ then $y = z$.

   b. If $x \neq 0$ and $xy = xz$ then $y = z$.

PROOF: Part b is an exercise. To do Part a, suppose $x, y, z \in E_1$ and further suppose $x + y = x + z$. Since $-x \in E_1$ by Axiom A6, we get

$$y = 0 + y = ((-x) + x) + y = (-x) + (x + y)$$

$$\stackrel{1}{=} (-x) + (x + z) = ((-x) + x) + z = 0 + z = z.$$

Is the step indicated by "1" begging the question? Look for the reasons for each step in the axioms above.

THEOREM 22.2:   a. The additive identity 0 is unique, i.e., if $w \in E_1$ and $x + w = x \ \forall x \in E_1$ then $w = 0$.

   b. The multiplicative identity 1 is unique. That is, if $u \in E_1$ and $xu = x \ \forall x \in E_1$ then $u = 1$.

   c. For each $x \in E_1$, $-x$ is unique.

   d. For each $x \in E_1$, if $x \neq 0$ then x^{-1} is unique.

PROOF PART a: Suppose $\exists w \in E_1 \ni x + w = x \ \forall x \in E_1$. Then, by Rule U.S. since $0 \in E_1$, $0 + w = 0$. But, since 0 has the property that $x + 0 = x \ \forall x \in E_1$, then $w + 0 = w$. Thus $0 = 0 + w = w + 0 = w$. That is $0 = w$. Hence 0 is unique. Parts b and c are Exercise 22.2 and 22.3.

PART d: Suppose $x \in E_1$ and $x \neq 0$ then $x^{-1} \in E_1$ and x^{-1} has the property that $xx^{-1} = 1$. To show x^{-1} is unique, suppose $\exists y \in E_1$ and $xy = 1$. Then $xy = xx^{-1}$. By Theorem 22.1b, $y = x^{-1}$. Hence x^{-1} is unique.

THEOREM 22.3: The following are true for all $x, y \in E_1$.

    a. $-(x + y) = (-x) + (-y)$

    b. $-(-x) = x$

    c. $x \cdot 0 = 0 = 0 \cdot x$

    d. $x(-y) = -(xy) = (-x)y$

    e. $xy = (-x)(-y)$

    f. $(xy = 0) \equiv (x = 0 \vee y = 0)$.

To do Parts a and d mimic the proof of Theorem 19.6 with a slight alteration. Part e is essentially Exercise 19.8. Parts b, c, and f are left to the exercises.

THEOREM 22.4: For all $a, b \in E_1$, if $a \neq 0$ then $\exists! x \in E_1 \ni ax = b$.

PROOF: There are two things to show–existence and uniqueness. Since $a \neq 0$, $a^{-1} \in E_1$ by A8. So by A1, since $b \in E_1$, $a^{-1}b \in E_1$. Let $x = a^{-1}b$, then $x \in E_1$ and

$$ax = a(a^{-1}b) = (aa^{-1})b = 1 \cdot b = b,$$

establishing existence. The uniqueness follows from Theorem 22.1b.

Several properties pertaining to the relations $<$ and $\leq$ will now be stated and verified. Most of these are familiar because they are the properties of the ordered field Q as well and so they require only ordered field axioms, i.e., with axioms A and B1 through B4. The LUB axiom, B5, is not needed in this theorem.

THEOREM 22.5: The following are true for all $x, y, z, t \in E_1$.

a. $x < y \Longrightarrow -y < -x$	h. $(0 \leq x^2 \wedge x \neq 0) \Longrightarrow 0 < x^2$
b. $0 < y \Longrightarrow -y < 0$	i. $0 < 1$
c. $x \leq y \Longrightarrow -y \leq -x$	j. $0 < x \Longrightarrow 0 < x^{-1}$
d. $(x < y \wedge 0 < t) \Longrightarrow xt < yt$	k. $0 < x < y \Longrightarrow y^{-1} < x^{-1}$
e. $(x \leq y \wedge 0 \leq t) \Longrightarrow xt \leq yt$	l. $0 < x \leq y \Longrightarrow y^{-1} \leq x^{-1}$
f. $(x < y \wedge t < 0) \Longrightarrow yt < xt$	m. $(xy < xz \wedge 0 < x) \Longrightarrow y < z$
g. $(x \leq y \wedge t \leq 0) \Longrightarrow yt \leq xt$	n. $(xy \leq xz \wedge 0 < x) \Longrightarrow y \leq z$

PROOF: a. Suppose $x < y$ then by axiom B3, $[(-x) + (-y)] + x < [(-x) + (-y)] + y$. By applying axioms A3, A2, A6, and A5, the result is $-y < -x$.

b. This follows from Part a, after observing the easily verified fact that $0 = -0$ (see Exercise 22.11).

c. This also follows easily from Part a.

d. In order to show $(x < y \wedge 0 < t) \implies xt < yt$, suppose $x < y$ and $0 < t$. Then using B3 and A6, $0 < y + (-x)$. So by B4, $0 < [y + (-x)]t$. Thus $0 < yt + (-x)t = yt + (-(xt))$ using A4 and Theorem 22.3d. Thus by B3 and A6, $xt < yt$.

e. This follows easily from Part d.

f. Suppose $x < y$ and $t < 0$ then $0 < y + (-x)$ and $0 < (-t)$. (Why?)   By B4, $0 < [y + (-x)](-t) = y(-t) + (-x)(-t)$. By Theorem 22.3d, e; $0 < -(yt) + xt$, so $yt < xt$.

g. This follows easily from Part f.

h. Two cases are considered in order to show $0 \leq x^2$. They are $0 < x$ and $x \leq 0$. If $0 < x$ then by B4, $0 = 0 \cdot x < x \cdot x = x^2$. And if $x \leq 0$ then by Part g one has $0 = 0 \cdot x \leq x \cdot x = x^2$. In any case, $0 \leq x^2$. The balance is left as an exercise. That is, show when $0 \neq x$ then $0 < x^2$.

i. The inequality $0 < 1$ follows from Part h and the fact that $0 \neq 1$ from A7. The details are left as an exercise.

j. To show $0 < x \implies 0 < x^{-1}$, use the trichotomy law and the preceeding Part i.

k. Parts k–n are left entirely to the student.

The systems Z and Q of integers and rational numbers are isomorphic to subsets of the set E_1 of real numbers because, first: $1 \in E_1$ by axiom A7. Then by closure of addition $\{1, 2, 3, \ldots, n, \ldots\} \subseteq E_1$. Since each positive integer n is in E_1, then by axiom A6, $-n$ is in E_1. With axiom A5, we have all of Z in E_1. Now by using Theorem 22.4, if $\frac{p}{q} \in Q$ then the solution $\frac{p}{q} = x$ to the equation $qx = p$, with $q \neq 0$, belongs to E_1. Hence, Z and Q are "subsets" of E_1 (technically isomorphic to subsets of E_1). Therefore properties such as closure for $+$ and $\cdot$ and well ordering for Z^+ hold for the subset $\{1, 2, 3, \ldots, n, \ldots\}$ of E_1. Similarly, Q^+, thought of as a subset of E_1 is closed with respect to addition and multiplication.

The next theorem is a very important order property called the *Archimedean order property* (AOP) of the real number system; however, in contrast to the properties just listed, its proof is our first one requiring the use of the LUB axiom B5.

THEOREM 22.6 (AOP): If $a, b \in E_1$ and $0 < a$ then $\exists n \in N \ni b < na$.

PROOF: Let $a, b \in E_1$ and $0 < a$. Note that either $b \leq 0$ or $b > 0$. If $b \leq 0$, then since $0 < a$, $b \leq 0 < a = 1 \cdot a$. So the assertion of the theorem is valid when b is not positive. On the other hand if $b > 0$ the objective is to show $\exists n \in N \ni b < na$. This is done by contradiction. Suppose $\sim [\exists n \in N \ni b < na]$, then $\forall n \in N$, $na \leq b$. Thus, b is an upper bound of the set $S = \{na \mid n \in N\}$. Now $S \neq \emptyset$, since $1 \cdot a \in S$, so by LUB $\exists c \in E_1$ such that c is the least upper bound of S. Since, $a > 0$, $c - a < c$. So $c - a$ is not an upper bound of S. Thus $\exists n_0 \in N \ni c - a < n_0 a \leq c$. Consequently, $c = c - a + a < n_0 a + a = (n_0 + 1)a$ and $(n_0 + 1) \cdot a \in S$, so c is less than the element $(n_0 + 1) \cdot a$, of S. This says c is not an upper bound of S, which is a contradiction. Hence $\exists n \in N \ni b < na$.

The next theorem asserts that given any pair of different real numbers (rational or irrational), there is a rational number between them. It is established after proving the following lemma.

LEMMA: If $a, b \in E_1$ and $0 \leq b < a$ then $\exists q \in Q \ni b < q < a$.

PROOF: Let $a, b \in E_1$ and $0 \leq b < a$. Then $0 < a - b$. By the Archimedean order property, since $0 < a - b$, $\exists n \in Z^+ \ni 1 < n(a - b)$, then $n^{-1} = \frac{1}{n} < a - b$. Thus $\frac{1}{n} + b < a$. Since $nb \in E_1$ and since $0 < 1$, $\exists m_1 \in Z^+ \ni nb < m_1 \cdot 1 = m_1$, again using Archimedean order. Thus the set $\{z \in Z^+ \mid nb < z\}$ is nonempty. Hence, by the well ordering property, there is a least positive integer m, such that $nb < m$. Therefore, $b < \frac{m}{n}$. Since m is the least integer greater than nb, $m - 1 \leq nb$ so $\frac{m}{n} - \frac{1}{n} = \frac{m-1}{n} \leq \frac{nb}{n} = b$, whence $\frac{m}{n} \leq b + \frac{1}{n}$. But we know $b < \frac{m}{n}$ so, combining the inequalities, $b < \frac{m}{n} \leq b + \frac{1}{n} < a$. Consequently, using $q = \frac{m}{n}$, $\exists q \in Q \ni b < q < a$, as we had sought to do.

Now this preliminary result is used to establish the theorem.

THEOREM 22.7: If $a, b \in E_1$ and $b < a$ then $\exists q \in Q$ such that $b < q < a$.

PROOF: Suppose $a, b \in E_1$ and $b < a$. Two cases will be considered: $0 \leq b$ and $b < 0$.

CASE 1: If $0 \leq b$ then the lemma applies and yields the desired result.

CASE 2: If $b < 0$ and $0 < a$ then $q = 0$ yields the desired conclusion, i.e., $b < q < a$. And if $a \leq 0$ then since $b < a \leq 0$, $0 \leq -a < -b$. Applying the lemma, $\exists q \in Q \ni -a < q < -b$, whence $b < -q < a$ where $-q \in Q$. Hence the result is valid in this case as well.

This theorem says, using other terminology, that the rational are <u>dense</u> in the reals. An important corollary result is obtained by assuming a and b are irrational real numbers. Thus between every pair of distinct irrational numbers there is a rational number. Now we "reverse these ideas."

EXAMPLE 22.1:   Between every pair of distinct rational numbers there is an irrational number.

PROOF: Let $a, b \in Q$ and $b < a$. Then $a - b > 0$. By the Archimedean order property, since $a - b > 0$ and since $\sqrt{2} > 0$ $\exists n \ni n \in Z^+ \wedge n(a-b) > \sqrt{2}$. Thus $na > \sqrt{2} + nb$. Since $n \in Z^+$, $\frac{1}{n} \in Q^+$, so $\frac{\sqrt{2}}{n} > 0$. Thus

$$a > \frac{\sqrt{2}}{n} + b > b.$$

Now $\sqrt{2}$ is irrational. (See Rudin's *Principles of Mathematical Analysis*, or Theorem 5.7 of this text.) Also since $\frac{1}{n} \in Q$, $\frac{\sqrt{2}}{n}$ is irrational.

(Why?) And consequently, $\frac{\sqrt{2}}{n} + b$ is irrational. (Why?) Thus between the rational numbers a and b, where $b < a$, there is an irrational number $\frac{\sqrt{2}}{n} + b$.

The preceding theorem and example establish that between every pair of distinct irrational numbers there is a rational number and between every pair of distinct rational numbers there is an irrational number. Nevertheless we shall see in the next section, that the infinite sets of rational and irrational numbers are NOT in $1 - 1$ correspondence. Initially this may seem to be a rather startling fact considering what we just learned about the denseness of each of these sets of numbers in the other. So chalk this up to another wonderfully unexpected result.

Exercises:

22.1   Answer the following questions by writing True or False.

   a. Every set of real numbers has an upper bound.

   b. Every nonempty set of real numbers which has an upper bound contains its least upper bound.

   c. The set of rational numbers is a complete ordered field.

   d. Between every pair of distinct rational numbers there is a unique irrational number.

    e. The set of irrational numbers is closed with respect to addition.

    f. The set of irrational numbers is closed with respect to multiplication.

    g. The systems Q and E_1 of rationals and reals are each ordered fields.

    h. Every set of rational numbers with rational number upper bound has a least upper bound in the system E_1.

    i. An important distinction between the ordered fields Q and E_1 is that E_1 is a complete ordered field, while Q is not.

22.2 Prove that 1 is the unique real number such that $x \cdot 1 = x \; \forall x \in E_1$. (That is, prove Theorem 22.2b.)

22.3 Prove Theorem 22.2c, i.e., $\forall x \in E_1, -x$ is unique.

22.4 Prove Theorem 22.3b, i.e., $\forall x \in E_1, -(-x) = x$.

22.5 Prove Theorem 22.3c, i.e., $\forall x \in E_1, x0 = 0 = 0x$.

22.6 Prove that if $y \in E_1$ and $0 < y$ then $-y < 0$. That is, prove Theorem 22.5b.

22.7 Prove Theorem 22.5c, i.e., $\forall x, y \in E_1, x \le y \Longrightarrow -y \le -x$.

22.8 Prove Theorem 22.5g, i.e., $\forall x, y, t \in E_1, (x \le y \wedge t \le 0) \Longrightarrow y \cdot t \le x \cdot t$.

22.9 Prove the rest of Theorem 22.5h, i.e., the $x \ne 0$ part.

22.10 Prove Theorem 22.5i, i.e., prove $0 < 1$.

22.11 Prove $0 = -0$.

22.12 Prove Theorem 22.5j, i.e., prove $0 < x \Longrightarrow 0 < x^{-1}$.

22.13 Prove $\forall x, y \in E_1, [0 < x < y \Longrightarrow y^{-1} < x^{-1}]$. That is prove Theorem 22.5k.

22.14 Prove $\forall x, y \in E_1, [0 < x \le y \Longrightarrow y^{-1} \le x^{-1}]$. That is, prove Theorem 22.5l.

22.15 Prove $\forall x, y, z \in E_1, [(xy \le xz \wedge 0 < x) \Longrightarrow y < z]$. That is, prove Theorem 22.5m.

22.16 Prove $\forall x, y, z \in E_1, (xy \le xz \wedge 0 \le x) \Longrightarrow y \le z$. That is, prove Theorem 22.5n.

22.17 Prove that 1 is the least positive integer in E_1. Hint: Use Theorem 22.5i and the fact that Z^+ is a well ordered subset of E_1.

22.18 Prove that if $a, b \in E_1$ and $a \in Q$ and $b \in E_1 - Q$ and $a \ne 0$ then $a + b$ and ab are in $E_1 - Q$.

22.19 Is the middle equality, indicated by 1, of the proof of Theorem 22.1 "begging the question?" (In other words, are we assuming what we are to prove?)

22.20 What does it mean in Theorem 22.2a, to say 1 is unique? Repeat this for Parts b, c, and d.

SECTION V

TRANSFINITE CARDINAL NUMBERS

CHAPTER 23

FINITE AND INFINITE SETS

Sets of many different sizes have been considered up to now; some have been finite and some infinite. In this section we will investigate what distinguishes finite from infinite and, in fact, because of the work of Georg Cantor (1845–1918), and others, we will even be able to draw distinctions between infinite sets on the basis of size. Then infinite sets of different sizes will be assigned different cardinal numbers. Finally, arithmetic and order properties of these numbers will be pursued.

No doubt the student already knows that the natural numbers form an infinite set and the alphabet forms a finite set. However, a finite set may be quite large. For example, at this moment the collection of all words in all books of all libraries on the earth is finite, although it would be physically impossible to count them. The two finite sets mentioned above, as well as any other finite set, share the property of "finiteness." Since the ability to physically "count" elements is certainly not the condition for determining finiteness, how may a set be determined to have a finite size?

Actually, the means for comparison of sets of any size depends upon the establishment of a 1 – 1 correspondence. The notion of 1 – 1 correspondence was introduced in Chapter 14 and it was a fundamental idea in the development of finite cardinal numbers in Chapter 18.

Cantor used the idea of one-to-one correspondence to serve as a sorting device to distinguish sets by size, and it is used every time a collection of objects is counted. This is done by pairing, in a 1 – 1 manner one of the elements in the collection with the number 1, another with 2 and another with 3, etc. until each of the objects in the collection have been "counted." The number that was paired with the last object is the number of elements in the set. Thus the notion of 1 – 1 correspondence is the keystone to the development of counting procedures as well as for categorizing sets by size.

Cardinal Equivalence:

The process of counting is the establishment of a 1 – 1 correspondence between the set to be counted and a set everyone carries around in his or her brain. The "standard" sets by which we draw the 1 – 1 correspondences are sets of the form $\{1, 2, 3, \ldots, k\}$ for some natural number *k*. The 1 – 1 correspondence is established by pointing to, or glancing toward, each object in turn and mentally or verbally reciting the numbers of the "standard" set, doing it in a one-to-one fashion.

Another common use of 1 – 1 correspondence was mentioned in Example 18.1. An observer standing in the front of an auditorium full of people can determine whether every seat is occupied or not and whether any persons are standing or not. The one-to-one correspondence is the pairing of each person with the chair he or she occupies, excluding exceptional cases such as two persons in one seat, etc. This procedure can accurately determine whether the set of persons and the set of chairs are of the same size, rendering the counting of each of these sets unnecessary. When such a one-to-one correspondence exists between two sets, the sets are said to be *cardinally equivalent*.

DEFINITION: Let A and B be sets then A is cardinally equivalent to B, denoted by $\cong$, if and only if there exists a function $f : A \xrightarrow{\text{1 – 1 onto}} B$.

THEOREM 23.1: If $\mathcal{P}$ is a nonempty set of sets then $\cong$ is an equivalence relation on $\mathcal{P}$.

PROOF: The proof is an exercise.

EXAMPLE 23.1: If N is the set of natural numbers, then $N \cong N$. Suppose V and D denote the sets of even and odd natural numbers, respectively, then $N \cong V$ and $N \cong D$. In other words, there are the same number of elements in N, in the sense of "is cardinally equivalent to," as there are even natural numbers. It is an exercise (Exercise 23.1) to supply actual functions f and g which are the one-to-one correspondences between N and V and between N and D.

Notice that V is a proper subset of N AND YET is cardinally equivalent to N. In other words, N is cardinally equivalent to a proper subset of itself. In fact, one of the ways to characterize an infinite set is exactly this property although we choose a different way to define finite and infinite.

DEFINITION: Let $k \in N$ and let N_k denote $\{1, 2, 3, \ldots, k\}$. The set A is finite if and only if $A = \emptyset$ or $\exists k \in N \ni A \cong N_k$. If A is not finite then A is infinite.

This definition says that a set is *finite* iff it is empty or is cardinally equivalent to one of the sets N_k. For example, if $A = \{a, b, c, d, \ldots, z\}$ is the alphabet, then $\exists k \in N \ni A \cong N_k$. That k is 26.

THEOREM 23.2: N is infinite.

PROOF: This is an indirect proof. If N were finite then

$$\exists k \in N \text{ and } \exists f \ni f : N \xrightarrow{\text{1-1 onto}} N_k.$$

Let $a_1 = f(1)$, $a_2 = f(2)$, ... , $a_k = f(k)$. Then since f is a one-to-one function from N ONTO N_k, the elements $a_1, a_2, a_3, \ldots, a_k$ are all distinct elements of N_k. Thus $N_k = \{a_1, a_2, a_3, \ldots, a_k\}$. Now $k+1 \in N$, and $f : N \longrightarrow N_k$ so $f(k+1)$ is an element of N_k. Thus $f(k+1) = a_i$ for some i such that $1 \leq i \leq k$. But $k+1 \neq i$ and $f(k+1) = a_i = f(i)$. This says f is NOT $1-1$, which is a contradiction. Consequently, it is false that N is finite, so N is infinite.

THEOREM 23.3: If A is finite and B is cardinally equivalent to A then B is finite.

PROOF: Suppose A is finite and $B \cong A$ then either $A = \emptyset$ or $\exists k \in N$ such that $A \cong N_k$. Since $\cong$ is an equivalence relation by Theorem 23.1, either B is empty or $\exists k \in N$ such that $B \cong N_k$. In both cases B is finite.

COROLLARY: If A is infinite and B is cardinally equivalent to A then B is infinite.

PROOF: If B were finite, then by the theorem, A would be finite. This is a contradiction, so B is infinite.

Now several properties of finite and infinite sets are explored with the objective of determining the effect of the subset relation and the binary operation of *union* upon those sets. Specifically, the objective is to establish that subsets of finite sets are finite. To do this, some preliminary results are needed.

LEMMA: If A is a finite set then $A \cup \{b\}$ is finite for any element h

PROOF: Suppose A is finite and b is an element.

CASE 1: If $b \in A$ then $A \cup \{b\} = A$, so $A \cup \{b\} \cong A$. Thus $A \cup \{b\}$ is finite.

CASE 2: If $b \notin A$ then since A is finite, $A = \emptyset$ or $A \cong N_k$ for some $k \in N$. If $A = \emptyset$ then $A \cup \{b\} = \{b\}$, which is finite. If $\exists k \in N \ni A \cong N_k$ then, $A \cup \{b\} \cong N_k \cup \{k+1\} = N_{k+1}$ (see Exercise 14.17). So in this case $A \cup \{b\}$ is finite. Consequently, $A \cup \{b\}$ is finite in <u>any</u> case.

THEOREM 23.4: For each $k \in N$, each subset of N_k is finite.

PROOF: This a proof by induction on k. We will show: $\forall k \in N$ if $A \subseteq N_k$ then A is finite using PMI.

CASE 1: (k=1). Let $A \subseteq N_1$ then either $A = \emptyset$, whence finite, or $A \neq \emptyset$. If $A \neq \emptyset$, since $N_1 = \{1\}$ and $A \subseteq N_1$ and $A \neq \emptyset$, then $A = N_1$. Thus A is finite. Case 1 of PMI has been established.

CASE 2: Suppose the implication $(i \in N \wedge A \subseteq N_i \implies A$ is finite$)$ is true. This is the induction hypothesis. The objective is to show

$$\left(i \in N \wedge A \subseteq N_{i+1} \implies A \text{ is finite} \right).$$

Let $i \in N$ and $A \subseteq N_{i+1}$. Now either $A \subseteq N_i$ OR $i + 1 \in A$. If $A \subseteq N_i$ then A is finite by the induction hypothesis. On the other hand, if $i + 1 \in A$ then the set $B = A - \{i+1\} \subseteq N_i$, whence B is finite by the induction hypothesis. But $A = B \cup \{i+1\}$, so by the preceeding lemma, A is finite. Thus Case 2 is established. By the PMI, it is true that for each $k \in N$ and each $A \subseteq N_k$, A is finite.

THEOREM 23.5: If A is finite and $B \subseteq A$ then B is finite.

PROOF: Suppose A is finite and $B \subseteq A$. Then either $A = \emptyset$ or $\exists k \in N \ni A \cong N_k$. If $A = \emptyset$ then $B = \emptyset$, hence B is finite. If $A \neq \emptyset$ then $\exists k \in N \ni A \cong N_k$. But then $\exists f$ such that

$$f : A \xrightarrow{\text{1-1 onto}} N_k.$$

Either $B = \emptyset$ or $B \neq \emptyset$. If $B = \emptyset$ then B is finite. If $B \neq \emptyset$ then the function $f : A \longrightarrow N_k$ restricted to B, i.e., $f|_B : B \longrightarrow N_k$, is $1-1$ by Theorem 15.3d. Let $T = f[B]$ then $T \subseteq N_k$, hence T is finite by Theorem 23.4. But then

$$f|_B : B \xrightarrow{\text{1-1 onto}} T$$

(by Exercise 23.5) so $B \cong T$. By Theorem 23.3, B is finite.

COROLLARY: If A is infinite and $A \subseteq B$ then B is infinite.

Do the proof of this as an exercise.

THEOREM 23.6: If A and B are finite then $A \cup B$ is finite.

PROOF: Suppose A and B are finite. If either A or B is empty the result is trivially true, so suppose neither is empty. Now either $A \cap B = \emptyset$ or $A \cap B \neq \emptyset$. We use this natural dichotomy to split the proof into two parts.

PART 1: Suppose $A \cap B = \emptyset$. Since A and B are finite, $\exists h, k \in N$ such that $A \cong N_h$ and $B \cong N_k$. That means there exists two functions f_1 and f_2 such that

$$f_1 : A \xrightarrow{\text{1-1 onto}} N_h \quad \text{and} \quad f_2 : B \xrightarrow{\text{1-1 onto}} N_k.$$

Let $T = \{1, 2, 3, \ldots, h, h+1, h+2, \ldots, h+k\}$ then $T = N_{h+k}$. Define f_3 by

$$f_3(x) = \begin{cases} f_1(x), & \text{if } x \in A; \\ f_2(x) + h, & \text{if } x \in B. \end{cases}$$

We wish to establish that f_3 is a function from $A \cup B$ to N_{h+k}. We will use the facts that f_1 and f_2 are functions from A and B to N_{h+k}, respectively, AND $A \cap B = \emptyset$. Now $\text{dom}(f_3) = A \cup B$, because if $x \in A$ then $f_3(x) = f_1(x) \in N_{h+k}$. And if $x \in B$ then $f_3(x) = f_2(x) + h$, so $1 + h \le f_3(x) = f_2(x) + h \le k + h$, which says $f_3(x) \in N_{h+k}$. Thus $f_3(x) \in N_{h+k}$ for all $x \in A \cup B$. Now we know the domain of f_3 is $A \cup B$ and the codomain is N_{h+k}. It remains to be shown that f_3 is single valued. That is an exercise (or one could use Exercise 14.17).

Now we need to establish that f_3 is $1-1$. This may be accomplished in the following four cases. Let $x_1, x_2 \in A \cup B$ and suppose $f_3(x_1) = f_3(x_2)$. Then x_1 will be shown to equal x_2 in each possible case.

CASE 1: If $x_1, x_2 \in A$ then $f_3(x_1) = f_3(x_2)$ implies $f_1(x_1) = f_1(x_2)$. Thus since f_1 is $1-1$, $x_1 = x_2$.

CASE 2: If $x_1, x_2 \in B$ then $f_3(x_1) = f_3(x_2)$ implies $f_2(x_1) + h = f_2(x_2) + h$. This implies $f_2(x_1) = f_2(x_2)$. Thus since f_2 is $1-1$, $x_1 = x_2$.

CASE 3: If $x_1 \in A$ and $x_2 \in B$ then $f_3(x_1) = f_1(x_1) \le h$ and $f_3(x_2) = f_2(x_2) + h > h$. Thus $f_3(x_1) \le h < 1 + h \le f_2(x_2) + h = f_3(x_2)$. Consequently, $f_3(x_1) \ne f_3(x_2)$. Thus this case is impossible.

CASE 4: If $x_1 \in B$ and $x_2 \in A$ then by an analogous argument to Case 3, this is seen to be impossible as well.

It is not difficult to show that f_3 is onto N_{h+k}, so $A \cup B$ is finite in the case where $A \cap B = \emptyset$.

PART 2: Suppose $A \cap B \ne \emptyset$. Let $C = B - A$ then by earlier properties (see Theorem 9.8). $A \cup C = A \cup B$ and $A \cap C = \emptyset$. After applying the result of Part 1 to the sets A and C, the result is $A \cup C$ is finite. But $A \cup C = A \cup B$ so $A \cup B$ is finite.

Exercises:

23.1 Prove that if V and D denote, respectively, the even and odd natural numbers then $N \cong V$ and $N \cong D$.

23.2 Prove that the relation $\cong$, of cardinal equivalence, is an equivalence relation. If S is a set of sets then how does $\cong$ partition the set S?

23.3 Prove the Corollary to Theorem 23.5. That is, if a subset of a set B is infinite then B is infinite.

23.4 Prove that if each A_i for $i = 1, 2, 3, \ldots, k$ is a finite set then $\displaystyle\bigcup_{i=1}^{n} A_i$ is a finite set. Hint: use mathematical induction.

23.5 Let $f : A \longrightarrow B$ be a function. Prove the following:

     a. If $f[A]$ denotes the image of A under f then $f : A \overset{\text{onto}}{\longrightarrow} f[A]$ (see Chapter 16).

     b. If $S \subseteq A$ then $f|_S : S \overset{\text{onto}}{\longrightarrow} f[S]$ (see Chapter 15).

     c. If $f : A \overset{1\text{-}1}{\longrightarrow} B$ and $S \subseteq A$ then $f|_S : S \xrightarrow{1\text{-}1 \text{ onto}} f[S]$.

23.6 Illustrate Theorem 23.6 using $A = \{a, b, c\}$ and $B = \{1, 2\}$. Show how the functions f_1, f_2, f_3 work for these sets A and B.

23.7 Prove. If A is an infinite set then $A - \{a\}$ is infinite.

CHAPTER 24

DENUMERABLE AND COUNTABLE SETS

The terms *denumerable* and *countable* may appear to be associated with a type of finiteness, however, as you have no doubt discovered by now, mathematical terms must be defined carefully. Such is the case with these words. In particular, *countable* does not necessarily mean the capacity to be physically counted and *denumerable* doesn't even refer to finite sets. Let's find out exactly what these terms do mean. Recall that the relation $\cong$ is an equivalence relation (see Exercise 23.2).

DEFINITION: If A is a set and N the set of natural numbers, then A is denumerable if and only if $A \cong N$.

It turns out that some infinite sets are not denumerable, so *denumerable* is NOT just another term for infinite. If this were the case, *denumerable* would be a redundant concept. There is one obvious example of a denumerable set: it is N. Since $\cong$ is an equivalence relation, $\cong$ is reflexive, so $N \cong N$. Thus we have proven our first theorem.

THEOREM 24.1: N is denumerable.

One of the special features that denumerable sets possess is the capacity to be written by "enumeration" (see Chapter 8, where enumeration of sets is first discussed).

THEOREM 24.2: If A is denumerable then A may be written as $A = \{a_1, a_2, a_3, \ldots, a_n, \ldots\}$ with $a_i = a_j$ iff $i = j$. Conversely, such an A is denumerable.

PROOF: If A is denumerable then $A \cong N$. Thus $\exists f \ni f : A \xrightarrow{1-1 \text{ onto}} N$. Define a_i for each $i \in N$ as follows

If $a \in A$ then $a = a_i$ if and only if $f(a) = i$.

Then $f(a_1) = 1$, $f(a_2) = 2$, $f(a_3) = 3$, ..., $f(a_n) = n$, ... The proof will be complete when the equality $A = \{a_i \mid i \in N\}$ is established.

It is clear from the equation $\text{dom}(f) = A$, that $\{a_i \mid i \in N\} \subseteq A$. If on the other hand, $a \in A$ then $f(a) \in N$, so $\exists i \ni i \in N \land f(a) = i$. But then $a = a_i$ so $a \in \{a_i \mid i \in N\}$. Thus $A = \{a_1, a_2, a_3, \ldots, a_n, \ldots\}$. The converse is obvious.

Since denumerable sets can be "enumerated" even though they are infinite and since finite sets can certainly be "enumerated" in some sense, we are ready to define *countable*.

DEFINITION: If A is a set then A is <u>countable</u> if and only if A is either finite or denumerable.

Since N is denumerable, N is countable and, in addition, every finite set is countable. Now obviously, even some finite sets are so large that it would be physically impossible to "count" them in the ordinary sense. However, that is NOT the criterion for "countability." Here are some examples.

EXAMPLE 24.1: N is countable, $\emptyset$ is countable, and $\forall k \in N$, N_k is countable. Refer to Theorem 24.1 and the definition of "finite" for justification.

Now suppose we have a denumerable set, such as N. Some subsets of N may be infinite and some may be finite or even empty. In each of these cases the subset is countable. That is what the next theorem says.

THEOREM 24.3:   If A is denumerable and $B \subseteq A$ then B is countable.

PROOF: Suppose A is denumerable and $B \subseteq A$. From Theorem 24.2, $A = \{a_1, a_2, a_3, \ldots, a_n, \ldots\}$. Now $B = \emptyset$ or $B \neq \emptyset$. If $B = \emptyset$ then B is finite, so B is countable. If $B \neq \emptyset$ then $\exists b \ni b \in B$. Since $B \subseteq A$, then $b \in A$, so $b = a_k$ for some $k \in N$. Since N is well ordered, select the least natural number k such that $a_k \in B$. Let this k be k_1. Now either $B - \{a_{k_1}\}$ is empty or nonempty. If $B - \{a_{k_1}\} = \emptyset$ then $B = \{a_{k_1}\}$, so B is finite and thus countable. If $B - \{a_{k_1}\}$ is nonempty then, since $B - \{a_{k_1}\} \subseteq A$, there exists $a_k \in B - \{a_{k_1}\}$. Let k_2 be the least such k. Now either $B - \{a_{k_1}, a_{k_2}\}$ is empty or not. If empty, then $B = \{a_{k_1}, a_{k_2}\}$ and thus B is countable. If $B - \{a_{k_1}, a_{k_2}\}$ is not empty then there is an element a_k of least index k, such that $a_k \in B - \{a_{k_1}, a_{k_2}\}$. Call it a_{k_3}. Repeating the process, either $B - \{a_{k_1}, a_{k_2}, a_{k_3}\}$ is empty or not.

Continuing this removal process will lead to one of two outcomes. Either the process will terminate at some step, say the mth step, or continue without stopping. In the former case, $B = \{a_{k_1}, a_{k_2}, a_{k_3}, \ldots, a_{k_m}\}$ and is therefore finite. In the latter case, we show that B must be $\{a_{k_1}, a_{k_2}, a_{k_3}, \ldots, a_{k_m}, \ldots\}$. Clearly B contains this infinite set of a_{k_i}'s since they were withdrawn from B. If B was not actually equal to the indicated set, then there would exist an element $b \in B$ such that

$b \notin \{a_{k_1}, a_{k_2}, a_{k_3}, a_{k_4}, \ldots, a_{k_m}, \ldots\}$. But that b is in B, so $b \in A$, which means $b = a_n$ for some $n \in N$. But B contains all a_{k_i} for $i \in N$, i.e., $a_{k_1}, a_{k_2}, a_{k_3}, \ldots, a_{k_i} \ldots$ are all in B. Because of the way the a_{k_i} are selected, that selection process will eventually get a_n into the set B. So the assumption that B would not equal $\{a_{k_1}, a_{k_2}, a_{k_3}, a_{k_4}, \ldots, a_{k_m}, \ldots\}$ is false. Thus in the latter case, $B \cong N$. So in any case B is denumerable and thus countable.

We now consider a remarkable theorem. The assertion of the theorem, at first glance, seems unquestionably false. It is easy to see why someone would want to reject the assertion as false. But such a rejection is likely based on intuition gathered from experience with finite sets. And finite based intuition is simply inadequate as a means of judging whether a conjecture about infinite sets is likely to be true or false. This is only the first of several properties that are true of infinite sets, but run counter to a corresponding type of property for finite sets.

THEOREM 24.4: If N denotes the natural numbers then $N \times N$ is denumerable. Or in other words, $N \times N \cong N$.

We outline the procedure for the proof of this theorem. The only parts left out are the details of establishing that a certain function is $1 - 1$ and onto N, which is slightly messy to do, but not hard to understand.

To prove this, one must show there is a function $f : N \times N \longrightarrow N$ such that f is $1 - 1$ and onto N. To arrive at this function, we employ a procedure pioneered by Cantor.

The elements of $N \times N$ may be arrayed in a row and column display with the first coordinate determining the row number and the second coordinate, the column number (see *Table 24.1*).

It should be clear that each $(p, q) \in N \times N$ is in this table. The pair (p, q) will be found in the pth row and q^{th} column. One can establish a pairing of $N \times N$ with N (in a $1 - 1$ fashion) by counting the elements, as indicated in *Table 24.1*, down the diagonals. If, on the first impulse, the counting were to proceed down a row or a column, not all elements of $N \times N$ would be involved in the pairing. This is because the elements of N would be exhausted in that row (or column) leaving no elements to be paired with the next row. However, by diagonalization, the pairing of elements of N with elements of $N \times N$ will eventually involve each element of $N \times N$. The only question to be settled is what natural number n will it be when the procedure of counting down the diagonals has arrived at the pair (p, q)? We will answer this question.

<u>Table 24.1</u>

1	2	4	7	11	16	
$(1,1)$	$(1,2)$	$(1,3)$	$(1,4)$	$(1,5)$	$(1,6)$	$(1,7)$ ⋯

3 ✓ 5 ✓ 8 ✓ 12 ✓ 17 ✓ ✓ ✓

$(2,1)$	$(2,2)$	$(2,3)$	$(2,4)$	$(2,5)$	$(2,6)$	$(2,7)$ ⋯

6 ✓ 9 ✓ 13 ✓ ✓ ✓ ✓ ✓

$(3,1)$	$(3,2)$	$(3,3)$	$(3,4)$	$(3,5)$	$(3,6)$	$(3,7)$ ⋯

10 ✓ 14 ✓ ✓ ✓ ✓ ✓ ✓

$(4,1)$	$(4,2)$	$(4,3)$	$(4,4)$	$(4,5)$	$(4,6)$	$(4,7)$ ⋯

15 ✓ ✓ ✓ ✓ ✓ ✓ ✓

$(5,1)$	$(5,2)$	$(5,3)$	$(5,4)$	$(5,5)$	$(5,6)$	$(5,7)$ ⋯

✓ ✓ ✓ ✓ ✓ ✓ ✓

$(6,1)$	$(6,2)$	$(6,3)$	$(6,4)$	$(6,5)$	$(6,6)$	$(6,7)$ ⋯

✓ ✓ ✓ ✓ ✓ ✓ ✓

$(7,1)$	$(7,2)$	$(7,3)$	$(7,4)$	$(7,5)$	$(7,6)$	$(7,7)$ ⋯

✓ ✓ ✓ ✓ ✓ ✓ ✓

$(8,1)$	$(8,2)$	$(8,3)$	$(8,4)$	$(8,5)$	$(8,6)$	$(8,7)$ ⋯

✓ ✓ ✓ ✓ ✓ ✓ ✓

⋮   ⋮   ⋮   ⋮   ⋮   ⋮   ⋮

Counting the elements of $N \times N$.
For example, the 17th ordered pair is (2,5)

The key to answering the question of what n is paired with a given (p,q) is based on several observations.

OBSERVATIONS:

1. On any diagonal, the sum of the coordinates is a constant, AND that constant is one more than the diagonal number. On the first diagonal the sum $p + q$, of the coordinates (p,q), is 2. On the second diagonal that sum is 3, on the third diagonal the sum is 4, etc. In general, the sum of p and q on the diagonal containing (p,q) is $p+q$ and the associated diagonal number is $p+q-1$. So $(4,2)$ is on the $(4+2-1)^{st} = 5^{th}$ diagonal.

2. A second observation is that on the kth diagonal, there are k pairs in that diagonal beginning with $(1,1)$ as the first diagonal. Thus on the diagonal containing (p,q), i.e., the $(p+q-1)^{\text{st}}$ diagonal, there are $p+q-1$ ordered pairs. For example, the pair $(3,4)$ is on

the $(3 + 4 - 1)^{\text{st}} = 6^{\text{th}}$ diagonal. That diagonal has $3 + 4 - 1 = 6$ pairs on it. Also the diagonal containing the pair $(4, 2)$ has 5 pairs on it, all of which sum to 6.

We now want to count the total number of pairs on all of the diagonals, beginning with the first diagonal, the second, etc. down to the position occupied by (p, q). Since (p, q) is on the $(p + q - 1)^{\text{st}}$ diagonal, we make the further observation that there are $p + q - 2$ diagonals PRECEDING the diagonal containing (p, q). We know how many elements there are in each of these diagonals – there are k elements in the kth diagonal. This sum is easy–there are a total of

$$1 + 2 + 3 + \cdots + (p + q - 2)$$

pairs in all of the diagonals preceding the diagonal containing (p, q). What is this sum? It is an arithmetic series beginning with 1 and having a common difference of 1 and $p + q - 2$ terms, so the sum is

$$\frac{(p + q - 2)(p + q - 1)}{2}.$$

Check this for the pair of choice in *Table 24.1*.

We have counted all pairs on the preceeding diagonals. We have not counted those pairs preceeding (p, q) on its own diagonal. So we need to count down the diagonal containing (p, q) to the pair (p, q). Since (p, q) is on the pth row, clearly (p, q) is the pth pair on the "final" diagonal. Thus our counting process has identified the natural number to be paired with (p, q), it is

$$\frac{(p + q - 2)(p + q - 1)}{2} + p. \tag{1}$$

These numbers appear in *Table 24.1*, above the pairs (p, q).

Define f to be $f((p, q))$ to be the number in Equation (1). Then certainly $f : N \times N \longrightarrow N$. It can be shown that f is a $1 - 1$ function from $N \times N$ onto N. Thus $N \times N \cong N$. Therefore N is denumerable.

Now that we know that the Cartesian product of two denumerable sets is a denumerable set, we next wish to focus on the union of denumerable sets. In fact, we consider the union of a denumerable number of denumerable sets. It turns out that such a union is still denumerable. To make sure the sets do not overlap, we define *pairwise disjoint*.

DEFINITION: If S is a collection of sets then that collection is said to be <u>pairwise disjoint</u> if and only if the implication $A, B \in S \land A \neq B \Longrightarrow A \cap B = \emptyset$.

THEOREM 24.5: If $S = \{A_1, A_2, \ldots, A_n, \ldots\}$ is a denumerable set of pairwise disjoint denumerable sets A_n then their union $\bigcup S = \bigcup_{n \in N} A_n$ is denumerable.

PROOF: Since each A_n is denumerable, it can be indexed by N, using Theorem 24.2. This is done as follows.

$$A_1 = \{a_{11}, a_{12}, a_{13}, a_{14}, \ldots, a_{1n}, \ldots\}$$
$$A_2 = \{a_{21}, a_{22}, a_{23}, a_{24}, \ldots, a_{2n}, \ldots\}$$
$$A_3 = \{a_{31}, a_{32}, a_{33}, a_{34}, \ldots, a_{3n}, \ldots\}$$
$$\vdots \qquad\qquad \vdots$$
$$A_n = \{a_{n1}, a_{n2}, a_{n3}, a_{n4}, \ldots, a_{nn}, \ldots\}$$
$$\vdots \qquad\qquad \vdots$$

Here it is assumed that in each A_i, $a_{ij} = a_{ik}$ if and only if $j = k$, so no elements are repeated in each of the sets. Notice also that the subscripts are similar to the entries in the table for $N \times N$ and that the first subscript n indicates which of the sets A_n an element belongs to, while the second subscript distinguishes between the elements in any given set A_n.

The union of the sets A_n is simply the set of all the elements that are in the sets A_n, so $\bigcup_{n \in N} A_n = \{a_{ij} \mid i, j \in N\}$. We want to show the union is denumerable, so we need a denumerable set and a function. Since there are double subscripts on the a_{ij}, it seems reasonable to pair the union with the denumerable set $N \times N$. Let us define

$$f : N \times N \longrightarrow \bigcup_{n \in N} A_n$$

by $\qquad\qquad f((p, q)) = a_{pq} \quad \forall\, (p, q) \in N \times N.$

Then f is a function because $(p, q) = (r, s) \Longrightarrow p = r \land q = s$ and therefore $a_{pq} = a_{rs}$. The domain of f is $N \times N$ because there is an element a_{pq} in the union, for each $(p, q) \in N \times N$. And clearly $f((p, q))$ is an element of $\bigcup_{n \in N} A_n$ (which element?), so $f : N \times N \longrightarrow \bigcup_{n \in N} A_n$ is a function.

Now we establish that f is $1-1$. If $a_{pq} = a_{rs}$ then, since $A_p \cap A_r = \emptyset$ for all $p \neq r$, we get $p = r$. Hence the elements a_{pq} and a_{rs} belong in the same row, namely A_p. Since $a_{pq} = a_{rs}$ and $p = r$ then q must be s. In other words $(p, q) = (r, s)$. Consequently, f is $1-1$.

It is an exercise to show that f is onto $\bigcup_{n \in N} A_n$, so the union $\bigcup_{n \in N} A_n$ is denumerable (see Exercise 24.1).

If the assumption that the sets be denumerable is relaxed appropriately, the conclusion still holds.

COROLLARY 1: If S is a denumerable set of pairwise disjoint countable sets B_n which are either all nonempty or at least one is denumerable, then $\bigcup S$, i.e., $\bigcup_{n \in N} B_n$, is denumerable.

PROOF: We break the proof into two cases.

CASE 1: Suppose $S = \{B_n \mid n \in N\}$ is such a set of pairwise disjoint countable sets B_n which are all nonempty. Then for all n, either B_n is finite or denumerable. Say

$$B_n = \begin{cases} \{b_{n1}, b_{n2}, b_{n3}, \ldots, b_{nm_n}\} & \text{if } B_n \text{ is finite;} \\ \{b_{n1}, b_{n2}, b_{n3}, \ldots, b_{nm_n}, \ldots\}, & \text{if } B_n \text{ is denumerable.} \end{cases}$$

Define C_n by $C_n = \{c_{n1}, c_{n2}, c_{n3}, \ldots, c_{ni}, \ldots\}$ where for each $i \in N$ the element c_{ni} is defined as follows

$$c_{ni} = \begin{cases} b_{ni}, & \text{if } B_n \text{ is denumerable;} \\ b_{ni}, & \text{if } i \leq m_n \text{ and if } B_n \text{ is finite;} \\ (n, i) & \text{if } i > m_n \text{ and if } B_n \text{ is finite.} \end{cases}$$

Then the following things are easy to see or to establish.

1. $C_n = B_n$ if B_n is denumerable
2. $B_n \subset C_n$ if B_n is finite
3. $(n, i) \in C_n$ if B_n is finite and $i > m_n$
4. $n_1 \neq n_2 \implies C_{n_1} \cap C_{n_2} = \emptyset$.

Clearly C_n is denumerable for all n and $\{C_n \mid n \in N\}$ is pairwise disjoint. As a consequence of this, Theorem 24.5 applies, so $\bigcup C_n$ is denumerable. But since $B_n \subseteq C_n$ for all n, $\bigcup B_n \subseteq \bigcup C_n$. (Why?) By Theorem 24.3, $\bigcup B_n$ is countable. Also, since each B_n is nonempty (by assumption in this case), then $b_{n1} \in B_n \ \forall n \in N$. Also $n_1 \neq n_2$ implies

$b_{n_11} \neq b_{n_21}$. Therefore the set $\{b_{n1} \mid n \in N\}$ is a denumerable subset of $\bigcup S$, so $\bigcup S$ is denumerable.

CASE 2:   This case occurs if not all B_n's are nonempty. Then at least one is denumerable; say it is B_{n_0} which is denumerable. Then $B_{n_0} \subseteq \bigcup S$. Thus $\bigcup S$ is not finite. But then an argument similar to Case 1 gets $\bigcup S$ to be a subset of a denumerable union of disjoint denumerable sets. So, by Theorem 24.3, $\bigcup S$ is countable and not finite, so it is denumerable.

Thus, in either case, the assertion $\bigcup S$ is denumerable has been established.

Now some other assumptions in the theorem can be relaxed so that we no longer need to assume that the sets in S be pairwise disjoint. The union is still denumerable.

COROLLARY 2:   If $S = \{A_1, A_2, A_3, \ldots, A_n, \ldots\}$ is a denumerable collection of denumerable sets then $\bigcup S$ is denumerable.

PROOF: The idea of the proof is this. A collection B_1, B_2, B_3, $\ldots$ of sets, which are pairwise disjoint, is formed from the A_i's and which satisfy the condition $\bigcup\limits_{n \in N} A_n = \bigcup\limits_{n \in N} B_n$. This is done as follows

$$\text{Let } B_1 = A_1, \quad B_2 = A_2 - A_1, \quad B_3 = A_3 - \bigcup_{i=1}^{2} A_i, \quad \cdots \ ,$$

and, in general

$$B_n = A_n - \bigcup_{i=1}^{n-1} A_i,$$

then it can be shown that the B_n's are pairwise disjoint and countable and that $\bigcup_{n \in N} B_n = \bigcup_{n \in N} A_n$. Now the hypothesis of Corollary 1 is satisfied by the B_n's. Thus $\bigcup_{n \in N} A_n$ is denumerable.

First, the pairwise disjointness is established. Since the sets A_i are all nonempty, then by the way the B_n's are defined, the B_n's are all different. Let B_i, B_j be any two unequal sets in the collection of B's. Since $B_i \neq B_j$ then $i \neq j$, so without loss of generality (WLOG), suppose $i < j$. If $x \in B_i$ then $x \in A_i$ and thus $x \in \bigcup_{m=1}^{j-1} A_m$, since $i \leq j - 1$. Therefore $x \notin B_j$. For if $x \in B_j$ then $x \in A_j$ and $x \notin A_m$ $\forall m < j$. But $x \in A_i$ and $i < j$ which is a contradiction. By conditional proof, we have established that when $i < j$ if $x \in B_i$ then $x \notin B_j$. On the other hand, if $x \in B_j$ then $x \in A_j - \bigcup_{m=1}^{j-1} A_m$ so $x \notin \bigcup_{m=1}^{j-1} A_m$. Thus $\forall m (1 \le m \le j - 1 \Longrightarrow x \notin A_m)$. Since $i < j$ and thus $1 \le i \le j - 1$, $x \notin A_i$ and so $x \notin B_i$, since $B_i \subseteq A_i$. Therefore if $x \in B_j$ then $x \notin B_i$. In any case, when $i < j$, $B_i \cap B_j = \emptyset$. Consequently, if $i \ne j$ then $B_i \cap B_j = \emptyset$.

The proof of the assertion that $\bigcup_{n \in N} B_n = \bigcup_{n \in N} A_n$ is left to the exercises.

COROLLARY 3: If $A_1, A_2, \ldots, A_k$ is a finite non–zero number k of denumerable sets A_i, then $\bigcup_{i=1}^{k} A_i$ is denumerable.

PROOF: Suppose $\{A_1, A_2, \cdots, A_k\}$ is such a finite non–zero set of denumerable sets. Let $A_{k+1} = A_{k+2} = A_{k+3} = \cdots = A_1$. Then the collection $S = \{A_1, A_2, A_3, \ldots, A_k, A_{k+1}, A_{k+2}, A_{k+3}, \ldots\}$ is a denumerable collection of denumerable (albeit not pairwise disjoint) sets. By Corollary 2, $\bigcup S$ is denumerable. But $\bigcup S = \bigcup_{i=1}^{k} A_i$, so $\bigcup_{i=1}^{k} A_i$ is denumerable.

COROLLARY 4: If $S = \{A_k \mid k \in N\}$ is a collection of pairwise disjoint finite and non–empty sets A_k then $\bigcup_{k \in N} A_k$ is denumerable.

The proof is left to the exercises.

THEOREM 24.6: a. The set Z of integers is a denumerable set.

b. The set Q of rational numbers is a denumerable set.

PROOF: The relation f defined by

$$f(n) = \begin{cases} \dfrac{n-1}{2}, & \text{if } n \in N \text{ and } n \text{ is odd;} \\ -\left(\dfrac{n}{2}\right), & \text{if } n \in N \text{ and } n \text{ is even.} \end{cases}$$

is a $1 - 1$ function from N onto Z. The verification that f is a function, f is $1 - 1$, and f is onto Z, is left to the exercises. Thus Z is denumerable.

The set $F = \{\frac{p}{q} \mid p, q \in N\}$ is cardinally equivalent to the set $N \times N$ using the $1 - 1$ correspondence $f\left(\frac{p}{q}\right) = (p, q)$. Now F has some repetitions, e.g., $\frac{6}{8}$ and $\frac{9}{12}$ are rational numbers that are somehow equivalent, but $(6, 8)$ and $(9, 12)$ are not equal ordered pairs. These repetitions must be deleted. Let F' be the subset of F, which results by deletion of all of the repetitions. Since $F' \subseteq F$ and F' contains the denumerable set

$\{\frac{1}{q} \mid q \in N\}$, F' is denumerable . But the set F' is exactly the set Q^+ of positive rational numbers, so the set Q^+ is denumerable. The set Q^- of negative rational numbers is cardinally equivalent to Q^+ by the $1-1$ correspondence $f(x) = -x \; \forall x \in Q^+$, so Q^- is also denumerable. Thus by Corollary 3 and Exercise 24.3, $Q = Q^+ \cup \{0\} \cup Q^-$ is denumerable.

Exercises:

24.1 Complete the proof of Theorem 24.5 by proving the f defined therein is onto the set $\bigcup_{n \in N} A_n$.

24.2 Prove that $\bigcup_{n \in N} B_n = \bigcup_{n \in N} A_n$ in Corollary 2 to Theorem 24.5.

24.3 Prove that if A is denumerable and B is finite then $A \cup B$ is denumerable.

24.4 Prove Corollary 4 of Theorem 24.5. In other words, prove if $S = \{A_k \mid k \in N\}$ is a denumerable collection of pairwise disjoint finite nonempty sets A_k then $\bigcup S$ is denumerable.

24.5 Prove: If A is a countable set and $B \subseteq A$ and B is denumerable then A is denumerable.

24.6 Prove that the function f defined in Theorem 24.6a is a function, is $1-1$, and is onto Z.

24.7 A real number r is an <u>algebraic number</u> if r satisfies an equation of the form $f(x) = 0$ where

$$f(x) = a_0 + a_1 x + a_2 x^2 + a_3 x^3 + a_4 x^4 + \cdots + a_n x^n$$

and where $a_i \in Z$. Examples of real algebraic numbers are any rational number, any real number of the form $\sqrt[m]{n}$ where $m, n \in N$, and, in fact, any number of the form $\sqrt[m]{q}$ where $q \in Q^+$ and $m \in N$. In other words, since it contains Q, the set A of algebraic numbers seems to be much larger than Q. Nevertheless A is denumerable (i.e., cardinally equivalent to N or to Q.) Find a proof in *A Survey of Modern Algebra* by Garret Birkhoff and Saunders Mac Lane or *Field Theory and its Classical Problems* by Charles Hadlock (MAA Carus Monograph), and then outline the proof. Or create a proof that A is denumerable based on ideas of diagonalization. (This problem is rather difficult.)

24.8 Explain why the number defined by Equation 1 in the proof of Theorem 24.4, is an integer.

CHAPTER 25

UNCOUNTABLE SETS

The sets discussed in the preceeding chapter are all countable, that is they are all finite or denumerable. Indeed, we have found that the sets of integers and rational numbers are denumerable. We also found that the rational numbers are densely spaced on the line, so it may be tempting to think that the set of all real numbers (i.e., rational and irrational) is also denumerable. To contribute to this misconception, see Example 22.1 and Theorem 22.7. But we shall soon see that this is false. In other words, we are going to show that the set of real numbers is not cardinally equivalent to N. Thus, as indicated previously, there are infinite sets which are not cardinally equivalent to some other infinite sets. Cantor was among the first to recognize this. In fact, we shall see later in this chapter that there are infinitely many infinite sets, all of different "sizes."

The sets we are now going to consider are those which are infinite and not denumerable. These sets are said to be uncountable, since they are not finite and not denumerable. As before, "uncountable" does not simply mean the set cannot be physically "counted." While it true that an uncountable set cannot be physically counted, it is also true that certain countable sets cannot be so counted. There is no contradiction here because "uncountable" has a precise technical meaning. A set is uncountable if and only if it is infinite and non denumerable, i.e., infinite and not cardinally equivalent to N.

Perhaps the diagram in *Figure 25.1* will aid in placing these concepts in perspective. All of the sets we discussed in Chapter 24 are in the two compartments on the left. The corollary to Theorems 25.1 and 25.3 and other properties in this section establish that E_1 and intervals of real numbers, as well as certain other sets, belong in the right compartment.

Now we are ready to justify the existence of sets of this latter type–
uncountable sets.

THEOREM 25.1: The interval $(0,1) = \big\{x \mid x \in E_1 \land 0 < x < 1\big\}$ of real numbers is uncountable, i.e., is non–denumerable and infinite.

The schema is to assume that $(0,1)$ is denumerable, hence $(0,1) \cong N$. Thus

$$\exists f \ni f : N \xrightarrow{\;1\,-\,1 \text{ onto}\;} (0,1).$$

Figure 25.1

All Sets

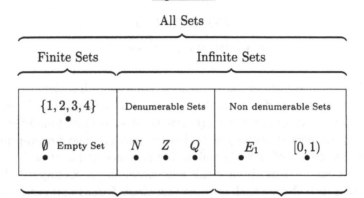

Countable Sets   Uncountable Sets

Our proof will be by contradiction and will consist of demonstrating that the f just mentioned is NOT onto $(0,1)$. To do this we will find a number r that belongs in the interval $(0,1)$, yet does not belong in the range of f. In other words, we will show $\exists r \ni r \in (0,1) \wedge r \notin f[N]$.

Before we show that f is not onto $(0,1)$, let us highlight a bit of notation regarding the significance of 10 in base 10 place value in the representation of real numbers. Every $x \in (0,1)$ has an infinite decimal representation of the form:

$$x = .d_1 d_2 d_3 \cdots d_n \cdots = \sum_{i \in N} d_i \cdot 10^{-i} \quad \text{where} \quad d_i \in \{0,1,2,3,\cdots,9\}.$$

This notation does, however, allow some numbers to be represented in more than one way as an infinite decimal. For example, it is not difficult to show that $0.1349999\cdots = .135 = .135000\cdots$ EXACTLY (see Exercise 25.5). To avoid difficulties that arise from the non-uniqueness of representation, we assume that all terminating decimals (such as 0.135) are expressed with infinitely many digits of 9 instead. Then each real number in the interval $(0,1)$ has one and only one representation, which is an infinite decimal. So two real numbers in this form are different if and only if their decimal representations differ in <u>any</u> place. In particular, since 0.135 is rejected in favor of $0.134999\cdots$, we have only one way to describe $\frac{135}{1000}$.

PROOF: This is an indirect proof. Suppose $(0,1)$ is denumerable then there exists an f such that f is a 1–1 function from N onto $(0,1)$ as mentioned previously. So let

$$
\begin{aligned}
f(1) = x_1 &= .d_{11}d_{12}d_{13}d_{14}\cdots d_{1n}\cdots \\
f(2) = x_2 &= .d_{21}d_{22}d_{23}d_{24}\cdots d_{2n}\cdots \\
f(3) = x_3 &= .d_{31}d_{32}d_{33}d_{34}\cdots d_{3n}\cdots \\
f(4) = x_4 &= .d_{41}d_{42}d_{43}d_{44}\cdots d_{4n}\cdots \\
&\vdots \qquad\qquad \vdots \\
f(n) = x_n &= .d_{n1}d_{n2}d_{n3}d_{n4}\cdots d_{nn}\cdots \\
&\vdots \qquad\qquad \vdots
\end{aligned}
\tag{1}
$$

Then the image of the set N under f, i.e., $f[N]$ is the following (denumerable) set

$$
\left\{ 0.d_{n1}d_{n2}d_{n3}d_{n4}\cdots d_{nm}\cdots \;\middle|\; n \in N \wedge d_{nm} \in \{0,1,2,3,\cdots,9\} \right\}
$$

which is the set $\{x_n \mid n \in N\}$. Cantor found that the set $f[N]$ is NOT equal to $(0,1)$. The way this is established is by constructing a number r that IS IN $(0,1)$, but is DIFFERENT from each number in $f[N]$. That is, r will be different from each number displayed in Equations (1).

Let r be defined by $r = 0.r_1r_2r_3\cdots r_n\cdots$ where each r_n is given by

$$
r_n = \begin{cases} 6, & \text{if } d_{nn} \neq 6; \\ 7, & \text{if } d_{nn} = 6. \end{cases}
\tag{2}
$$

(For example, if $x_1 = .4137699\cdots$ and $x_2 = .873999\cdots$ and $x_3 = .5261999\cdots$ then r will have 6 in its first decimal place, 6 in the second decimal place, and 7 in the third, etc.) Notice $r \neq x_1$ because $r_1 \neq d_{11}$. In other words, since r and x_1 differ in the first decimal place, they are unequal. Also, $r \neq x_2$ because r_2, the second decimal place digit of r, is unequal to d_{22}, the second decimal place of x_2. Similarly $r \neq x_3$ because r and x_3 differ in the third decimal place. In general, Equation 2 assures us that $r \neq x_n$ because $r_n \neq d_{nn}$, i.e., r and x_n differ in the nth decimal place. Thus we know that r is unequal to every element of $f[N]$. But since $r = 0.r_1r_2r_3\cdots$, where $r_1 = 6$ or $r_1 = 7$ then $0.6 < r < 0.8$. Since r is an infinite decimal representation of a number between 0 and 1, $r \in (0,1)$. But since r is not in $f[N]$, then $f[N] \neq (0,1)$. But this says f is NOT onto $(0,1)$, which is a contradiction. Thus $(0,1)$ is non-denumerable. Is it finite? Since $\{\frac{1}{2}, \frac{1}{3}, \frac{1}{4}, \ldots, \frac{1}{n} \ldots\} \subseteq (0,1)$, $(0,1)$ is infinite. Thus $(0,1)$ is non–denumerable and infinite, hence uncountable, as was to be shown.

An immediate corollary to this theorem establishes that E_1 is uncountable. By the corollary to Theorem 23.5, since $(0,1) \subseteq E_1$ and $(0,1)$ is infinite, E_1 is infinite. If E_1 were denumerable, then since $(0,1) \subseteq E_1$, $(0,1)$ would be countable. (Why?) But this is contrary to the theorem. So we have proven the corollary below.

COROLLARY: E_1 is uncountable.

Refer to *Figure 25.1* to see how E_1 fits into the diagram.

THEOREM 25.2:   If A is an infinite set then there exists a set B such that $B \subseteq A$ and B is denumerable.

PROOF: If A is infinite then A is nonempty. Thus $\exists a \in A$, say $a = a_1$. Then $A - \{a_1\} \neq \emptyset$. For if $A - \{a_1\} = \emptyset$ then $A = \{a_1\}$ would be finite. Therefore $\exists\, a_2 \in A - \{a_1\}$. Then $A - \{a_1, a_2\} \neq \emptyset$. For if $A - \{a_1, a_2\} = \emptyset$ then $A = \{a_1, a_2\}$ would be finite. Continuing, we get $A - \{a_1, a_2, a_3, \ldots, a_n\} \neq \emptyset$ and thus $\exists a_{n+1} \in A - \{a_1, a_2, a_3, \ldots, a_n\}$. The process does NOT terminate. For if the process terminated at some $k \in N$ then $A - \{a_1, a_2, a_3, \ldots, a_k\} = \emptyset$. This says A would be finite. Let $B = \{a_1, a_2, a_3, \ldots, a_k, \ldots\}$ then B is a denumerable subset of A.

An example of this occurs in the set of real numbers. As subsets of the uncountable set E_1, one can find denumerable sets Q, Z, and N. Also, if A were taken to be Q, the rational numbers, then Z would be an example of a denumerable subset of Q.

In the corollary just mentioned note that E_1 is uncountable. It turns out that not only are E_1 and $(0,1)$ uncountable, but $(0,1)$ is actually cardinally equivalent to the entire set E_1. Before reading on, imagine what this entails. We must find a $1-1$ function from $(0,1)$ onto E_1.

THEOREM 25.3: $(0,1) \cong E_1$.

PROOF: Let $f : (0,1) \longrightarrow (-\frac{\pi}{2}, \frac{\pi}{2})$ be given by: $f(x) = \pi(x - \frac{1}{2})$ $\forall x \in (0,1)$. Then f is a $1-1$ function, since its graph is a non–horizontal line segment. It is an exercise to show f is onto $(-\frac{\pi}{2}, \frac{\pi}{2})$. Thus $(0,1) \cong (-\frac{\pi}{2}, \frac{\pi}{2})$. Now let $g : (-\frac{\pi}{2}, \frac{\pi}{2}) \longrightarrow E_1$ be given by $g(x) = \tan x$ $\forall x \in (-\frac{\pi}{2}, \frac{\pi}{2})$. From calculus you should recall that g is a $1-1$ function from $(-\frac{\pi}{2}, \frac{\pi}{2})$ onto E_1. Finally, by transitivity of $\cong$, $(0,1) \cong E_1$.

The next theorem will verify that a certain property we have observed about the infinite set N, is in fact true of any infinite set. In Exercise 23.1 we found a proper subset of N that is cardinally equivalent to N. This is true of any infinite set, and not true of any finite set.

THEOREM 25.4: If A is infinite then there is a proper subset B of A such that B is cardinally equivalent to A.

PROOF: Suppose A is an infinite set. Then by Theorem 25.2, there is a set C such that $C \subseteq A$ and C is denumerable. Since C is denumerable, it may be written as $C = \{c_1, c_2, c_3, \ldots, c_k, \ldots\}$ by Theorem 24.2, where these elements are all distinct. Now let $D = A - C$ then $A = D \cup C$ and $D \cap C = \emptyset$. (What do we have that justifies this?)
Define $f : A \longrightarrow A$ by

$$f(x) = \begin{cases} x, & \text{if } x \in D \\ c_{i+1}, & \text{if } x = c_i \text{ (where } c_i \in C). \end{cases}$$

Then $\operatorname{dom}(f) = C \cup D = A$ and $\operatorname{ran}(f) = D \cup \{c_2, c_3, c_4, \ldots, c_k, \ldots\} = A - \{c_1\} \subset A$. Now f is $1-1$ because if $x, y \in A$ and $f(x) = f(y)$ then either $x \in D$ and $y \in D$ or $x \in C$ and $y \in C$. Note that $x \in C$ and $y \in D$ or vice-versa is impossible, since D is disjoint from $\{c_2, c_3, \ldots, c_k, \ldots\}$. If $x, y \in D$ then $x = f(x) = f(y) = y$ and if $x, y \in \{c_2, c_3, \ldots, c_k, \ldots\}$ then, e.g., $x = c_{i+1}$ and $y = c_{j+1}$, so $c_{i+1} = f(c_i) = f(x) = f(y) = f(c_j) = c_{j+1}$. Thus $c_{i+1} = c_{j+1}$ so $i + 1 = j + 1$, hence $i = j$. But then $c_i = c_j$. Consequently, f is $1-1$. Exercise 25.8 asks you to show f is onto the set $A - \{c_1\}$. Let $B = A - \{c_1\}$ then $\exists B \ni [B \subset A \wedge A \cong B]$. In summary, A has a proper subset B with which it is cardinally equivalent.

THEOREM 25.5: If A is infinite, B is countable, $B \subseteq A$ and $A - B$ is infinite then $A \cong A - B$.

COROLLARY: If A is infinite and B is countable then $A \cup B \cong A$.

The proofs to the theorem and corollary will not be given; however, an outline of the proof of the theorem is given in Exercise 25.6.

Exercises:

25.1 Prove that the closed interval $[0, 1]$ is an infinite set.

25.2 Find a set (guaranteed by Theorem 25.2) that is a denumerable subset of $\mathcal{P}(N)$.

25.3 Let A be, in turn, each of the following sets. Illustrate Theorem 25.4, by finding proper subsets B of A such that $A \cong B$.

    a. Let $A = N$.

    b. Let $A = Z$.

    c. Let $A = Q$.

    d. Let $A = (0, 1)$.

    e. Let $A = E_1$.

    f. Let $A = \{f \mid f : E_1 \longrightarrow E_1 \wedge f \text{ is differentiable}\}$. The student need not verify directly that $A \cong B$.

25.4 Let A be each of the sets in Exercise 25.3 d,e,f, but find a proper denumerable subset B of A so that $A - B \cong A$.

25.5 Justify that $0.135 = .1349999 \cdots$ EXACTLY.

25.6 Prove Theorem 25.5 by verifying each of the following steps.

    a. Let $C = A - B$. Show $\exists D \subseteq C \ni D$ is denumerable.

    b. Let $E = C - D$. Show $E \cup D = C$ and $E \cap D = \emptyset$.

    c. Show $B \cup D$ is denumerable.

    d. Show $\exists f \ni f : B \cup D \longrightarrow D$, and f is 1–1 and onto D.

    e. Show $A = B \cup D \cup E$.

    f. Show $(B \cup D) \cap E = \emptyset$.

    g. Show $g(x) = \begin{cases} x, & \text{if } x \in E; \\ f(x), & \text{if } x \in B \cup D, \end{cases}$ is a function.

    h. Show g is $1 - 1$. See the proof of Theorem 23.6 for some ideas.

    i. Show g is ONTO and draw the appropriate conclusion.

25.7 Prove the Corollary to Theorem 25.5.

25.8 Prove the f of Theorem 25.4 is onto $A - \{c_1\}$.

25.9 The set $E_1 - Q$ denotes the set of irrational numbers. Prove that $E_1 - Q$ is uncountable.

25.10 Referring to Exercise 24.7, prove that if A denotes the set of algebraic numbers then $E_1 - A$ is uncountable. The set $E_1 - A$ is called the set of *real transcendental* numbers.

25.11 Prove that for no finite set A is it true that A is cardinally equivalent to a proper subset B of A.

CHAPTER 26

TRANSFINITE CARDINAL NUMBERS

In Chapter 18 a study was made of finite cardinal numbers. In that section it was assumed that finite sets have cardinal numbers and in particular the cardinal number of each set N_k is k. We chose to symbolize the expression, "the cardinal number of A," by $\#(A)$. In Chapter 18, after defining the finite cardinal numbers, operations for addition and multiplication were imposed on those numbers and properties such as commutativity and associativity were explored. Then the relation "$<$" was defined and some of its properties were considered. Now we will make a somewhat parallel investigation for infinite sets and their cardinal numbers. The reason for the word "somewhat" is that not all properties true of "$=$" and "$<$" in the finite context are true for cardinal numbers of infinite sets.

In this section, it is assumed that every set (in particular, every infinite set) has a cardinal number. So if A is a set, its cardinal number is $\#(A)$. To distinguish between the cardinal numbers of finite sets and infinite sets, we shall call the cardinal number of a finite set a <u>finite cardinal number</u> and the cardinal number of an infinite set a <u>transfinite cardinal number</u>. As before, if $\exists\, f : A \longrightarrow B \ni f$ is 1–1 and onto B then A and B are cardinally equivalent, i.e., $A \cong B$.

DEFINITION: Let A and B be any sets then
1. $\#(A) = \#(B)$ if and only if $A \cong B$.
2. If $A \cap B = \emptyset$ then $\#(A) + \#(B) = \#(A \cup B)$.
3. $\big(\#(A)\big) \cdot \big(\#(B)\big) = \#(A \times B)$.

Notice that the first item tells us that two sets are cardinally equivalent if and only if they have the same cardinal number. Also notice that in 2, we add cardinal numbers just as we did in Chapter 18, and just as you did when you first learned to add, paying attention to the assumption that A and B are disjoint? Finally, although you didn't learn to multiply numbers in elementary school like Part 3 says to do, it is nevertheless a very reasonable way to define multiplication and it agrees with what you should expect for products.

As a result of these definitions the following properties are true. The proofs are virtually the same as the proofs of the corresponding properties for finite sets in Chapter 18.

THEOREM 26.1:   If A, B, and C are any sets (finite or infinite) and A, B, C are pairwise disjoint and if $a = \#(A)$, $b = \#(B)$, and $c = \#(C)$ are their respective cardinal numbers, then

    a. $a + b = b + a$ and $ab = ba$

    b. $a + (b + c) = (a + b) + c$ and $a(bc) = (ab)c$

    c. $a(b + c) = ab + ac$

    d. $a \cdot 0 = 0$ and $a + 0 = a$

    e. $a \cdot 1 = a$ where $1 = \#(\{\emptyset\})$

Because this theorem holds in the context of finite as well as transfinite cardinal numbers, there is nothing surprising in it, regarding transfinite cardinal numbers. However, one cannot infer that all properties of finite cardinal numbers carry over to transfinite cardinal numbers. To make such an assumption would be reckless, indeed, for we already know that infinite sets are cardinally equivalent to proper subsets of themselves, while this is not so for finite sets. This dichotomy is only the first clue of what is to be seen in the strange new world of *transfinite arithmetic*.

In the next definition, symbols will be assigned for transfinite cardinal numbers in much the same way that the numerals $0, 1, 2, 3, \ldots$ were defined in Chapter 18 to be the cardinal numbers of certain finite sets. For example, $0 = \#(\emptyset)$, $1 = \#(\{0\})$, $2 = \#(\{0, 1\})$, $3 = \#(\{0, 1, 2\})$, etc.

DEFINITION: Let $\aleph_0 = \#(N)$, then $\aleph_0$ is called <u>aleph null</u>. And let c be defined by $c = \#(E_1)$. Then c is called the <u>cardinal number of the continuum</u>.

Notice that $\aleph_0$ is the cardinal number, not only of N, but of any denumerable set, such as Z or Q. This is because $Q \cong N$ means that $\#(Q) = \#(N) = \aleph_0$. In addition, notice that c is the cardinal number of any interval of real numbers, such as $(0, 1)$ or $(-\frac{\pi}{2}, \frac{\pi}{2})$ or even the whole real line (see Exercise 26.19 as well as Theorem 25.3).

We now prove two properties, not true of any whole number greater than one and simultaneously reinforce the previous assertion, that there are unusual properties which hold for transfinite cardinal numbers.

THEOREM 26.2:

    a. $\aleph_0 + \aleph_0 = \aleph_0$

    b. $\aleph_0 \cdot \aleph_0 = \aleph_0$

PROOF: Let $V = \{2n \mid n \in N\}$ and $D = \{2n - 1 \mid n \in N\}$ be the sets of even and odd natural numbers, respectively. Then $N = V \cup D$ and $V \cap D = \emptyset$. So using Exercise 23.1 and previous results, $\#(N) = \#(V) = \#(D) = \aleph_0$. Thus we get

$$\aleph_0 + \aleph_0 = \#(V) + \#(D) = \#(V \cup D) = \#(N) = \aleph_0.$$

This justifies Part a.

By Theorem 24.4, $N \times N \cong N$. So $\aleph_0\aleph_0 = \#(N \times N) = \#(N) = \aleph_0$. So Part b is true too.

THEOREM 26.3: If n is any finite cardinal number, $\aleph_0 + n = \aleph_0$ and if $n \neq 0$ then $n\aleph_0 = \aleph_0$. (Why must n be non–zero?)

PROOF: The proof of the first part is an exercise (see Exercise 26.7). Since $n = \#(N_n) = \#(\{1, 2, 3, \cdots, n\})$ and (by Exercise 11.8) since

$$N_n \times N = \{1, 2, 3, \ldots, n\} \times N = (\{1\} \cup \{2\} \cup \{3\} \cup \cdots \cup \{n\}) \times N$$
$$= (\{1\} \times N) \cup (\{2\} \times N) \cup (\{3\} \times N) \cup \cdots \cup (\{n\} \times N)$$

then, according to the definition of product of cardinal numbers,

$$n\aleph_0 \overset{1}{=} \#(N_n \times N)$$

$$\overset{2}{=} \#\Big((\{1\} \times N) \cup (\{2\} \times N) \cup \cdots \cup (\{n\} \times N)\Big)$$

$$\overset{3}{=} \#(\{1\} \times N) + \#(\{2\} \times N) + \#(\{3\} \times N) + \cdots + \#(\{n\} \times N)$$

$$\overset{4}{=} \#(N) + \#(N) + \#(N) + \cdots + \#(N)$$

$$= \aleph_0 + \aleph_0 + \aleph_0 + \cdots + \aleph_0 \overset{5}{=} \aleph_0$$

where there are n summands. Step 3 is because $\{1\}, \{2\}, \ldots, \{n\}$ are pairwise disjoint. Step 4 follows by use of Exercise 14.15. You might want to find the reasons for the other steps. The preceding two theorems lead us to still another property of transfinite numbers that illustrates distinctions between finite and transfinite cardinal numbers.

EXAMPLE 26.1: The cancelation laws of addition and multiplication do not hold for transfinite cardinal numbers. To see this, consider the equations arising from Theorems 26.2 and 26.3.

$$\aleph_0 + \aleph_0 = \aleph_0 = \aleph_0 + n \quad \text{and} \quad \aleph_0\aleph_0 = \aleph_0 = n\aleph_0.$$

If the same member $\aleph_0$ were "canceled" from each side of either equation, the result would be $\aleph_0 = n$. But $\aleph_0 \neq n$ for any finite cardinal number n. (Why? Cite a specific theorem or theorems.)

Order for Cardinal Numbers:

Before pursuing additional properties of addition and multiplication, the relation "$<$" for transfinite cardinal numbers is defined. It is desirable to have the relation "$<$" be applicable to finite as well as transfinite cardinal numbers. The definition should agree with the notion of "$<$" for natural numbers in the finite case, which was discussed in Chapter 18. Furthermore, the definition must lead to a distinction between two infinite sets that are not cardinally equivalent, for otherwise such properties as the trichotomy law may not hold.

DEFINITION: If A and B are sets then $\#(A) < \#(B)$ if and only if

1. $\exists f \ni f : A \xrightarrow{1\text{-}1} B$, and
2. $A \not\cong B$.

EXAMPLE 26.2:   The definition yields the correct inequality for the cardinal numbers of the finite sets $A = \{1,2,3\}$ and $B = \{a,b,c,d\}$ because $f = \{(1,a),(2,b),(3,c)\}$ is a $1-1$ function from A into B, BUT, clearly $\{1,2,3\} \not\cong \{a,b,c,d\}$, so $\#(\{1,2,3\}) < \#(\{a,b,c,d\})$.

If one were asked for a definition of "$<$" for cardinal numbers, a first impulse might be to say $\#(A) < \#(B)$ iff A is a proper subset of B. But consider Example 23.1. In this example, N is the set of natural numbers and V the set of even natural numbers. Now V is a proper subset of N, but $V \cong N$; hence $\#(V) = \aleph_0 = \#(N)$. Thus it would be impossible to have $\#(V) < \#(N)$, as would result from defining "$<$" for cardinal numbers in terms of proper subset. As a consequence, the definition of "$<$" must exclude such cases as $V \cong N$. This is why Part b of the definition of "$<$" is essential. After considering the incorrectness of saying $\#(A) < \#(B)$ iff A is actually a proper subset of B, one may be tempted to say A must be cardinally equivalent to a proper subset of B. But this is still inadequate as we shall see.

The next example illustrates a use of the definition to establish an inequality between transfinite cardinal numbers.

EXAMPLE 26.3: $\aleph_0 < c = \#(E_1)$. This is because $\exists f \ni f : N \xrightarrow{1\text{-}1} E_1$ namely $f(n) = n.000\cdots \ \forall n \in N$. Also $N \not\cong E_1$ by Theorem 25.1, or its corollary. Thus $\#(N) < \#(E_1)$.

Cantor's Theorem and the Continuum Hypothesis:

We now consider another important theorem due to Cantor. Later we shall see that this theorem is related to the property just considered in Example 26.3. This theorem applies to all sets, finite and infinite.

Recall that if A is a set then $\mathcal{P}(A)$, the power set of A, is given by $\mathcal{P}(A) = \{S \mid S \subseteq A\}$. Here is Cantor's theorem.

THEOREM 26.4 (Cantor): If A is ANY set then $\#(A) < \#(\mathcal{P}(A))$.

PROOF: If $A = \emptyset$ then $\#(A) = 0$ and $\#(\mathcal{P}(A)) = \#(\mathcal{P}(\emptyset)) = \#(\{0\}) = 1$. So $\#(A) < \#(\mathcal{P}(A))$. Now suppose $A \neq \emptyset$. In order to establish $\#(A) < \#(\mathcal{P}(A))$, there are two parts to show:

1. $\exists f \ni f : A \overset{1\text{-}1}{\longrightarrow} \mathcal{P}(A)$ and
2. $A \not\cong \mathcal{P}(A)$.

PART 1: Let f be given by $f(a) = \{a\}$ for all $a \in A$ then $\mathrm{dom}(f) = A$ and $\mathrm{ran}(f) = \{\{a\} \mid a \in A\} \subseteq \mathcal{P}(A)$. If $a = b$ then $\{a\} = \{b\}$, so $f : A \longrightarrow \mathcal{P}(A)$ is a function. Now f is also $1 - 1$, for if $\{a\} = \{b\}$ then $a = b$.

PART 2: To do this part, we suppose the contrary, i.e., suppose $A \cong \mathcal{P}(A)$. Then

$$\exists g \ni g : A \xrightarrow{\;\;1-1\ \text{onto}\;\;} \mathcal{P}(A).$$

Now we know that for each $a \in A$, $g(a) \in \mathcal{P}(A)$, so $g(a) \subseteq A$. If $a_0 \in A$ then either $a_0 \in g(a_0)$ or $a_0 \notin g(a_0)$. Consider the set S given by $S = \{a \mid a \in A \wedge a \notin g(a)\}$. Then $S \subseteq A$. Therefore $S \in \mathcal{P}(A)$. Since g is onto $\mathcal{P}(A)$, $\exists s \in A \ni g(s) = S$. Then either $s \in S$ or $s \notin S$. If $s \in S$, then since $S = \{a \mid a \in A \wedge a \notin g(a)\}$ then $s \in A$ and $s \notin g(s)$. By simplification, $s \notin g(s)$, but $g(s) = S$, so $s \notin S$. Consequently, $s \in S \implies s \notin S$. On the other hand, if $s \notin S$ then $s \notin g(s)$, so since $s \in A$, $s \in \{a \mid a \in A \wedge a \notin g(a)\} = S$, hence $s \in S$. Thus $s \notin S \equiv s \in S$, which is a contradiction. Therefore $A \not\cong \mathcal{P}(A)$. Thus $\#(A) < \#(\mathcal{P}(A))$ as we sought to do.

This theorem assures us that there are infinitely many infinite sets of ever increasing cardinal number. Here is how. For each set S, $\mathcal{P}(S)$ is a set of greater cardinal number, so if we begin with N then we get

$$\#(N) < \#(\mathcal{P}(N)) < \#(\mathcal{P}(\mathcal{P}(N))) < \cdots$$

$$< \#\Big(\mathcal{P}(\mathcal{P}(\cdots(\mathcal{P}(N))\cdots))\Big) < \cdots . \tag{1}$$

Therefore, not only are there infinite sets of different "sizes," there are infinitely many different infinite sets of different infinite "sizes," a truly remarkable fact.

Several questions arise at this juncture. Cognizance of the string of inequalities in (1) above, and of the cardinal numbers $c = \#(E_1)$, $\aleph_0 = \#(N)$ and of $\#(\mathcal{P}(N))$ leads to the first question.

QUESTION 1: "Where does $\#(E_1)$ fit, if anywhere, in the string of inequalities in (1)?"

This question should be asked because of the known results $\#(N) < \#(E_1)$ and $\#(N) < \#(\mathcal{P}(N))$ (see Example 26.3 and Theorem 26.4). Clearly there is the possibility that $\#(E_1)$ may be comparable with $\#(\mathcal{P}(N))$.

QUESTION 2: "Is there any set S such that $\#(N) < \#(S) < \#\mathcal{P}(N)$?"

This related question is natural to ask, since at this point E_1 is a candidate for S. Even if E_1 is ruled out for S, it is natural to ask if anything else would serve for S.

QUESTION 3: "How does $\aleph_0$ [i.e., $\#(N)$] compare with other numbers, both finite and infinite?"

The first question will be answered in Theorem 26.9 because it requires a bit more mathematical development. The third question will be answered in Theorem 26.5. The second question has an interesting history and will be answered next.

Cantor expended much effort to prove that there is no set S such that $\#(N) < \#(S) < \#(\mathcal{P}(N))$, but alas, was unsuccessful. In fact, it is believed that his health suffered greatly as a result of this attempt. He died without settling his conjecture. It became known as the *continuum hypothesis*, and for many years, other mathematicians tried unsuccessfully to prove or disprove it.

The conjecture was generalized to, "For each infinite set B there exists no set A such that $\#(B) < \#(A) < \#(\mathcal{P}(B))$." This was called the *generalized continuum hypothesis*. Attempts to prove or disprove this conjecture met the same fate: failure. However, with this dilemma, a somewhat unexpected pathway out began to be discovered through a series of theorems of profound mathematical significance.

In 1940, the Austrian logician, Kurt Gödel (1906–1978), published *The Consistency of the Continuum Hypothesis*, which appeared in <u>Annals of Mathematical Studies</u> (Princeton University Press, 1940), establishing that the generalized continuum hypothesis is not inconsistent with the usual axioms of set theory, i.e., it cannot be disproved. Consistency is an important feature of axiomatic systems. For a good discussion of this, and other fundamental issues, see *Introduction to the Foundations of Mathematics* by Raymond Wilder. Gödel's result was a big leap forward toward obtaining an answer concerning the validity or invalidity of the continuum hypothesis. In 1963, Paul Cohen of Stanford University proved that the continuum hypothesis is independent of the axioms of set theory (i.e., now known to be neither provable, nor disprovable in the framework of set theory). So an answer to the second question is that there is no answer in the context of the "usual" axioms of set theory. In other words, the generalized continuum hypothesis could be taken as an axiom itself, because it can neither be established nor can its denial established in the framework of that "usual" set of axioms for set theory.

The answer to Question 3 is in the following theorem. Once again, Question 1 will be answered later, after a suitable preparation is made.

THEOREM 26.5: If α is any finite cardinal number then $\alpha < \aleph_0$, and if α is a transfinite cardinal number then $\aleph_0 \leq \alpha$.

PROOF: If α is a finite cardinal number then $\alpha = \#(N_\alpha)$. Now if f is the function defined by $f(x) = x$ for all $x \in N_\alpha$ then $f : N_\alpha \longrightarrow N$ is a $1-1$ function. If $N_\alpha \cong N$ then by Theorem 23.3, N is finite, but this is contrary to Theorem 23.2, hence $N_\alpha \not\cong N$. Thus $\alpha = \#(N_\alpha) < \#(N) = \aleph_0$ by the definition of $<$ for cardinal numbers.

If α is transfinite, then there is an infinite set A such that $\alpha = \#(A)$. Thus A has a denumerable subset B. (Why?) Now either $A \cong B$ or not. If $A \cong B$ then $\aleph_0 = \#(B) = \#(A) = \alpha$. And if $A \not\cong B$, then still

$$\exists f \ni f : B \xrightarrow{1\text{-}1} A$$

where f is given by $f(x) = x$ for all $x \in B$. Thus $\aleph_0 = \#(B) < \#(A) = \alpha$. In any case $\aleph_0 \leq \alpha$.

Thus $\aleph_0$ is greater than any finite cardinal number and, at the same time, is less than or equal to each transfinite cardinal number.

The trichotomy law, which might be expected at this point, cannot be proven here because it requires, and is, in fact, logically equivalent to another axiom. That axiom, the *axiom of choice*, has not been mentioned yet. It is an extremely important and useful mathematical statement and it will be considered in the next chapter, after a suitable framework is established. Stated without proof at this time, is the *trichotomy law*, which holds for infinite sets as well as finite sets (see Theorem 29.2 and its lemma, as well as *Equivalents of the Axiom of Choice* by Rubin and Rubin).

TRICHOTOMY LAW: If A and B are any sets then one and only one of the following is true:

$$\#(A) < \#(B) \quad \text{or} \quad \#(A) = \#(B) \quad \text{or} \quad \#(B) < \#(A).$$

Powers of Cardinal Numbers:

Now we set up the machinery to answer Question 1 of this section. The question is, "Where does $\#(E_1)$ fit in relation to the members of the inequality $\#(N) < \#(\mathcal{P}(N))$?" The answer, to be found in Theorem 26.9, is that $\#(E_1) = \#(\mathcal{P}(N))$. But along the way we get a definition and some properties of exponents for cardinal numbers.

DEFINITION: Suppose $\mu = \#(A)$ and $\nu = \#(B)$ then

$$\mu^\nu = \#(\{f \mid f \text{ is a function from } B \text{ to } A \}).$$

Up to now we know that the notion of product of cardinal numbers pertains to the Cartesian product of sets and the notion of addition of cardinal numbers pertains to the union of disjoint sets. Now we know how to deal with exponents. It pertains to the number of functions from one set to another. To find μ^ν by the definition requires first finding sets A and B with $\mu = \#(A)$ and $\nu = \#(B)$. And second to determine the number of functions from B to A. The latter number will be μ^ν.

EXAMPLE 26.4:   In this example we compute a few exponents in the finite case to familiarize you with the process of calculating exponents.

$$2^1 = \#(\{f \mid f : \{a\} \longrightarrow \{a, b\}\}) = \#(\{\{(a, a)\}, \{(a, b)\}\}) = 2$$
$$1^2 = \#(\{f \mid f : \{b, c\} \longrightarrow \{a\}\}) = \#(\{\{(b, a), (c, a)\}\}) = 1$$
$$2^2 = \#(\{f \mid f : \{a, b\} \longrightarrow \{c, d\}\})$$
$$= \#(\{\{(a, c), (b, c)\}, \{(a, c), (b, d)\}, \{(a, d), (b, c)\}, \{(a, d), (b, d)\}\}) = 4$$

Many, but not all, properties which are true for exponents in the finite setting, are true in the more general setting of all cardinal numbers. Some of the ones that are true in the general setting follow.

THEOREM 26.6: Let μ be any cardinal number.

    a. $\mu^1 = \mu$

    b. $\mu^0 = 1$

    c. $\mu \neq 0 \Longrightarrow 0^\mu = 0$

    d. $1\mu = \mu$

PROOF PART a: Let $\mu = \#(A)$ for some set A and $1 = \#(\{0\})$ then $\mu^1 = \#(\{f \mid f : \{0\} \longrightarrow A\})$. Now for each $a \in A$ there exists a unique function $f_a : \{0\} \longrightarrow A$, satisfying $f_a(0) = a$. Also $a_1 \neq a_2$ implies $f_{a_1} \neq f_{a_2}$. Thus $\#(A) = \#(\{f \mid f : \{0\} \longrightarrow A\}) = \mu^1$. That is $\mu^1 = \mu$.

PART b: Let $\mu = \#(A)$, then $\mu^0 = \#(\{f \mid f : \emptyset \longrightarrow A\})$. Since the empty function $\emptyset$ is a function from $\emptyset$ to A and since it is the only function $f : \emptyset \longrightarrow A$, then $\mu^0 = 1$.

Parts c and d are exercises.

Some additional familiar properties of exponents are true for the expanded notion of exponents of cardinal numbers. Here is a theorem, without proof, exhibiting some of the more important ones.

THEOREM 26.7: The following are true for all μ, ν, τ, and ρ where $\tau\mu \neq 0$.

    a. $\mu^{\nu+\rho} = \mu^\nu \mu^\rho$

    b. $(\mu\tau)^\rho = \mu^\rho \tau^\rho$

    c. $(\mu^\nu)^\rho = \mu^{\nu\rho}$

Notice that if $\nu = \#(A)$ and $2 = \#(\{0,1\})$ then from the definition of exponents

$$2^\nu = \#(\{f \mid f : A \longrightarrow \{0,1\}\}).$$

(See Example 14.13.) With this equation, we are able to generalize to transfinite numbers the fact that if a set A has a finite number n of elements, then there are 2^n subsets of A, i.e., $2^n = \#(\mathcal{P}(A))$. This is accomplished by use of the following theorem.

THEOREM 26.8: If $\nu = \#(A)$ then $2^\nu = \#(\mathcal{P}(A))$.

PROOF: Let ν be a cardinal number and A a set such that $\nu = \#(A)$. The theorem will be established if we show that

$$\mathcal{P}(A) \cong \{f \mid f : A \longrightarrow \{0,1\}\}. \tag{2}$$

Toward that end, define: For each $C \subseteq A$

$$\Psi_C(x) = \begin{cases} 1, & \text{if } x \in C; \\ 0, & \text{if } x \in A - C. \end{cases}$$

The verification that Ψ_C is a function is left to the exercises (see Exercise 26.10). Thus $\forall C \big($if $C \subseteq A$ then $\Psi_C : A \longrightarrow \{0,1\}.\big)$ See Example 14.13 for information regarding the *characteristic* function.

Now notice from Equation (2) that we seek a 1–1 correspondence between $\mathcal{P}(A)$ and $\{f \mid f : A \longrightarrow \{0,1\}\}$. So let Θ be defined by

$$\Theta(C) = \Psi_C \ \forall C \in \mathcal{P}(A).$$

The remainder of the proof is a matter of showing that $\Theta : \mathcal{P}(A) \longrightarrow \{f \mid f : A \longrightarrow \{0,1\}\}$ is a function from $\mathcal{P}(A)$ and is $1 - 1$ and is onto the set $\{f \mid f : A \longrightarrow \{0,1\}\}$. It should be clear from the definition of Θ that $\mathrm{dom}(\Theta) = \mathcal{P}(A)$. Also, since each Ψ_C is a function from A to $\{0,1\}$ then $\mathrm{ran}(\Theta) \subseteq \{f \mid f : A \longrightarrow \{0,1\}\}$. To show Θ is a function, suppose $C_1 = C_2$ then $\Psi_{C_1}(x) = \Psi_{C_2}(x) \ \forall x \in A$. So $\Psi_{C_1} = \Psi_{C_2}$. (Why? See Exercise 26.14.) Thus Θ is a function. Now if $C_1 \neq C_2$ then WLOG $\exists c \big[c \in C_1 - C_2 \big]$. Thus $\Psi_{C_1}(c) = 1 \neq 0 = \Psi_{C_2}(c)$. Thus $\Psi_{C_1} \neq \Psi_{C_2}$. So Θ is $1 - 1$.

To show Θ is onto, let $f_0 \in \{f \mid f : A \longrightarrow \{0,1\}\}$. We seek $C \in \mathcal{P}(A)$ such that $\Theta(C) = f_0$. So let $C = \{a \in A \mid f_0(a) = 1\}$. Then $C \in \mathcal{P}(A)$ and $\Psi_C = f_0$. (This is Exercise 26.15.) Thus $\Theta(C) = f_0$. Thus Θ is onto the set $\{f \mid f : A \longrightarrow \{0,1\}\}$. Thus we have finally shown $\mathcal{P}(A) \cong \{f \mid f : A \longrightarrow \{0,1\}\}$. Recall that 2^ν is defined to be $\#\big(\{f \mid f : A \longrightarrow \{0,1\}\}\big)$ so we have the result we sought. That is

$$2^\nu = \#\big(\{f \mid f : A \longrightarrow \{0,1\}\}\big) = \#\big(\mathcal{P}(A)\big).$$

COROLLARY 1: $2^\nu > \nu$ for all cardinal numbers ν.

Notice that this result generalizes the well–known inequality that $2^n > n$ for all natural numbers n. The proof of the corollary is left to the student. Hint, see Theorem 26.4.

COROLLARY 2: $2^{\aleph_0} = \#\big(\mathcal{P}(N)\big).$

This corollary is a trivial consequence of the theorem. Also think about what the corollary says and how it is a result which generalizes the concept that 2^n is the "number of subsets of a set with n elements."

The answer to Question 1 from earlier in this section is that $\#(\mathcal{P}(N))$ is equal to $\#(E_1) = c$. In the next theorem we get a result that gets us closer to this answer. The final piece needed to conclude $\#(\mathcal{P}(N)) = c$ will occur after Schröder-Bernstein is established in Theorem 29.3. It says if A is cardinally equivalent to a subset of B, and B is cardinally equivalent to a subset of A, then $A \cong B$.

Suppose that counting is done in base two, then the first ten whole numbers in base 2 (which are 0 through 9 in base 10) are 0, 1, 10, 11, 100, 101, 110, 111, 1000, 1001. The rational numbers $\frac{1}{2}, \frac{1}{4}, \frac{1}{8}, \frac{2}{3}, \frac{1}{5}$ are $0.1, 0.01, 0.001, 0.10101\cdots, 0.00110011\cdots$ respectively, in binary. In fact, each $r \in [0,1]$ has a binary expansion of the form

$$r = 0.b_1 b_2 b_3 \cdots b_n \cdots = \sum_{n \in N} b_n \cdot 2^{-n} \quad \text{where } b_n \in \{0,1\}.$$

Now representations in binary are not unique. For example $0.0111\cdots$ and 0.1 are equal. To maintain uniqueness of representation, all expansions with infinite repeating 1's are excluded except that the representation of 1 itself is taken as $0.11111\cdots$ (see the proof of Theorem 25.1 for a somewhat similar situation with decimal notation).

THEOREM 26.9: $c \leq \#(\mathcal{P}(N))$ and $\#(\mathcal{P}(N)) \leq c$.

The schema is to show that $\exists f \ni f : [0,1] \xrightarrow{\ 1-1\ } \mathcal{P}(N)$. Then $c = \#(E_1) \leq \#(\mathcal{P}(N))$. Then show $\exists g \ni g : \mathcal{P}(N) \xrightarrow{\ 1-1\ } [0,1]$. Then $\#(\mathcal{P}(N)) \leq c$.

PROOF: Let $r \in [0,1]$ then r is denoted by $0.b_1 b_2 b_3 \cdots b_n \cdots$ in binary notation, where $b_n \in \{0,1\}$ for all $n \in N$ and subject to the conditions of uniqueness mentioned above. For each r, as above, let A_r be given by $A_r = \{n \in N \mid b_n = 1\}$ then $A_r \in \mathcal{P}(N)$. Now define f by $f(r) = A_r$ for each $r \in [0,1]$. We seek to show this f is the desired $1-1$ function: $f : [0,1] \longrightarrow \mathcal{P}(N)$.

To see how f is applied to various numbers, notice the following: $f(0.1) = \{1\}$ and $f(0.011) = \{2,3\}$ and $f(0.111\cdots) = N$.

First notice that $f \subseteq [0,1] \times \mathcal{P}(A)$. Next, it is necessary to show that f is a function. To do this suppose $r_1, r_2 \in [0,1]$ and $r_1 = r_2$. Then the binary expansions of r_1 and r_2 are identical. In other words, for each $n \in N$ the following is true. In the nth place in the expansions of r_1 and r_2 either both of those n^{th} places are 0 or both are 1. If both are 0, then

n is in neither A_{r_1} nor A_{r_2}. If both are 1, then n is in both A_{r_1} and A_{r_2}. Thus $A_{r_1} = A_{r_2}$, i.e., $f(r_1) = f(r_2)$, so f is indeed a function with domain $[0, 1]$.

On the other hand, if r_1 and r_2 are two real numbers in $[0, 1]$, expressed in binary, and if $A_{r_1} = A_{r_2}$ then $n \in A_{r_1}$ if and only if $n \in A_{r_2}$. We are now ready to show f is $1 - 1$.

To do this, suppose

$$r_1 = 0.b_1 b_2 b_3 \cdots b_n \cdots \quad \text{and} \quad r_2 = 0.c_1 c_2 c_3 \cdots c_n \cdots \quad \text{and} \quad A_{r_1} = A_{r_2}.$$

Now let $n \in N$ then $n \in A_{r_1} = A_{r_2}$ or $n \notin A_{r_1} = A_{r_2}$. If $n \in A_{r_1} = A_{r_2}$ then $b_n = c_n = 1$ and if $n \notin A_{r_1} = A_{r_2}$ then $b_n = c_n = 0$. Thus for each $n \in N$ the nth place of r_1 is the same as the nth place of r_2. Thus $r_1 = r_2$. Consequently, f is $1 - 1$. Therefore $c \leq \#(\mathcal{P}(N))$.

To get the other inequality, define $g : \mathcal{P}(N) \longrightarrow [0, 1]$ as follows. If $A_r \subseteq N$ and $i \in N$ then $d_i = 1$ if $i \in A_r$, otherwise $d_i = 0$. Let $r = 0.d_1 d_2 \cdots d_i \cdots$ be a <u>decimal</u> expansion and define $g(A_r) = r$ $\forall A_r \subseteq N$. The domain of g is $\mathcal{P}(N)$ and codomain is $[0, 1]$. Clearly g is not onto $[0, 1]$ since $0.3 \in [0, 1]$ but $0.3 \notin \text{ran}(g)$. Also g is $1 - 1$, so $\#(\mathcal{P}(N)) \leq c$.

To see how g is applied to various subsets of N notice the following: $g(N) = 0.111 \cdots$ and $g(\{1, 3, 5, \ldots\}) = 0.10101 \cdots$.

Why didn't we write $\#\mathcal{P}(N)) = c$? Actually, it is, but we have not established the necessary things to conclude that $\leq$ is a partial order relation yet. That is where Schröder-Bernstein comes in. We choose to defer Schröder-Bernstein until Theorem 29.3, although it could have been done at this point (see Birkhoff and Mac Lane). Instead, let us assume it is already done. Then we have the following corollary.

COROLLARY: $2^{\aleph_0} = c$.

PROOF: Do this as an exercise using the theorem above and the second corollary to Theorem 26.8.

THEOREM 26.10: If α is transfinite then

    a. $\aleph_0 + \alpha = \alpha$

    b. $\aleph_0 \alpha = \alpha$

PROOF: Do this as an exercise using Theorem 25.5.

THEOREM 26.11: $c + c = c$.

PROOF: By use of Theorem 25.5, $[0, 1) \cong [0, 1]$, so $\#([0, 1)) = c$. Also the function f given by $f(x) = x + 1$ $\forall x \in [0, 1)$ establishes that $[0, 1) \cong$

$[1, 2)$. Thus $\#([1, 2)) = c$. Similarly, the function g given by $g(x) = 2x \ \forall x \in [0, 1)$ is a 1–1 function from $[0, 1)$ onto $[0, 2)$, then $\#([0, 2)) = c$ as well.

By the definition of addition of cardinal numbers, since $[0, 1) \cap [1, 2) = \emptyset$ and $[0, 1) \cup [1, 2) = [0, 2)$ we get

$$c + c = \#([0, 1)) + \#([1, 2)) = \#([0, 1) \cup [1, 2)) = \#([0, 2)) = c.$$

Thus $c + c = c$.

The next theorem is remarkable. It is reputed that Cantor said, "I see it, but I don't believe it," after he proved a version of it.

THEOREM 26.12: The real Cartesian plane has cardinal number c.

PROOF: This can be established by an interesting technique. The real plane is $E_1 \times E_1$, so

$$\#(E_1 \times E_1) = c \cdot c = c^2 = \left(2^{\aleph_0}\right)^2 = 2^{\aleph_0 \cdot 2} = 2^{2\aleph_0} = 2^{\aleph_0} = c.$$

Some additional results whose proofs are similar to the proof above are stated in the next theorem.

THEOREM 26.13: If n is a finite cardinal number and c the cardinal number of the continuum, then

a. $c^n = \left(2^{\aleph_0}\right)^n = 2^{n\aleph_0} = 2^{\aleph_0} = c$

b. $c^{\aleph_0} = \left(2^{\aleph_0}\right)^{\aleph_0} = 2^{\aleph_0 \aleph_0} = 2^{\aleph_0} = c$

Exercises:

26.1 Prove that 1 has binary expansion $0.1111 \cdots$ and $\frac{2}{3}$ has expansion $0.10101 \cdots$.

26.2 Prove: If $A \subseteq B$ then $\#(A) \le \#(B)$.

26.3 Prove that $\#([0, 1]) = \#([0, 1)) = \#((0, 1)) = c$.

26.4 Prove Part c of Theorem 26.1. That is, prove that $a(b+c) = ab+ac$ for all cardinals a, b, c.

26.5 Prove Part d of Theorem 26.1. That is, prove that $a \cdot 0 = 0$ for all cardinal numbers a. Hint: $\#(\{0\}) = 1$ (see Chapter 18).

26.6 Prove Theorem 26.1e. That is, prove that $a \cdot 1 = a$ for all cardinals a where $1 = \#(\{0\})$.

26.7 Prove the first part of Theorem 26.3. That is prove $\aleph_0 + n = \aleph_0$.

26.8   Prove that $n \neq \aleph_0$ where n is a finite cardinal number (see Example 26.1).

26.9   Prove that $\mu \neq 0 \Longrightarrow 0^\mu = 0$, i.e., Part c of Theorem 26.6.

26.10   Prove that the Ψ_C of Theorem 26.8 is indeed a function.

26.11   In a similar way to Theorems 26.12 or 26.13, prove: $\aleph_0^{\aleph_0} = c$. (Hint: split $\aleph_0$ as $2\aleph_0$ and expand.)

26.12   Prove that $1^\nu = 1$ for all cardinal numbers ν.

26.13   Prove that $c^c = 2^c$.

26.14   Referring to the proof of Theorem 26.8, show that if $C_1 = C_2$ then $\Psi_{C_1} = \Psi_{C_2}$.

26.15   See the "onto" part of the proof of Theorem 26.8. Show that $\Psi_C = f_0$.

26.16   Prove $2^\nu > \nu$ for all cardinals ν i.e., Corollary 1 to Theorem 26.8.

26.17   Prove Corollary 2 to Theorem 26.8.

26.18   Show that $2^3 = 8$ following Example 26.4.

26.19   Let $a, b \in E_1$. If $a < b$ show that the cardinal number of the interval (a, b) is c.

26.20   Explain the reasons for steps 1, 2, $\cdots$, 5 in Theorem 26.3.

26.21   How many functions are there from E_1 to E_1?

26.22   Show that $<$ for transfinite numbers is transitive.

26.23   Show by induction $\aleph_0 + \aleph_0 + \cdots + \aleph_0 = \aleph_0$ where there are the finite number n terms in the sum on the left (see Step 5 in the proof of Theorem 26.3).

SECTION VI

AXIOM OF CHOICE AND ORDINAL NUMBERS

CHAPTER 27

PARTIALLY ORDERED SETS

In this section, our attention is directed toward some of the deeper concepts of set theory. The topics to be considered will be treated in a fairly elementary, non-rigorous manner, yet some essential ideas will be developed and some major results will be established. The ideas to be considered are the fundamental tools for work in a very broad spectrum of mathematics. Without these tools, much of mathematics today simply could not exist in its present form. Much of the later material in this chapter is 20th century mathematics and as a result of its comparative newness, this body of information is growing and changing as new results are established. Interested readers are encouraged to consult the reading list to gain a wider perspective and learn some of the exciting history associated with these topics.

Partially Ordered Sets:

The launching point for the notions of this section is the concept of a partial order relation. This type of relation was introduced in Chapter 12 (see Exercises 12.18–12.20 and 12.23). The definition of a partial order relation is recorded here again for convenience.

DEFINITION: A <u>partial</u> <u>order</u> <u>relation</u> <u>on</u> <u>a</u> <u>set</u> S is a relation r on S such that

1. $(x \, r \, x) \; \forall x \in S$ Reflexive
2. $(x \, r \, y \,\wedge\, y \, r \, x \Longrightarrow y = x) \; \forall x, y \in S$ Antisymmetric
3. $(x \, r \, y \,\wedge\, y \, r \, z \Longrightarrow x \, r \, z) \; \forall x, y, z \in S$ Transitive

EXAMPLE 27.1:   The relation $\leq$ on E_1 is a partial order relation on E_1, since it satisfies the three required conditions. In particular,

$$\left[(x \leq x) \;\wedge\; (x \leq y \wedge y \leq x \Longrightarrow x = y) \;\wedge\; (x \leq y \wedge y \leq z \Longrightarrow x \leq z) \right]$$

holds for all $x, y, z \in E_1$.

EXAMPLE 27.2:   If S is a set of sets and $\subseteq$ is the usual "is a subset of" relation then $\subseteq$ is a partial order relation on S. The justification for the three conditions is in the solution to Exercise 12.18.

EXAMPLE 27.3:   If the relation "divides" is defined on N by

$$\text{If } x, y \in N \text{ then } x \mid y \text{ if and only if } \exists\, k \in N \ni x \cdot k = y,$$

then " $\mid$ " (read as "divides") is a partial order relation on Z.

PROOF: Since $x \cdot 1 = x \ \forall x \in N$ then $x \mid x \ \forall x \in N$. So $\mid$ is reflexive. If $x \mid y$ and $y \mid x$ then $\exists\, h, k \in N \ni [xh = y] \wedge [yk = x]$. Thus $(xh)k = yk = x$, hence $hk = 1$. By use of order properties in N (or in Z^+) the cases $h < 1$ and $k < 1$ can both be eliminated. Thus $h = k = 1$. Consequently, $x = y$. So $\mid$ is antisymmetric. Finally, if $x \mid y$ and $y \mid z$ then $\exists\, h, k \in N \ni xh = y$ and $yk = z$. Thus $x(hk) = (xh)k = yk = z$. Since $hk \in N$, $x \mid z$, so $\mid$ is transitive. Since it is reflexive, antisymmetric and transitive, $\mid$ is a partial order relation on N (also see Exercise 12.23).

The following property was posed as Exercise 12.20. It indicates that the only relation that is simultaneously a partial order relation and an equivalence relation on a set A is the diagonal relation I_A of A.

THEOREM 27.1:   If A is a nonempty set and r is a relation on A then r is simultaneously a partial order relation on A and an equivalence relation on A if and only if $r = I_A$.

PROOF: We will establish the part of the biconditional in the direction $\Longrightarrow$, since the $\Longleftarrow$ part is trivial.

Suppose r is a partial order relation on A and an equivalence relation on A. Then $I_A \subseteq r$. (Why?) The objective is to show $r = I_A$. So, if $(x, y) \in r$ then, since r is an equivalence relation on A, $(y, x) \in r$. Since r is an antisymmetric relation, and since $(x, y) \in r$ and $(y, x) \in r$, then $x = y$. Thus $(x, y) = (x, x) \in I_A$. Hence $r \subseteq I_A$. Therefore $r = I_A$.

DEFINITION: A partially ordered set (abbreviated *poset*) is an ordered pair (S, r) where S is a set and r is a partial order relation on S.

EXAMPLE 27.4:   The pairs $(E_1, \leq)$, $(Z, \geq)$, $(Q, =)$ and $(\mathcal{P}(\{a, b, c\}), \subseteq)$ are examples of partially ordered sets with respect to the usual $\leq$, $\geq$, $=$ and $\subseteq$ relations, respectively.

In the remainder of this chapter when a poset in general is considered, it will be convenient to use the symbol $\leq$ or a variation of it for the partial order relation. This usage is not meant to imply that the relation is necessarily ordinary, "less than or equal to." In specific examples, such as $(\mathcal{P}(S), \subseteq)$ or $(E_1, \geq)$, the actual relation is given. In case several sets or several partial orderings on a set are considered, it may be convenient to modify the notation slightly to distinguish between the different posets. This may be accomplished by subscripting the partial order relation as in $(A, \leq_A)$ or $(A, \leq_1)$ or $(A, \leq_2)$.

Notice also that $(A, \leq_A)$ and $(B, \leq_B)$ are different posets if either $A \neq B$ or $(A = B$ and $\leq_A$ is different from $\leq_B)$. This follows from the fact that a poset is an ordered pair.

DEFINITION: If $(S, \leq)$ is a poset and $a, b \in S$ then a and b are <u>comparable</u> iff either $a \leq b$ or $b \leq a$.

This is an important definition because not all posets have the property that all pairs are comparable. The following illustration demonstrates this.

EXAMPLE 27.5: In the poset $(\mathcal{P}(\{a, b, c\}), \subseteq)$, the elements $\{a, b\}$ and $\{b\}$ are comparable, since $\{b\} \subseteq \{a, b\}$. However, $\{a, b\}$ and $\{b, c\}$ are NOT comparable since $\{a, b\} \not\subseteq \{b, c\}$ and $\{b, c\} \not\subseteq \{a, b\}$. It cannot be overemphasized that in a poset $(A, \leq)$, one cannot assume for $a, b \in A$ that necessarily $a \leq b$ or $b \leq a$.

A poset can sometimes be depicted as a lattice diagram. The elements of the set are the points (or vertices or nodes). Two elements x and y are comparable if and only if they are joined by an edge (i.e., line segment with x and y as endpoints) or by a series of edges connected end-to-end that do not contain "vertices" pointing up or down.

A lattice diagram of the poset $(\mathcal{P}(\{a, b, c\}), \subseteq)$ of Example 27.5 is given in *Figure 27.1*.

Notice that all elements of $\mathcal{P}(\{a, b, c\})$ are vertices in the diagram and that all comparable elements of $\mathcal{P}(\{a, b, c\})$ are joined by an edge or series of end-to-end connected edges that do not form a vertex pointing up or down. For example, $\{c\}$ and $\{a, c\}$ are comparable as are $\{a, b, c\}$ and $\{a\}$. However, $\{a, b\}$ and $\{c\}$ are not comparable since each series of end-to-end connected edges joining them will have a vertex that points down at $\emptyset$, or up at $\{a, b, c\}$. Notice also that the segments from $\{a, c\}$ to $\{a\}$ or to $\{c\}$ are edges that do not intersect the edges from $\{b\}$ to $\{a, b\}$ or $\{b, c\}$. This is suggested by the "interruption" in the segments. You might visualize this as a cube standing on one vertex, $\emptyset$, with the opposite vertex $\{a, b, c\}$ at the top. Then two vertices are comparable if it is possible to move from one to the other by traveling along edges either

Figure 27.1

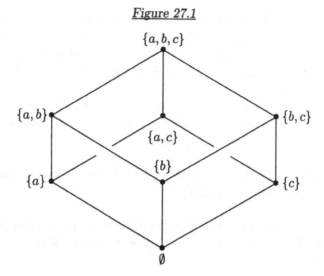

by going steadily upward or steadily downward, but not doing both on the journey from one vertex to another.

EXAMPLE 27.6: In Example 27.5, if

$$B = \mathcal{P}(\{a,b,c\}) - \{\{a,b,c\}\} = \{\{a,b\},\{a,c\},\{b,c\},\{a\},\{b\},\{c\},\emptyset\}$$

then $(B,\subseteq)$ is a poset too. This illustrates the more general principle described in the following theorem. The order relation on B is determined by the order relation on $\mathcal{P}(\{a,b,c\})$. Use *Figure 27.1* with $\{a,b,c\}$ removed as well as the edges joining $\{a,b,c\}$ to adjacent vertices to see the order relation *restricted to B*.

THEOREM 27.2: If $(S,\leq_S)$ is a poset and $A \subseteq S$ and $\leq_A$ denotes the *restriction* of $\leq_S$ to A, (i.e., $\leq_A$ is $\{(x,y) \mid x \leq_S y \wedge x \in A \wedge y \in A\}$), then $(A,\leq_A)$ is a poset.

PROOF: Suppose $(S,\leq_S)$ is a poset and $A \subseteq S$ and $\leq_A$ is the restriction of $\leq$ to A. Then $\leq_A = \{(x,y) \mid (x,y) \in \leq_S \wedge x,y \in A\}$, to employ the notation mentioned in Example 12.2.

SHOW: $\leq_A$ is reflexive. To do this, let $x \in A$ then $x \in S$ and so $x \leq_S x$. Thus, since $x \in A$, $x \leq_A x$. Consequently $\leq_A$ is reflexive.

SHOW: $\leq_A$ is antisymmetric. To do this, suppose $x,y \in A$ and $x \leq_A y$ and $y \leq_A x$ then, since $x,y \in S$ $x \leq_S y$ and $y \leq_S x$. Thus $x = y$, so $\leq_A$ is antisymmetric.

Figure 27.2

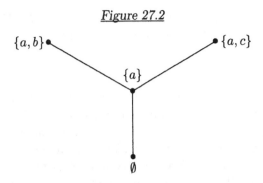

SHOW: $\leq_A$ is transitive. Suppose $x, y, z \in A$ and $x \leq_A y$ and $y \leq_A z$ then it is an exercise to establish that $x \leq_A z$.

Consequently, $(A, \leq_A)$ is a poset.

Another observation that can be made from Example 27.6 is that the element $\{a, b\}$ of B has the property that no element of B is strictly "larger" with respect to $\subseteq$. In other words, there is no element $T \in B$ such that $\{a, b\} \subset T$. This does <u>not</u> say that every $T \in B$ satisfies $T \subseteq \{a, b\}$ because, for example, $\{a, c\} \not\subseteq \{a, b\}$ even though $\{a, c\} \in B$. This state of affairs is described by saying that $\{a, b\}$ is a *maximal* element in the poset of Example 27.6. ("Maximal" is not the same as "maximum." See the next two definitions.)

DEFINITION: Let $(A, \leq)$ be a poset, $B \subseteq A$ and $b \in B$. Then b is <u>maximal</u> in B with respect to $\leq$ if and only if

$$\forall y \big[(y \in B \wedge b \leq y) \Longrightarrow b = y\big].$$

Thus b is maximal in B iff no element in B is "greater" than b. The definition does <u>not</u> say every element of B is "less than or equal to" b because, as mentioned before, not every pair of elements in a poset are comparable

EXAMPLE 27.7: The subset $B = \{\{a, b\}, \{a, c\}, \{a\}, \emptyset\}$ of $\mathcal{P}(\{a, b, c\})$ has two maximal elements $\{a, b\}$ and $\{a, c\}$ with respect to $\subseteq$. (See *Figure 27.2* for a lattice diagram of the set B.) The maximal elements are, in a sense, at the top of the diagram. Since there are two maximal elements in B, there is no element $T \in B$ such that $S \subseteq T \ \forall S \in B$. Because of this, there is no "maximum" element in B in the sense of the following definition.

DEFINITION: Let $(A, \leq)$ be a poset, $B \subseteq A$ and $b \in B$. Then b is "the" <u>maximum</u> element of B with respect to $\leq$ if and only if

$$\forall y [y \in B \Longrightarrow y \leq b].$$

The word "the" in the definition asserts that if a maximum element exists, it is unique. The verification of uniqueness is left as an exercise.

A set may have several maximal elements, but no maximum as was observed in Example 27.7. It should be clear in the general case that if a set S in a poset $(A, \leq)$ has a maximum element m, that m is also maximal. The following illustration points out that there are posets $(A, \leq)$ which have subsets B possessing neither maximal, nor maximum elements.

EXAMPLE 27.8:   Let $B = \{ 2n \mid n \in Z^+ \}$ and $A = Z$, partially ordered by usual $\leq$. Then clearly B has neither maximal, nor a maximum element with respect to $\leq$.

The notions of minimal and minimum are defined in a manner similar to maximal and maximum. For the record, the definitions follow.

DEFINITION: Let $(A, \leq)$ be a poset, $B \subseteq A$ and $b \in B$. Then b is <u>minimal</u> in B with respect to $\leq$ if and only if $\forall y [(y \in B \wedge y \leq b) \Longrightarrow y = b]$. And b is the <u>minimum</u> element of B with respect to $\leq$ if and only if $\forall y [y \in B \Longrightarrow b \leq y]$.

In a partially ordered set, those assertions regarding maximum and maximal have analogous counterparts for minimum and minimal. A minimum element is always minimal, but the converse is false. Also, there may be many minimal elements or none. And if a set has a minimum element, it is unique, hence the term "the" minimum.

EXAMPLE 27.9:   Consider the pair $(N, \mid)$ ordered by "divides" where $a \mid b$ iff $\exists k \in N \ni ak = b$. As observed earlier in Exercise 12.23 the relation $\mid$ really is a partial order relation on N. The number 1 is the "minimum" element of N with respect to $\mid$, NOT because it is the smallest element sizewise, BUT because $1 \mid n$ for all $n \in N$. If $B = \{ x \in N \mid x \neq 1 \}$ then 2 is a minimal element of B because, if $y \in B$ and $y \mid 2$ then $y = 2$. Similarly 3 is a minimal element of B. However, 4 is not a minimal element of B, because $2 \in B$ and $2 \mid 4$ but $2 \neq 4$. Hence $[y \in B \wedge y \mid 4] \not\Longrightarrow y = 4$. In Exercise 27.3 you are asked to pursue this line of questions to find all minimal elements of B.

EXAMPLE 27.10: Let $S = \{a, b, c\}$ then $(\mathcal{P}(S), \subseteq)$ is a partially ordered set. The element $\emptyset \in \mathcal{P}(S)$ is a minimal element in $\mathcal{P}(S)$ and is in fact a minimum element. If $B = \mathcal{P}(S) - \{\emptyset\}$ then B has three minimal elements $\{a\}, \{b\}$, and $\{c\}$. It might help to see the lattice diagram in *Figure 27.1*. In that figure, remove the element $\emptyset$ and the segments connecting it to the rest of the partially ordered set. In particular, $\{a\}$ is minimal because

$$\{a\} \in B \quad \wedge \quad \left(T \in B \quad \wedge \quad T \subseteq \{a\}\right) \quad \Longrightarrow \quad T = \{a\}.$$

In Chapter 21, the notions of upper and lower bounds and least upper and greatest lower bounds with respect to usual $\leq$ were explored in the system of real numbers. It is possible to generalize these concepts to any partial order relation.

DEFINITION: Let $(A, \leq)$ be a poset, $\emptyset \neq B \subseteq A$ and $a \in A$. Then
1. a is an <u>upper</u> <u>bound</u> <u>of</u> B <u>in</u> A <u>with</u> <u>respect</u> <u>to</u> <u>the</u> <u>partial</u> <u>order</u> <u>relation</u> $\leq$ if and only if
$$\forall x \left[x \in B \Longrightarrow x \leq a\right].$$

2. a is an <u>lower</u> <u>bound</u> <u>of</u> B <u>in</u> A <u>with</u> <u>respect</u> <u>to</u> <u>the</u> <u>partial</u> <u>order</u> <u>relation</u> $\leq$ if and only if
$$\forall x \left[x \in B \Longrightarrow a \leq x\right].$$

To see what these concepts mean in a particular situation, consider the following examples.

EXAMPLE 27.11: Let $A = \mathcal{P}(\{a, b, c\})$ and $B = \{\{a, b\}, \{b, c\}\}$ then $\{b\}$ and $\emptyset$ are lower bounds of B in A with respect to $\subseteq$ since $\{b\}$ and $\emptyset$ are in A and $x \in B \Longrightarrow \{b\} \subseteq x$ and $x \in B \Longrightarrow \emptyset \subseteq x$. Notice that B does not have to contain its lower bounds. On the other hand $\{a, b, c\}$ in A is an upper bound of B in A with respect to $\subseteq$ because $\{a, b, c\} \in A$ and $x \in B \Longrightarrow x \subseteq \{a, b, c\}$. It is also true that B does not have to contain its upper bounds.

Example 27.11 deals with a particular type of partial order relation and the notions of upper and lower bounds for that relation. However, an additional example follows to emphasize an important point. Always carefully follow the definitions pertaining to the given partial order relation and do not be influenced by other partial order relations that may make sense for that given set.

EXAMPLE 27.12: Consider the poset $(N, \mid)$ with $B = \{8, 12\}$ partially ordered by "divides", i.e., $\mid$. Then the elements $1, 2, 4$ of N are lower bounds of B not because they are less than 8, but because they divide 8 and 12. In particular, 4 is a lower bound because $4 \mid x \; \forall x \in B$. Clearly a number such as 6 is not a lower bound for B, even though it is smaller sizewise than 8 and 12. This is because it does not divide both 8 and 12.

The element 24 is an upper bound of B in N with respect to $\mid$ because $x \in B \Longrightarrow x \mid 24$. Clearly 16 is not an upper bound of B even though it is "larger" sizewise. What is the "least" upper bound of B in N? What is the "greatest" lower bound of B in N? Be careful. The answer is found in the next chapter.

Exercises:

27.1 Is $r = \{(a, a), (b, b), (c, c), (a, b), (a, c), (b, c)\}$ a partial order relation on $A = \{a, b, c\}$? What is r^{-1}? Is r^{-1} a partial order relation on A?

27.2 Complete the proof of Theorem 27.2 by showing that the restriction $\leq_A$ of $\leq_S$ to A is transitive.

27.3 Let $(N, \mid)$ be a poset and $B = \{x \in N \mid x \neq 1\}$. Find all minimal elements of B with respect to the partial order relation "divides." Also find all minimum, maximal, and maximum elements, if they exist, of B with respect to $\mid$. What is another name for the minimal elements of B? See Example 27.9.

27.4 Let $S \neq \emptyset$ and $\mathcal{P}(S)$ the power set of S.
   a. Verify that $\subseteq$ is a partial order relation on $\mathcal{P}(S)$.
   b. Verify that $(\mathcal{P}(S), \subseteq)$ is a poset.
   c. Let $B = \mathcal{P}(S) - \{\emptyset\}$ and find all minimal elements of B. Are there any minimum elements? (Is this certain? What if S is a singleton?)
   d. Repeat Exercise 27.4c, as is, for maximal and maximum elements.

27.5 Prove: If $(A, \leq)$ is a poset, $B \subseteq A$ and B has a maximum element b then b is unique. If B has a unique maximal element, is it necessarily a maximum element? (The student need not justify.)

27.6 Prove that a maximum element in a poset B is also a maximal element.

27.7 Prove that: r is a partial order relation on A iff r^{-1} is a partial order relation on A. In other words, (A, r) is a poset iff (A, r^{-1}) is a poset.

27.8 If $(A, \leq_1)$ and $(B, \leq_2)$ are posets and $A \cap B = \emptyset$, then show there exists a partial order relation $\leq$ on $A \cup B$ whose restriction to A, i.e., $\leq_A$ is equal to $\leq_1$ and whose restriction to B, i.e., $\leq_B$ is equal to $\leq_2$.

27.9 Determine whether or not the relation, "is comparable to," on a poset $(A, \leq)$ is an equivalence relation. Prove your conjecture is correct.

CHAPTER 28

LEAST UPPER BOUND AND GREATEST LOWER BOUND

In Chapters 21 and 22 the concepts of upper bounds and lower bounds were considered for the particular posets $(E_1, \leq)$ where $\leq$ is the usual "less than or equal to" relation. In Chapter 27, generalizations of these concepts were investigated for any set A and any partial order relation on A. Likewise, in Chapters 21 and 22 the notions of least upper bound and greatest lower bound were considered for the particular poset $(E_1, \leq)$. These concepts are now generalized for any partially ordered set. As before, the symbol $\leq$ is sometimes used for the partial order relation. But $\leq$ is NOT necessarily the ordinary "less than or equal to." At times the actual partial order relation will be used, e.g., $\subseteq$ in $\big(\mathcal{P}(\{a, b, c\}), \subseteq\big)$.

DEFINITION: Let $(A, \leq)$ be a poset and $\emptyset \neq B \subseteq A$.

    i. The element $a \in A$ is "the" <u>least</u> <u>upper</u> <u>bound</u> <u>of</u> $\underline{B}$ <u>in</u> $\underline{A}$ <u>with</u> <u>respect</u> <u>to</u> $\leq$ if and only if

        1. a is an upper bound of B in A with respect to $\leq$, and

        2. $a \leq u$ for all upper bounds u of B in A with respect to $\leq$ (not necessarily ordinary $\leq$).

    ii. The element $a \in A$ is "the" <u>greatest</u> <u>lower</u> <u>bound</u> <u>of</u> $\underline{B}$ <u>in</u> $\underline{A}$ <u>with</u> <u>respect</u> <u>to</u> $\leq$ if and only if

        1. a is a lower bound of B in A with respect to $\leq$, and

        2. $k \leq a$ for all upper bounds k of B in A with respect to $\leq$ (not necessarily ordinary $\leq$).

The inclusion of the article "the" in the two definitions above, is due to the fact that if a set B has a least upper bound b (or a greatest lower bound b), then that element b is unique. Verification is left to Exercise 28.4. Also note, condition 1 in Part i says a is an upper bound and condition 2 makes it the "least" upper bound.

In addition, the terminology can be simplified when the underlying partial ordering is clear. For example, the sentence, "The element a in A is the least upper bound of B in A with respect to $\leq$," may be relaxed to read, "The element a in A is the least upper bound of B," by deletion of, "in A with respect to $\leq$." Similar condensations can be made for the greatest lower bound, when the underlying poset is clear.

For simplicity, let the least upper bound of B and the greatest lower bound of B be denoted by $\text{LUB}(B)$ and $\text{GLB}(B)$ respectively. As before, the underlying poset $(A, \leq)$ must be clearly understood.

EXAMPLE 28.1:   Let $A = \mathcal{P}(\{a, b, c\})$ be partially ordered by $\subseteq$ and $B = \{\{a, b\}, \{b, c\}\}$, then the $\text{LUB}(B) = \{a, b, c\}$. This is true because of the following: $\{a, b, c\} \in A$ and $\forall T \in B, T \subseteq \{a, b, c\}$, hence $\{a, b, c\}$ is an upper bound of B. AND furthermore, if $U \in A$ and U is an upper bound of B then $U \supseteq \{a, b\}$ and $U \supseteq \{b, c\}$. Thus $U \supseteq \{a, b, c\}$ by Theorem 9.13. (Look it up.) Consequently, $\text{LUB}(B) = \{a, b, c\}$.

On the other hand, $\text{GLB}(B) = \{b\}$ because $\{b\} \in A$ and $\{b\} \subseteq T \,\forall T \in B$, so $\{b\}$ is a lower bound of B. In addition, if U is any lower bound of B then $U \subseteq \{a, b\}$ and $U \subseteq \{b, c\}$, hence $U \subseteq \{b\}$ by Theorem 9.14. (Check it out.) Therefore, $\text{GLB}(B) = \{b\}$.

EXAMPLE 28.2:   Consider the poset $(N, \,|\,)$ partially ordered by "divides." Let $B = \{8, 12\}$. Since $24 \in N$, $8 \mid 24$, and $12 \mid 24$, 24 is an upper bound of B. Similarly 48 and 72 are each upper bounds of B. Condition i, Part 2 remains to be established for some upper bound a.

So let u be an "upper bound" of B. Then $8 \mid u$ and $12 \mid u$. Thus $2^3 \mid u$ and $2^2 \cdot 3 \mid u$. Therefore $2^3 \cdot 3 \mid u$. In other words, $24 \mid u$. Thus $\text{LUB}(B) = 24$.

In an analogous way 1, 2, and 4 are the lower bounds of $B = \{8, 12\}$. Since $1 \mid 4$ and $2 \mid 4$ and $4 \mid 4$ then $k \mid 4$ for each lower bound k of B. Thus $\text{GLB}(B) = 4$.

Notice that in Example 28.1 not every pair of elements in A satisfies the relation, "is comparable to." In particular, the pair of sets $\{a, b\}$ and $\{a, c\}$ are not comparable. In Example 28.2, 8 and 12 are not comparable. On the other hand, the subset $B_1 = \{\{a, b, c\}, \{b, c\}, \{b\}\}$ in $(A, \subseteq)$ is a subset of $A = \mathcal{P}(\{a, b, c\})$ in which each pair of elements is comparable. In particular $\{b, c\} \subseteq \{a, b, c\}$, $\{b\} \subseteq \{a, b, c\}$, $\{b\} \subseteq \{b, c\}$, as well as the containments $T \subseteq T \,\forall T \in B_1$. Thus if $C, D \in B_1$ then $C \subseteq D$ or $D \subseteq C$, i.e., C and D are comparable.

Likewise, the subset $B_2 = \{4, 12, 24, 72\}$ in the poset $(N, \,|\,)$ is a set in which each pair of elements is comparable with respect to the partial order relation "divides." In other words, if $x, y \in B_2$ then either $x \mid y$ or $y \mid x$.

Each of the two sets B_1 and B_2 in Example 28.2 is an example of what is called a *linearly ordered set* or *chain*, or *totally ordered set*. Here is the official definition.

DEFINITION: Let $(A, \leq)$ be a poset and $C \subseteq A$. Then C is a <u>chain</u> (or <u>linearly ordered subset</u> or <u>totally ordered subset</u>) of A with respect to $\leq$ iff $\forall c_1, c_2 \in C$, either $c_1 \leq c_2$ or $c_2 \leq c_1$.

In other words, every pair of elements of a chain C is comparable with respect to the partial order relation. In the cases cited previously, $B_1 = \{\{a, b, c\}, \{b, c\}, \{b\}\}$ and $B_2 = \{4, 12, 24, 72\}$ are chains with respect to the partial order relations $\subseteq$ and $\mid$, respectively. On the other hand, $B_3 = \{\{a, b, c\}, \{b, c\}, \{a\}\}$ is not a chain with respect to $\subseteq$. Why not?

EXAMPLE 28.3: The poset $(Q, \leq)$ with usual $\leq$ is a chain. Likewise, $(Q, \geq)$ is a chain, where $\geq$ is the usual, "greater than or equal to." This suggests a more general principle whose verification is left as an exercise. That is, if C is a chain with respect to a partial order relation $\leq$ then C is also a chain with respect to $\leq^{-1}$.

EXAMPLE 28.4: In $(\mathcal{P}(\{a, b, c\}), \subseteq)$ the set $C = \{\{a, b, c\}, \emptyset, \{a, b\}\}$ is a chain, but it is not a "maximal" chain. This is because the set $C' = \{\{a, b, c\}, \{a, b\}, \{b\}, \emptyset\}$ is a chain in $(\mathcal{P}(\{a, b, c\}), \subseteq)$ such that $C \subset C'$.

Each of the notions of chains and maximality have been considered. We put these concepts together. To say that a chain C is maximal in a poset $(A, \leq)$ means there is no chain C' in $(A, \leq)$ such that C' properly contains C.

DEFINITION: A chain C in a poset $(A, \leq)$ is a <u>maximal chain</u> iff

$$\forall C' [C \subseteq C' \wedge C' \text{ is a chain in } (A, \leq) \implies C = C'].$$

EXAMPLE 28.5: If $A = \{1, 2, 3, 4, \ldots, 20\}$ partially ordered by "divides" then the set $B = \{2, 8, 16\}$ is a chain in A, but B is not a maximal chain in $(A, \mid)$ since $C = \{1, 2, 4, 8, 16\}$ is a chain in $(A, \mid)$ and $B \subset C$. However, C is a maximal chain in $(A, \mid)$. Likewise $\{1, 3, 6, 12\}$ and $\{1, 5, 10, 20\}$ are other maximal chains in $(A, \mid)$

Exercises:

28.1 Consider the poset $(N, \mid)$ and let $B = \{12, 15\}$.

    a. Find all maximal and maximum elements of B, if any, with respect to the partial order relation $\mid$.

    b. Find all minimal and minimum elements of B, if any, with respect to the partial order relation $\mid$.

c. List three upper bounds and two lower bounds of B in N with respect to $|$.

d. Find $\mathrm{GLB}(B)$ and $\mathrm{LUB}(B)$ with respect to $|$ in N.

e. Is B a chain or subset of a chain in N with respect to $|$?

f. With the poset $(\{12,15\}, |)$ in $(N, |)$, what is its greatest lower bound by another name?

g. With the poset $(\{12,15\}, |)$, what is its least upper bound by another name?

h. Is $\mathrm{LUB}(B) \in B$? Is $\mathrm{GLB}(B) \in B$?

28.2 Let $S = \{a,b,c\}$ and $A = \mathcal{P}(S)$ (the power set of S). Suppose $\subseteq$ is the partial order on A.

a. If $B = \{\{a\},\{a,c\}\}$ then find $\mathrm{GLB}(B)$ and $\mathrm{LUB}(B)$ in A.

b. Is B a chain with respect to $\subseteq$ in A?

c. Find a maximal chain containing B.

28.3 Let $(A, \leq)$ be a poset.

a. Prove that if $a \in A$ then $\{a\}$ is a chain.

b. Prove that if $C \subseteq A$ and C is a chain and $B \subseteq C$ then B is a chain.

c. Prove that if C_1 and C_2 are chains in A then $C_1 \cap C_2$ is a chain in A.

d. Prove that if C is a chain in $(A, \leq)$ then C is also a chain with respect to $\leq^{-1}$.

e. If $C = N$ prove that if $(N, \leq)$ is a poset with usual $\leq$ then C is a chain. What is $\leq^{-1}$? Is $(C, \leq^{-1})$ a chain?

28.4 Prove that if $(A, \leq)$ is a poset, $\emptyset \neq B \subseteq A$, and if B has a least upper bound b then b is unique. Consequently, "the" is an appropriate article in the definition.

28.5 Repeat Exercise 28.4 for greatest lower bound.

28.6 Let $S \neq \emptyset$, $A = \mathcal{P}(S)$ and $\subseteq$ be the usual partial order relation. If $\emptyset \neq B \subseteq A$ then carefully prove each of the following.

a. $\bigcup_{T \in B} T = \mathrm{LUB}(B)$.   b. $\bigcap_{T \in B} T = \mathrm{GLB}(B)$.

28.7 Let $S = \{a,b,c,d\}$ and $A = \mathcal{P}(S) - \{a,b,c,d\}$ and let $B = \{\{a\},\{a,c\}\}$. Is B a chain with respect to $\subseteq$? Find a maximal chain in A containing B. Is there more than one maximal chain containing B?

CHAPTER 29

AXIOM OF CHOICE

Introduction:

Suppose $\{a, b, c\}$, $\{n, e\}$ and $\{y\}$ are subsets of the alphabet and S is the set $\{\{a, b, c\}, \{n, e\}, \{y\}, Q\}$ of nonempty sets. Is there a set H consisting of exactly one element from each member of S? In this case the answer seems obvious. Since each $T \in S$ is nonempty, an element may be selected from it. Then the set H is formed from the selected elements. For example, a possible H would be $H = \{a, n, y, 0.5\}$.

But what if S were an infinite set of nonempty sets? In particular, suppose $S = \{T_i \mid i \in N\}$ where each $T_i \neq \emptyset$ and the T_i's are pairwise disjoint. To make an infinite number of selections would be physically impossible if each selection depends in some way on the previous selection or if the selections are simply made one at a time. This presumes that there is some minimum amount of time to make an individual selection or some means by which the selection process takes place. Yet it may still seem reasonable that a set H does exist such that H consists of exactly one element t_i from each T_i in S. The justification for the existence of H is the *axiom of choice*.

AXIOM OF CHOICE: Suppose S is a nonempty set of nonempty sets S_λ indexed by Λ. Then there exists a set H such that $\forall S_\lambda$, if $\lambda \in \Lambda$ then $S_\lambda \cap H$ is a singleton, i.e., there is a set H such that H consists of exactly one element from each S_λ, where $\lambda \in \Lambda$.

This principle not only handles situations of the finite and denumerable types mentioned in the introduction, but it applies in cases where S and perhaps some or all of the S_λ are uncountable as well. This, in fact, is where its real power comes into play. The knowledge that each $S \in S$ is nonempty is sufficient to invoke the axiom of choice (hereinafter denoted by A.C.) and voilà the set H is there, ready to be used.

The need for A.C. may not be apparent. Indeed, it is such a reasonable and natural rule it was not apparent to mathematicians and was not recognized as a principle of set theory until approximately 1900 even though it had been used unknowingly prior to that time.

In case this may seem like too much ado about a triviality, don't prematurely judge it so. In fact, nothing can be further from the truth. In 1904, the mathematician Ernst Zermelo stated the axiom of choice and proved that it implied the astounding *well ordering theorem* (to be discussed in Chapter 30). Zermelo's well ordering theorem was not well received at first because, among other things, it said that the real numbers could be "well" ordered, but no one could find such a well ordering of the real numbers. A huge debate among mathematicians ensued. (For some interesting historical comments, as well as other mathematical developments pertaining to these ideas, consult the Reading List.)

The reluctance to accept the well ordering theorem left two alternatives. Accept A.C. and thus the well ordering theorem or reject the well ordering theorem and thus A.C. The alternative of rejecting A.C. discards important mathematical results, and then without A.C., any mathematics having set theoretic foundations, would be more limited.

This unsettled state of affairs characterized the status of the A.C. for a number of years. There was no convincing reason to accept it as an axiom or reject it as an axiom, because no one had found a way to establish it as consistent with the other axioms of set theory or independent from them. In 1938, Kurt Gödel proved that A.C. was consistent with the then existing axioms for set theory. Though this did not totally clarify the status of A.C., it was a big step toward establishing that A.C. could be taken as an axiom of set theory. The next breakthrough came in 1963 when Paul Cohen of Stanford University proved, in effect, that A.C. is independent of the other axioms of set theory. The result is that A.C. may be taken as an axiom of set theory.

Part of the reason for the importance of the A.C. is that quite a large number of results are equivalent to it and many results require it for substantiation. Some of these will be considered after we review a few terms that we will be using in this section.

Recall that a chain in a partially ordered set $(A, \leq)$ is a subset C, of A, partially ordered with respect to $\leq_C$ and having the property that each pair c_1, c_2 of elements of C is comparable, i.e., $c_1 \leq_C c_2$ or $c_2 \leq_C c_1$. Also recall that an upper bound for a subset S of A is an element a of A such that $s \leq a \ \forall s \in S$. And a maximal element for a subset S of A is an element $m \in S$ such that $\forall s \in S(m \leq s \implies m = s)$. As previously mentioned, $\leq$ does not necessarily mean ordinary less than or equal.

The following principles are equivalent to the A.C.

ZORN'S LEMMA (Z.L.): If $(A, \leq)$ is a partially ordered set such that every chain C in A has an upper bound in A then A has a maximal element.

HAUSDORFF MAXIMALITY PRINCIPLE (H.M.P.): Every partially ordered set has a maximal chain.

For convenience, we use Z.L. and H.M.P. to represent, respectively, Zorn's lemma and Hausdorff maximality principle. The following theorem asserts the equivalence of the three principles.

THEOREM 29.1: A.C. $\Longleftrightarrow$ H.M.P. $\Longleftrightarrow$ Z.L.

To prove the equivalence of the three statements A.C., H.M.P., and Z.L., it would be sufficient to prove the three implications

$$Z.L. \stackrel{1}{\Longrightarrow} H.M.P. \stackrel{2}{\Longrightarrow} A.C. \stackrel{3}{\Longrightarrow} Z.L.$$

PROOF PART 1: We do the first of these parts, i.e., Z.L. $\Longrightarrow$ H.M.P. Suppose Z.L. and let $(A, \leq)$ be a partially ordered set. Also let

$$S = \{C \mid C \text{ is a chain w.r.t. } \leq \text{ in } A\}.$$

From Example 27.2 or Exercise 12.18, $\subseteq$ is a partial order relation on any set S of sets, so $(S, \subseteq)$ is also a partially ordered set. In order to conclude H.M.P., it is sufficient to show that S has a maximal element with respect to $\subseteq$, i.e., show S has a maximal chain. (This is our objective because $(A, \leq)$ was an arbitrary poset and S is the set of all chains, so a maximal chain in A is a maximal element in S.)

Keep in mind that Z.L. is our assumption so to use Z.L. we must show every chain in S has an upper bound in S. To do this, let C be a chain in S. (Don't confuse the chain C, in S with the chains C in A. C is, in fact, a chain in S of chains C in A.)

First, two items need to be shown with regard to the chain C in S.

a. $\bigcup_{C \in C} C$ is a chain in A, and therefore it is an element of S.

b. $\bigcup_{C \in C} C$ is an upper bound for C in S.

To do Item a, let $\bigcup C = \bigcup_{C \in C} C$ for the sake of brevity. In order for $\bigcup C$ to be a chain, each pair in $\bigcup C$ must be comparable, so let $T_1, T_2 \in \bigcup C$. Then from the definition of union, $\exists C_1 \in C \ni T_1 \in C_1$ and $\exists C_2 \in C \ni T_2 \in C_2$. Since C is a chain in $(S, \subseteq)$ and $C_1, C_2 \in C$ then either $C_1 \subseteq C_2$ or $C_2 \subseteq C_1$. In either case, T_1 and T_2 are both in C_1 or T_1 and T_2 are both in C_2. But both C_1 and C_2 are chains in $(A, \leq)$ so T_1 and T_2 are comparable with respect to $\leq$. That is either $T_1 \leq T_2$ or $T_2 \leq T_1$. Thus every pair in $\bigcup C$ is comparable, hence $\bigcup C$ is a chain.

To do Item b, refer to Exercise 10.6, which says, for each $C \in \mathcal{C}$, $C \subseteq \bigcup \mathcal{C}$. With this result, it is immediate that $\bigcup \mathcal{C}$ is an upper bound for $\mathcal{C}$, so Item b is done.

To complete the proof that Z.L. $\implies$ H.M.P., we now know that $(\mathcal{S}, \subseteq)$ is a partially ordered set such that each chain $\mathcal{C}$ in $\mathcal{S}$ has an upper bound $\bigcup \mathcal{C}$ in $\mathcal{S}$. By using the hypothesis, i.e., Z.L., $\mathcal{S}$ has a maximal element. That maximal element is a chain (because everything in $\mathcal{S}$ is a chain) in $(A, \leq)$ (see the definition of $\mathcal{S}$). So A has a maximal chain. Thus Z.L. $\implies$ H.M.P. is proven.

PART 2: The next major part of the proof of this theorem is to show the next implication: H.M.P. $\implies$ A.C. To do this, suppose H.M.P. is true and S is a nonempty set of nonempty sets S. To establish A.C. as the conclusion, it is necessary to show that $\exists \, H \ni H$ consists of exactly one element from each $S \in S$. Toward that end, define $\mathcal{F}$ by

$$\mathcal{F} = \left\{ f \ \Big| \ \Big(\exists \, S' \subseteq S \ni f : S' \longrightarrow \bigcup_{S \in S} S \Big) \text{ and } \big(f(S) \in S \ \forall S \in S' \big) \right\}$$

and let $S_0 \in S$. Then $\{S_0\} \subseteq S$ and since $S_0 \neq \emptyset$, $\exists \, s_0 \in S_0$, so $\{(S_0, s_0)\}$ is a function between (not from) S and $\bigcup S$ (albeit with just one ordered pair in it). Furthermore, if $S' = \{S_0\}$ and $f = \{(S_0, s_0)\}$ then $f : S' \longrightarrow \bigcup S$ where $S' \subseteq S$ and $f(S) \in S$ $\forall S \in S'$. (The latter statement holds since $f(S_0) = s_0 \in S_0$ and S_0 is the only element of S'.) As a consequence $\{(S_0, s_0)\}$ is an element of the set $\mathcal{F}$. Thus $\mathcal{F} \neq \emptyset$. And since $\mathcal{F}$ is a set of functions, $\subseteq$ is a partial order relation on $\mathcal{F}$ so $(\mathcal{F}, \subseteq)$ is a partially ordered set and since H.M.P. is the assumption, $\mathcal{F}$ has a maximal chain which we call $\mathcal{C}$.

At this point, it might help to point out where we are going. We are showing that A.C. holds for the arbitrarily selected nonempty set S of nonempty sets S. To do this, we need a function which selects exactly one element from each $S \in S$. That is the next objective.

Let $F = \bigcup_{f \in \mathcal{C}} f$ i.e., F is the union of the maximal chain $\mathcal{C}$ of functions f. There are three items to show with regard to F:

   a. F is a function.

   b. $\text{dom}(F) = S$.

   c. $\text{ran}(F) \subseteq \bigcup_{S \in S} S$.

Items a and c are left as exercises, while Item b will now be done. The first thing in Item b to be shown is $\text{dom}(F) \subseteq S$. Before doing this, we make the observation that since $F = \bigcup_{f \in C} f$ then, from the definition of generalized union, $t \in F \iff \exists f \in C \ni t \in f$. (Note this says t is an ordered pair.) Let $x \in \text{dom}(F)$ then $\exists y \ni (x,y) \in F$. Therefore $\exists f \in C \ni (x,y) \in f$. But then from the definition of $\mathcal{F}$, $\exists S \in \mathcal{S} \ni (x,y) = (S, f(S))$. So $x = S \in \mathcal{S}$. Consequently, $\text{dom}(F) \subseteq \mathcal{S}$.

To complete the proof of Item b in Part 2, suppose, on the contrary, that $\text{dom}(F) \subset \mathcal{S}$. Then $\exists S_1 \in \mathcal{S} \ni S_1 \notin \text{dom}(F)$. Since $S_1 \neq \emptyset$, $\exists s_1 \in S_1$. Therefore $\{(S_1, s_1)\}$ is a function in $\mathcal{F}$ and $F_1 = F \cup \{(S_1, s_1)\}$ is a function in $\mathcal{F}$ properly containing F. It is an exercise at the end of this section to show $C \cup \{F_1\}$ is a chain in $\mathcal{F}$ properly containing C. This is contrary to the maximality of C. Hence $\text{dom}(F) = \mathcal{S}$. Thus when Items a, b, and c are completed, there is a function F from $\mathcal{S}$ to $\bigcup \mathcal{S}$ such that $F(S) = s$ where $s \in S$. This holds for all $S \in \mathcal{S}$. Let $H = F[\mathcal{S}] = \{F(S) \mid S \in \mathcal{S}\}$ be the image of $\mathcal{S}$ under F, then H consists of exactly one s from each $S \in \mathcal{S}$. Thus the Hausdorff maximality principle implies the axiom of choice.

PART 3: We will not do the third part of the equivalence, $A.C. \implies Z.L.$

The following example illustrates a technique commonly used to prove assertions using Zorn's lemma.

DEFINITION: Let $U \neq \emptyset$ and let A be a subset of the power set $\mathcal{P}(U)$. The set A has the finite intersection property (f.i.p.) if and only if it is true that for each finite subset A' of A, $\bigcap \{S \mid S \in A'\} \neq \emptyset$. In other words, A has f.i.p. iff the intersection of each finite collection of elements of A is nonempty.

The next example says a subset A of $\mathcal{P}(U)$ having f.i.p. is contained in a maximal subset B of $\mathcal{P}(U)$ having f.i.p..

EXAMPLE 29.1: Let $U \neq \emptyset$, and $A \subseteq \mathcal{P}(U)$, and A have f.i.p. Then $\exists B \in \mathcal{P}(U) \ni A \subseteq B$ and B has f.i.p. and B is a maximal subset of $\mathcal{P}(U)$ having f.i.p.

PROOF: Let $U \neq \emptyset$ and $A \subseteq \mathcal{P}(U)$ and suppose A has f.i.p. The objective is to show that A is contained in some maximal subset B of $\mathcal{P} = \mathcal{P}(\mathcal{U})$ where B has f.i.p. In order to show this, let

$$\mathcal{F} = \{B \mid A \subseteq B \subseteq \mathcal{P}(U) \text{ and } B \text{ has f.i.p.}\}.$$

Then $\mathcal{F} \neq \emptyset$ (Why?) and $(\mathcal{F}, \subseteq)$ is a partially ordered set. (Why?) (Expressed in terms of $\mathcal{F}$, our objective is to find a maximal B_0 in $\mathcal{F}$.

The machinery for doing this is Z.L. which says, in reference to $\mathcal{F}$, if every chain in $\mathcal{F}$ has an upper bound in $\mathcal{F}$ then $\mathcal{F}$ has a maximal element B_0. The set B_0 would be the set we are looking for.)

In order to use Z.L., its hypothesis must be met. In other words, we must show that every chain in $\mathcal{F}$ has an upper bound in $\mathcal{F}$. So let C be a chain in $\mathcal{F}$. Then the goal is to show $\bigcup_{C \in \mathcal{C}} C$, i.e., $\bigcup \mathcal{C}$ is an upper bound of C in $\mathcal{F}$. This objective can be accomplished by establishing the following parts:

a. $B \subseteq \bigcup_{C \in \mathcal{C}} C$ for each $B \in C$. (This will make $\bigcup_{C \in \mathcal{C}} C$ an upper bound w.r.t. $\subseteq$ for the chain C.)

b. $\bigcup_{C \in \mathcal{C}} C$ has f.i.p.

c. $A \subseteq \bigcup_{C \in \mathcal{C}} C$.

d. $\bigcup_{C \in \mathcal{C}} C \subseteq \mathcal{P}(U)$ (Parts b, c, d ensure $\bigcup_{C \in \mathcal{C}} C$ in $\mathcal{F}$).

(After these parts are done, the arbitrary chain C in $\mathcal{F}$ has an upper bound in $\mathcal{F}$. Thus the hypothesis of Z. L. is met. The conclusion is that $\mathcal{F}$ has a maximal element.)

(Parts c and d are exercises. Don't despair there are previous properties or exercises which give these results almost trivially; just find them and reference them.)

PART a: Suppose $B \in C$, then by a property of generalized unions $B \subseteq \bigcup_{C \in \mathcal{C}} C$. Thus Part a is true. That is, $\bigcup \mathcal{C}$ is an upper bound for C.

PART b: Let F be a finite subset of $\bigcup_{C \in \mathcal{C}} C$, say $F = \{B_1, B_2, \ldots, B_n\}$ then each $B_i \in \bigcup_{C \in \mathcal{C}} C$. Thus for each $B_i \in F$ $\exists C_i \in C \ni B_i \in C_i$. Since C is a chain and F is a finite set $\exists C_{i_o} \in C \ni B_1 \in C_{i_o}$, $B_2 \in C_{i_o}, \ldots, B_n \in C_{i_o}$. In other words, each element in the finite set F is contained in the same C_{i_o}. But recall that $C_{i_o} \in C$, so $C_{i_o} \in \mathcal{F}$. Thus C_{i_o} has the finite intersection property. So the intersection of the finite collection F in C_{i_o} is nonempty. In other words, $\bigcap_{B_i \in F} B_i \neq \emptyset$. Since F was arbitrarily selected in $\bigcup \mathcal{C} = \bigcup_{C \in \mathcal{C}} C$, $\bigcup \mathcal{C}$ has the f.i.p.

After establishing Parts c and d, the conclusion is that for an arbitrary chain C in $\mathcal{F}$, $\bigcup_{C \in \mathcal{C}} C \in \mathcal{F}$ and $\bigcup_{C \in \mathcal{C}} C$ is an upper bound of C in $\mathcal{F}$. Thus, the hypotheses of Zorn's lemma have been satisfied and so $\mathcal{F}$ has a maximal element. In other words, if A is a set in $\mathcal{P}(U)$ with f.i.p. then A is contained in a maximal set, in $\mathcal{P}(U)$, with finite intersection property.

The next major result of this section is the trichotomy law for cardinal numbers. But in order to establish it, a lemma is needed. This lemma depends upon Zorn's lemma for its justification.

LEMMA: If A and B are sets then either $\exists\, B_0 \subseteq B \ni B_0 \cong A$ or $\exists\, A_0 \subseteq A \ni A_0 \cong B$.

PROOF: If $A = \emptyset$ or $B = \emptyset$ the assertion is easily verified, so suppose $A \neq \emptyset \land B \neq \emptyset$. Let

$$\mathcal{F} = \{f \mid f : A_0 \xrightarrow{\;1-1\ \text{onto}\;} B_0 \land A_0 \subseteq A \land B_0 \subseteq B\}.$$

Since $A \times B \neq \emptyset$ $\exists\,(a,b) \in A \times B$. So $A_0 = \{a\} \subseteq A$ and $B_0 = \{b\} \subseteq B$ and $f = (a,b)$ is a $1-1$ function from A_0 to B_0. Therefore $\mathcal{F} \neq \emptyset$. Now $\mathcal{F}$ is partially ordered by $\subseteq$. Let $\mathcal{C}$ be an arbitrary nonempty chain in the poset $(\mathcal{F}, \subseteq)$. If $\mathcal{C}$ were to have an upper bound in $\mathcal{F}$ then by Z.L., $\mathcal{F}$ would have a maximal element, f_0 where $f_0 : A_0 \longrightarrow B_0$ for some $A_0 \subseteq A$ and $B_0 \subseteq B$. Then after verifying that either $A_0 = A$ or $B_0 = B$ the conclusion would follow.

So there are two major things to do. a: We need to be sure the hypothesis of Z.L. is satisfied, then invoke Z.L., and b: we need to show $A_0 = A$ or $B_0 = B$.

PART a: The objective is to show the nonempty chain $\mathcal{C}$ in $\mathcal{F}$ has an upper bound in $\mathcal{F}$. To do this, we verify that $\bigcup \mathcal{C}$ i.e., $\bigcup_{f \in \mathcal{C}} f$, is an upper bound of $\mathcal{C}$ in $\mathcal{F}$. Let $f_1 \in \mathcal{C}$ then, by Exercise 10.6 (a property of generalized union), $f_1 \subseteq \bigcup_{f \in \mathcal{C}} f$. Thus $\bigcup \mathcal{C}$ will be an upper bound of $\mathcal{C}$ in $\mathcal{F}$ if it is established that $\bigcup \mathcal{C}$ is a function from a subset of A to a subset of B. This is what we do next.

SHOW: $\bigcup \mathcal{C} \in \mathcal{F}$. Let $(a,b_1), (a,b_2) \in \bigcup \mathcal{C}$ then $\exists\, f_1, f_2 \in \mathcal{C} \ni (a,b_1) \in f_1 \land (a,b_2) \in f_2$. Since $\mathcal{C}$ is a chain, either $f_1 \subseteq f_2$ or $f_2 \subseteq f_1$. Suppose, WLOG, $f_1 \subseteq f_2$ then $(a,b_1) \in f_2 \land (a,b_2) \in f_2$ hence $b_1 = b_2$. Thus $\bigcup \mathcal{C}$ is a function. Now $(a,b) \in \bigcup \mathcal{C}$ iff $\exists\, f \in \mathcal{C} \ni (a,b) \in f$. Thus $a \in \mathrm{dom}(\bigcup \mathcal{C}) \implies \exists\, f \in \mathcal{C} \land \exists\, b \in B \ni (a,b) \in f$. Since $f \in \mathcal{C}$, $f \in \mathcal{F}$, so f has some domain A_1 and range $\mathrm{ran}(f) \subseteq B$. Therefore $a \in \mathrm{dom} f = A_1 \subseteq A$. Thus $\mathrm{dom}(\bigcup \mathcal{C}) \subseteq A$ and $\mathrm{ran}(\bigcup \mathcal{C}) \subseteq B$. Consequently, $\bigcup \mathcal{C}$ is a function from the subset, say A_1, of A onto a subset, B_1 of B. If $(a_1,b), (a_2,b) \in \bigcup \mathcal{C}$ then, as done previously, since $\mathcal{C}$ is a chain, $\exists\, f \in \mathcal{C} \ni (a_1,b),(a_2,b) \in f$. But f is $1-1$ so $a_1 = a_2$, therefore $\bigcup \mathcal{C}$ is $1-1$. So we now know $\bigcup \mathcal{C} \in \mathcal{F}$.

Now since $\bigcup \mathcal{C}$ has been shown to be an upper bound of $\mathcal{C}$ in $\mathcal{F}$, $\mathcal{F}$ has a maximal element, say f_0, by Zorn's lemma. And f_0 has a domain, say A_0 and a range B_0. This completes Part a.

SHOW: PART b: Either $A_0 = A$ or $B_0 = B$.

To do this, suppose on the contrary that $A_0 \neq A \wedge B_0 \neq B$ then $\exists\, a \in A - A_0$ and $\exists\, b \in B - B_0$; so $a \notin \text{dom}\,(f_0)$ and $b \notin \text{ran}\,(f_0)$. Thus $f_0 \cup \{(a, b)\}$ is a function from $A_0 \cup \{a\}$ to $B_0 \cup \{b\}$. It is easy to verify that $f_0 \cup \{(a, b)\}$ is a $1 - 1$ function from $A_0 \cup \{a\}$ onto $B_0 \cup \{b\}$. Thus $f_0 \cup \{(a, b)\} \in \mathcal{F}$. But $f_0 \subset f_0 \cup \{(a, b)\}$ contrary to the maximality of f_0. Thus $A_0 = A$ or $B_0 = B$. If $A_0 = A$ then $f_0 : A \longrightarrow B_0$ is $1 - 1$ and onto B_0, so $A \cong B_0$. The other case is analogous. So either $\exists\, B_0 \subseteq B \ni A \cong B_0$ or $\exists\, A_0 \subseteq A \ni A_0 \cong B$, as desired.

THEOREM 29.2 (The trichotomy law for cardinal numbers):
Let α, β be cardinal numbers and A, B be sets such that $\alpha = \#(A)$ and $\beta = \#(B)$. Then $\alpha < \beta$ or $\alpha = \beta$ or $\beta < \alpha$. And only one is true.

PROOF: The proof is outlined as follows, with the details left as an exercise. Let A and B be sets. Next we apply the lemma above.

If $\exists\, A_0 \subseteq A \ni A_0 \cong B$ then $\exists\, f : A_0 \longrightarrow B$ such that f is 1–1 and onto B. Then f^{-1} is a 1–1 function from B onto $A_0 \subseteq A$. Therefore $\#(B) \leq \#(A)$. If $\sim \exists\, g : B \longrightarrow A$ such that g is 1–1 and onto A then $\#(B) < \#(A)$. On the other hand, if $\exists\, g : B \longrightarrow A$ such that g is 1–1 and onto A then $\#(B) = \#(A)$. The rest is left as an exercise.

The final theorem of this section is the famous Schröder–Bernstein theorem. The proof given here depends heavily on the trichotomy law which is a consequence of the axiom of choice. However, the Schröder–Bernstein theorem is provable with far less powerful assumptions. In fact, several references in the reading list give very ingenious proofs which do not assume A.C. So the proof given here hits the problem with a sledgehammer rather than with the finesse of the more standard proofs.

THEOREM 29.3   (Schröder–Bernstein): If A and B are sets and there exists B_1 such that $B_1 \subseteq B$ and $A \cong B_1$ and there exists A_1 such that $A_1 \subseteq A$ and $A_1 \cong B$ then $A \cong B$.

PROOF: This theorem says if each of two sets is cardinally equivalent to a subset of the other, then the two sets are cardinally equivalent. This proof employs the earlier mentioned property, Exercise 26.2, that, "If $A \subseteq B$ then $\#(A) \leq \#(B)$."

If $A \cong B_1$ where $B_1 \subseteq B$ then $\#(A) = \#(B_1) \leq \#(B)$, and if $A_1 \cong B$ where $A_1 \subseteq A$ then $\#(B) = \#(A_1) \leq \#(A)$. Thus $\#(A) \leq \#(B)$ and $\#(B) \leq \#(A)$. By the trichotomy law $\#(A) = \#(B)$. Thus $A \cong B$.

Exercises:

29.1 From Part 2 of the proof of Theorem 29.1:

    a. Argue that $\{(S_0, s_0)\}$ is a function in $\mathcal{F}$.

    b. Argue that $(\mathcal{F}, \subseteq)$ is indeed a poset.

    c. Prove that F from Theorem 29.1 is a function.

    d. Prove that $\mathrm{ran}\,(F) \subseteq \bigcup_{S \in \mathcal{S}} S$.

    e. Show that if $\mathcal{C}$ is the maximal chain described in Theorem 29.1 Part 2 and F_1 is as described there, then $\mathcal{C} \cup \{F_1\}$ is a chain in $\mathcal{F}$ properly containing the chain $\mathcal{C}$.

29.2 a. Argue that the pair $(\mathcal{F}, \subseteq)$ in Example 29.1 is, indeed, a partially ordered set (poset).

    b. Prove that $A \subseteq \bigcup_{C \in \mathcal{C}} C$ in the proof in Example 29.1. In other words, establish Part c of that property.

    c. Prove that $\bigcup_{C \in \mathcal{C}} C \subseteq \mathcal{P}(U)$ in Example 29.1, i.e., Part d.

29.3 a. Verify that the lemma preceeding the trichotomy law is valid in the case $A = \emptyset$ or $B = \emptyset$.

    b. Verify that $\subseteq$ is a partial order relation on $\mathcal{F}$ in the lemma.

    c. Verify, in the lemma, that $f_0 \cup \{(a, b)\}$ is a $1-1$ function from $A_0 \cup \{a\}$ onto $B_0 \cup \{b\}$.

    d. Fill in the details left out in the last two sentences of the proof of the lemma.

29.4 Complete the proof of Theorem 29.2, i.e., the trichotomy law for cardinal numbers.

CHAPTER 30

WELL ORDERED SETS

In an earlier chapter the notion of well ordering was discussed in the more limited setting of usual $\leq$ in the number system N. This notion is now considered in relation to a more general setting of a linearly ordered set or a chain. As before $\leq$ will not necessarily mean ordinary $\leq$.

DEFINITION: Let $(A, \leq)$ be a linearly ordered set (i.e., a chain). Then A is <u>well ordered with respect to</u> $\leq$ iff for each B, if $\emptyset \neq B \subseteq A$ then B has a <u>first element</u>, (i.e., an element a such that $a \leq x$ for all $x \in B$). In this case $(A, \leq)$ is called a <u>well ordered set</u>.

EXAMPLE 30.1: a. The set N of natural numbers is well ordered with respect to usual $\leq$.

  b. Z is not well ordered with respect to usual $\leq$ because the set $B = \{-2, -4, -6, \ldots\}$ is a nonempty subset of Z having no least element.

  c. Q is not well ordered with respect to usual $\leq$ because the set $B = \{1/2, 1/4, 1/8, \ldots\}$ is a nonempty subset of Q having no least element.

  d. The set E_1 is not well ordered with respect to usual $\leq$ since Q is a nonempty subset of E_1 having no least element.

The fact that usual $\leq$ on N well orders N is important. Indeed, the principle of mathematical induction is equivalent to the statement that N is well ordered by usual $\leq$, although we won't verify this.

Perhaps the most famous theorem pertaining to well ordering is the well ordering theorem of Zermelo. This consequence of the axiom of choice sparked heated debate as mentioned in Chapter 20. The theorem is quite simply stated, but the consequences stretch intuition to the breaking point.

THEOREM 30.1 (*Zermelo*): If S is any set then there is an order relation $\leq$ on S such that $\leq$ is a well ordering of S. (In other words, for every set S, there is a well ordering of S.)

The proof will only be sketched to give a flavor of the procedure.

"PROOF:" Let S be a set and let $\mathcal{F}$ be the set of ordered pairs given by

$$\mathcal{F} = \left\{(A, \leq_A) \mid A \subseteq S \text{ and } \leq_A \text{ well orders } A\right\}.$$

Partially order $\mathcal{F}$ by $(A, \leq_A) \preceq (B, \leq_B)$ iff $A \subseteq B$ and $\leq_A \subseteq \leq_B$. The latter containment makes sense because $\leq_A$ and $\leq_B$ are relations, hence they are sets of ordered pairs and for sets, containment makes sense. Then it is an exercise to show $\preceq$ is a partial order relation on $\mathcal{F}$, hence $(\mathcal{F}, \preceq)$ is a poset. Let $\mathcal{C}$ be a chain in $\mathcal{F}$. To use Z.L., one needs to know $\mathcal{C}$ has an upper bound in $\mathcal{F}$. So let $D = \bigcup \{A \mid (A, \leq_A) \in \mathcal{C}\}$ and let $\ll = \bigcup \{\leq_A \mid (A, \leq_A) \in \mathcal{C}\}$. Although we will not do it, the proof will be complete when $\ll$ is shown to well order D.

The interested student may consult references in the reading list. Some of them have clear and somewhat elementary proofs of this important theorem (see Raymond Wilder or Rubin and Rubin).

DEFINITION: Let $(A, \leq)$ be a poset and $a \in A$ then the <u>initial segment of A determined by a</u> is given by $S_a = \{x \mid x \in A \wedge x < a\}$. [$x < a$ means $x \leq a \wedge x \neq a$]. Sometimes S_a is simply called an *initial segment*.

EXAMPLE 30.2: If $A = \{\emptyset, \{1\}, \{2\}, \{3\}, \{1, 2\}, \{1, 3\}, \{2, 3\}\}$ and the partial ordering of A is usual $\subseteq$ and $a = \{1, 3\}$ then

$$S_a = \{x \mid x \in A \wedge x < a\} = \{x \mid x \in A \wedge x \subset \{1, 3\}\} = \{\emptyset, \{1\}, \{3\}\}.$$

Notice that S_a is poset itself, but S_a may or may not be a linearly ordered set. In Example 30.2, we see that S_a is not linearly ordered; however, initial segments of initial segments are initial segments. This is the essence of the next theorem whose proof is an exercise.

THEOREM 30.2: If $(A, \leq)$ be a poset, P an initial segment of A, and Q an initial segment of P, then Q is an initial segment of A.

DEFINITION: Let $(A, \leq_A)$ and $(B, \leq_B)$ be posets.

  1. A function $f : A \longrightarrow B$ is <u>order preserving</u> iff

$$\forall x, y \in A, \left(x \leq_A y \implies f(x) \leq_B f(y)\right).$$

(Assuming there is no confusion, the subscripts on $\leq$ will usually be dropped.)

Figure 30.1

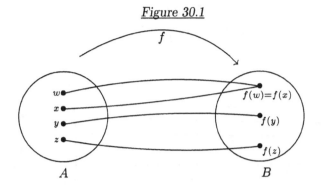

Figure 30.2

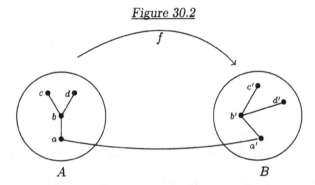

2. A function $f : A \longrightarrow B$ is an <u>order isomorphism</u> iff f is $1 - 1$, and $\forall x, y \in A (x \le y \implies f(x) \le f(y))$. If $f : A \longrightarrow B$ is an order isomorphism *onto* B then A and B are said to be <u>order isomorphic</u>. In this case the notation is $A \approx B$.

EXAMPLE 30.3:

a. In the lattice diagram of *Figure 30.1*, $f : A \longrightarrow B$ is a function, but is not $1 - 1$. Think of "larger" as being "higher" in the diagram. Since $a_1 \le a_2 \implies f(a_1) \le f(a_2) \; \forall a_1, a_2 \in S$, f is order preserving. That is, whenever a_1 is below a_2 in A then $f(a_1)$ is below $f(a_2)$ in B on the right. (Notice that $A = \{w, x, y, z\}$ and $D = \{f(w), f(y), f(z)\}$ are posets with the order as described in the discussion surrounding *Figure 27.1*, i.e., "larger" elements are "higher" in the diagram.)

b. Consider the posets $A = \{a, b, c, d\}$ and $B = \{a', b', c', d'\}$ with order as indicated in the lattice diagram in *Figure 30.2* ("larger" is "higher"). If $f(x) = x'$ for each $x \in A$, e.g., $f(a) = a'$, then f is an order isomorphism of A onto B. That is $A \approx B$.

Figure 30.3

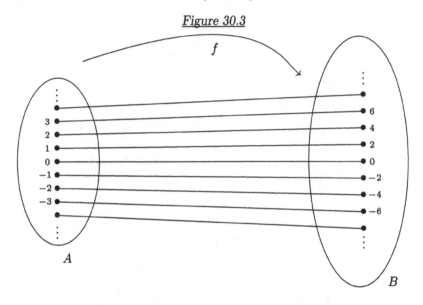

c. If $A = Z$, $B = \{2n \mid n \in Z\} = V$, and $\forall a \in Z, f(a) = 2a$ then f is an order isomorphism of Z onto V. A diagram for this order isomorphism of infinite sets is illustrated in *Figure 30.3*.

DEFINITION: Let A and B be well ordered sets. Then A and B have the <u>same ordinality</u> (denoted $A \approx B$) iff A and B are order isomorphic.

This definition is the analogue of the relation, "have the same cardinal number," of Chapter 18. An alternate terminology for $A \approx B$ is A and B are <u>ordinally equivalent</u>. Thus, in Part c of Example 30.3, the set of all integers is ordinally equivalent to the set of all even integers. The effect that $\approx$ has on A and B is that A and B are indistinguishable except for the symbols used for the elements and the symbol for the order relation.

EXAMPLE 30.4:   Let $A = \{2, 6, 12, 48\}$ be partially ordered by $\mid$ and $B = \{\{a\}, \{a, b\}, \{a, b, c\}, \{a, b, c, d\}\}$ be partially ordered by $\subseteq$. Define

$$f = \{(2, \{a\}), (6, \{a, b\}), (12, \{a, b, c\}), (48, \{a, b, c, d\})\}.$$

Then f is a $1 - 1$ function from A onto B and f is order isomorphic because if $x, y \in A$ then $x \mid y$ iff $f(x) \subseteq f(y)$. Therefore $A \approx B$.

DEFINITION: Let A and B be well ordered sets.   Then A has <u>less ordinality</u> than B (denoted $A \prec B$) iff A is order isomorphic to an initial segment of B.

EXAMPLE 30.5: $\{1,2,3\} \prec \{1,2,3,4,5\}$ because first, these sets are well ordered by their usual order, and second, $\{1,2,3\} \approx S_4$ where S_4 is the initial segment $S_4 = \{x \mid x \in N \wedge x < 4\} = \{1,2,3\}$ of $\{1,2,3,4,5\}$. The function $f(x) = x \ \forall x \in \{1,2,3\}$ is the order isomorphism since $x \leq y \Longrightarrow f(x) \leq f(y) \ \forall x, y \in \{1,2,3\}$.

We know that $(N, \leq)$ is a well ordered set. If a set B is ordinally equivalent to N then perhaps the easiest way to show that is to use the following theorem. Of course, this may be done with any pair of order isomorphic sets.

THEOREM 30.3: Suppose $(A, \leq_A)$ and $(B, \leq_B)$ are partially ordered sets. If $f : A \longrightarrow B$ is an order isomorphism of A onto B and A is well ordered with respect to $\leq_A$ then B is well ordered with respect to $\leq_B$.

PROOF: Suppose $(A, \leq_A)$ and $(B, \leq_B)$ are partially ordered sets and $f : A \longrightarrow B$ is an order isomorphism of A onto B and A is well ordered. To show B is well ordered, suppose $\emptyset \neq S \subseteq B$. Then $\emptyset \neq f^{-1}[S] \subseteq A$. Thus $f^{-1}[S]$ has a least element, say a. Then $f(a) \in S$ and $f(a)$ is the least element of S because if $s \in S$ then $a \leq_A f^{-1}(s) \in f^{-1}[S]$. Thus $f(a) \leq_B s$.

Exercises:

30.1 Let $A = N$ and suppose A is partially ordered by $\mid$, i.e., if $x, y \in N$ then x is related to y iff $x \mid y$. Find the initial segment of A determined by 24.

30.2 Verify that $\preceq$ defined in the proof of Theorem 30.1 is a partial order relation.

30.3 Thinking of functions $f : E_1 \longrightarrow E_1$, with both the domain and codomain ordered by usual $\leq$, which of the following functions are order preserving functions?

    a. $f(x) = x^3$.

    b. $f(x) = \sin x$.

    c. $f(x) = e^x$.

    d. Is E_1 order isomorphic to E_1^+?

30.4 Prove Theorem 30.2.

30.5 Referring to Example 30.4 let $A' = \{2, 6, 12\}$ be partially ordered by $\mid$ and $B = \{\{a\}, \{a, b\}, \{a, b, c\}, \{a, b, c, d\}\}$ be partially ordered by $\subseteq$. Determine if $A' \prec B$. Then verify your conjecture.

30.6 Let N be partially ordered by usual $\leq$ and $B = \{10^n \mid n \in N\}$ be partially ordered by $\mid$.

a. Verify $N \approx B$.

b. We know N is well ordered by usual $\leq$. Is B well ordered with respect to $\mid$?

30.7 Use the fact that N is well ordered with respect to $\leq$ to show $\{1, 2, 4, 8, 16, \ldots\}$ is well ordered with respect to $\mid$.

30.8 Although the student does not have to prove Theorem 30.1, outline the things that are to be done to complete the proof that $\ll$ well orders D.

30.9 Describe well orderings of the set Z of integers and the set Q^+ of positive rational numbers.

READING LIST

Bell, E.T., *Men of Mathematics*, Simon & Schuster, New York, 1937.

Birkhoff, G.D. and Mac Lane, S., *A Survey of Modern Algebra*, 4th ed., MacMillan, New York, 1977.

Cohen, P.J., *Set Theory and the Continuum Hypothesis*, W.A. Benjamin, New York, 1966.

Dunham, W.W., *Journey Through Genius*, John Wiley & Sons, New York, 1990.

Eves, H.W., *Great Moments in Mathematics (After 1650)*, Mathematical Association of America Press, 1981.

Gödel, K., *The Consistency of the Continuum Hypothesis*, Princeton University Press, Princeton, NJ, 1940.

Landau, E.G.H., *Grundlagen der Analysis*, 4th ed., Chelsea Publishing, New York, 1957.

Morash, R.P., *Bridge to Abstract Mathematics*, Random House/Birkhäuser, New York, 1987.

Pinter, C.C., *Set Theory*, Addison–Wesley, Reading, MA, 1971.

Rubin, H. and Rubin, J.E., *Equivalents of the Axiom of Choice*, North–Holland, Amsterdam, Holland, 1985.

Smith, D., Eggen, M., and St. Andre, R., *A Transition to Advanced Mathematics*, Brooks/Cole, Monterey, CA, 1986.

Wilder, R.L., *Introduction to the Foundations of Mathematics*, Krieger, Huntington, NY, 1980.

HINTS AND SOLUTIONS TO SELECTED EXERCISES

Chapter 1, p. 12:

1.1   a. This is ambiguous, not a statement.
     b. This is a (false) statement.
     c. This is not a statement.
     d. This is a (false) statement.
     e. This is a (true) statement.
     f. This is not a statement; it's paradoxical.
     g. This is a statement, but its truth value may be hard to decide.

1.3   e.

P	Q	S	$P \to Q$	$\sim(\sim Q \vee S)$	$(P \to Q) \to \sim(\sim Q \vee S)$
T	T	T	T	F	F
T	T	F	T	T	T
T	F	T	F	F	T
T	F	F	F	F	T
F	T	T	T	F	F
F	T	F	T	T	T
F	F	T	T	F	F
F	F	F	T	F	F

1.5   a. $\sim P \vee Q$.
     b. $\sim(P \wedge Q) \wedge R \longrightarrow Q \vee R$.
     c. All parentheses are necessary.
     d. $P \wedge R \longrightarrow \sim Q$.

1.7   a. The contrapositive is: "If F is not an integral domain then F is not a field." The converse is: "If F is an integral domain then F is a field."
     d. The contrapositive is: "If S does not have an upper bound then S is not bounded." The converse is: "If S has an upper bound then S is bounded."
     e. The contrapositive is: "If n is prime then n is not an even integer greater than 2." The converse is: "If n is not prime then n is an even integer greater than 2."
     f. The contrapositive is: "If $\sim(|f(x) - L| < \epsilon)$ then $\sim(0 < |x - a| < \delta)$," which is, "If $|f(x) - L| \geq \epsilon$ then $|x - a| \geq \delta$ or $x = a$." The converse is: "If $|f(x) - L| < \epsilon$ then $0 < |x - a| < \delta$."

1.9   b. "P is necessary and sufficient for Q."
     d. "P is necessary for Q," or, "Q is sufficient for P."
     e. "P is necessary and sufficient for Q."

Chapter 2, p. 25:

2.1 Tautologies 8a, 22 are verified below by abbreviated truth tables.

$(P$	$\longrightarrow$	$Q)$	$\equiv$	$(\sim Q)$	$\longrightarrow$	$\sim P$
T	T	T	T	F	T	F
T	F	F	T	T	F	F
F	T	T	T	F	T	T
F	T	F	T	T	T	T
1	4	2	6	3	5	2

$\sim Q$	$\wedge$	$(P$	$\longrightarrow$	$Q)$	$\Longrightarrow$	$\sim P$
F	F	T	T	T	T	F
T	F	T	F	F	T	F
F	F	F	T	T	T	T
T	T	F	T	F	T	T
1	4	1	3	2	5	2

2.9 a. False, b. True, c. True, d. True.

2.11 b. The form of the argument is: "If $(P \longrightarrow Q) \wedge (\sim P)$ then $\sim Q$" which is denying the antecedent. It is invalid.

c. The form is "If $(P \longrightarrow Q) \wedge (\sim Q)$ then $\sim P$," which is valid. See rule 22.

2.13 $P \wedge Q \wedge R \Longleftrightarrow P \wedge (Q \wedge R) \Longleftrightarrow P \wedge (R \wedge Q) \Longleftrightarrow P \wedge R \wedge Q.$

2.15 Let P be, "T is a topological subspace of the reals," Q be, "T is closed," R be, "T is bounded," and S be, "T is compact." The form is $P \wedge Q \wedge R \Longrightarrow S.$

2.17 a. If a^{-1} is not an element of E_1 then either a is not an element of E_1 or $a = 0$.

b. The negation of Part a is $\sim [a^{-1} \notin E_1 \Longrightarrow a \notin E_1 \vee a = 0]$. That is $a^{-1} \notin E_1 \wedge [a \in E_1 \wedge a \neq 0]$.

2.18 e. $\sim (P \wedge \sim Q)$ is true if P is __True__ and Q is __True__ .

2.19 Proof:

1.	$J \Longrightarrow N$	Rule P
2.	$\sim J \Longrightarrow D$	Rule P
3.	$D \Longrightarrow \sim A$	Rule P
4.	$\sim J \Longrightarrow \sim A$	2,3 Transitivity
5.	$A \Longrightarrow J$	4 D.N. and contrapositive
6.	$A \Longrightarrow N$	5, 1 Transitivity
7.	$\sim A \vee N$	Step 6 and Rule 13

2.21 Rule 21b.

Chapter 3, p. 39:

3.3 $\forall x[x$ is an integer $\wedge\ x \geq 0 \Longrightarrow x^2 \geq x]$.

3.5 $\exists x$ such that x in Z and x^3 is even.

3.6 $\forall x[x$ is real $\Longrightarrow x^2 + 6x + 10 > 0]$.

3.7 $\sim\exists x[x$ is real $\wedge\ x^2 + 1 = 0]$.

3.9 Let $Q(p)$ denote, "p is prime." The given statement becomes
$(Q(p) \wedge p \mid a \cdot b) \Longrightarrow (p \mid a \vee p \mid b)$.

3.10 $\forall x[x$ is a student $\wedge\ x$ is clever] says that, "Every person x is a student AND is clever." So every person is clever. This is not the same as, "Every person who is a student is a clever person."

3.12 b. $[\forall x\ \exists\ y, (x+y=2)]$ is true, while $[\exists y\ \forall x, (x+y=2)]$ is false.

3.13 b. $\forall x\big[\exists y[S(x) \wedge R(y) \longrightarrow T(x)] \longrightarrow T(y)\big]$.

3.14 c. False, choose $x = -4$.

3.15 a. $\exists x \in Z, [x^2 > 0]$; b. $\forall x \in Z, [x^2 > 0]$; c. $\exists x \in Z[x < 3 \wedge (x^2 \geq 0) \wedge (x \neq 0)]$; f. $(\forall x \in R)[x \leq 0 \vee 3x^3 - 5x^2 + 3x - 5 \neq 0]$.

Chapter 4, p. 49:

4.1 Proof:

1. $M \longrightarrow J$	Rule P
2. $J \longrightarrow \sim H$	Rule P
3. $\sim H \longrightarrow \sim J$	Rule P
4. $H \longrightarrow M$	Rule P
5. $H \longrightarrow \sim J$	1,2,3,4 Transitivity
6. J	Rule P
7. $\sim H$	5,6 & D.N. and Rule 22

4.2 $\sim(P \wedge P) \overset{\text{DeM}}{\Longrightarrow} \sim P \vee \sim P \overset{2.8c}{\Longrightarrow} \sim P$. By Transitivity, $\sim(P \wedge P) \Longrightarrow \sim P$. By the contrapositive, $P \longrightarrow (P \wedge P)$.

4.5 Invalid.

4.6 Valid.

4.8 Proof:

1. $\forall x[A(x) \Longrightarrow B(x)]$	Rule P.
2. $\exists x[P(x) \wedge A(x)]$	Rule P.
3. $P(y) \wedge A(y)$	Rule E.S.
4. $A(y) \longrightarrow B(y)$	Rule U.S.
5. $A(y)$	3, Simplification
6. $P(y)$	3, Simplification
7. $B(y)$	4, 5 Detachment
8. $P(y) \wedge B(y)$	6, 7, Conjunction
9. $\exists x[P(x) \wedge B(x)]$	8, Rule E.G.

Rule E.S. must appear before Rule U.S. or else E.S. is violated.

4.11 Existence: Now $f(x) = x^3 + x^2 + 5x - 15$ is continuous on any interval. And $f(0) = -15 < 0$ and $f(2) = 8 + 4 + 10 - 15 > 0$. By the intermediate value theorem $\exists\, c[0 < c < 2 \wedge f(c) = 0]$. Therefore $\exists\, x[x^3 + x^2 + 5x - 15 = 0]$. Uniqueness: $f'(x) = 3x^2 + 2x + 5 = 3(x^2 + \frac{2}{3}x) + 5 = 3(x^2 + \frac{2}{3}x + \frac{1}{9}) + 5 - \frac{3}{9} = 3(x + \frac{1}{3})^2 + \frac{42}{9} > 0$, so $f'(x) > 0$ for all x. If there were also a $d \neq c$ such that $f(d) = 0$ then by Rolle's theorem there would be a number e such that e is between c and d and $f'(e) = 0$. But $f'(x) > 0$ for all x, so this is a contradiction. Hence there is at most one value of x making $f(x) = 0$. Combining the two parts, $\exists!\, x \in E_1, (f(x) = 0)$.

4.13 At step 3, y was freed by the use of Rule E.S. So at step 5, when dequantifying $\exists\, x(B(x))$ at step 4, we are prevented by Rule E.S. from using y again. So we must "free" a different variable.

4.15 Using Rules 18a, E.S., and E.G the given proposition is valid.

Chapter 5, p. 66:

5.1 b. Solution: Let P be, "p is a prime," Q be, "p is a factor of ab," R be, "p is a factor of a," and S be, "p is a factor of b." Then the argument becomes $P \wedge Q \Longrightarrow R \vee S$.

e. Solution: Let P_1, P_2, and P_3 be the three premises, and Q_1 be, "$f'(c) = 0$," and Q_2 be, "$f'(c)$ does not exist." Then the argument has the form: $P_1 \wedge P_2 \wedge P_3 \Longrightarrow Q_1 \vee Q_2$.

5.3 a. Suppose $f'(x)$ exists at x_0 then show f is continuous at x_0.

c. Suppose a is between $|x|$ and 0. Either $x > a$ or not. If $x > a$ the assertion is valid. If $x \not> a$ then show $x < -a$.

e. Suppose all premises. A strategy: observe that $f'(c)$ exists or $f'(c)$ does not exist. In the former case show $f'(c) = 0$.

5.6 Suppose x is real and $0 < x$ and $x < 1$. Then by Assumed Property C, $x \cdot x < 1 \cdot x$ which says $x^2 < x$.

5.11 False. If $a = 0$, $b = 1$, $c = 2$, then $ab = 0 \cdot 1 = 0 = 0 \cdot 2 = ac$, but $b \neq c$. Thus the conjecture is false.

5.13 a. If $x \geq 0$ then $|x| = x$ and $-x < 0$, so $|-x| = -(-x) = x$. Thus $|x| = |-x|$. If $x < 0$ then $|x| = -x$ and $-x > 0$. Therefore $|-x| = -x = |x|$. Thus in either case $|x| = |-x|$.

b. Suppose $|x| < b$. Either $x \geq 0$ or $x < 0$. If $x \geq 0$ then $|x| = x$. So $-b < 0 \leq x = |x| < b$. If $x < 0$ then $|x| = -x$. So $-b < 0 < -x = |x| < b$. In any case $-b < x < b$. Conversely, if $-b < x < b$ then show $|x| < b$ which may done by cases.

c. By Theorem 5.9, $|a| - |b| \leq |a - b| \ \forall a, b$. By analogy, $|b| - |a| \leq |b - a|$. Thus $|a| - |b| \geq -|b - a| = -|a - b|$, by 5.13a. Combining the inequalities we get $-|a-b| \le |a|-|b| \le |a-b|$. By Exercise 5.13b, $||a| - |b|| \le |a - b|$.

Chapter 6, p. 77:

6.2 Suppose $x < 5$. If $x \le 0$ then either $x^2 < 25$ or $x \le 0$. If $x \not\le 0$ then $x \ne 0$ and $x \not< 0$. So, by the trichotomy law, $x > 0$. Now we know $0 < x < 5$, so $x^2 = x \cdot x < 5 \cdot x < 5 \cdot 5 = 25$.

6.5 Use the definition of $|x|$ given in Chapter 5.

6.7 As cited, the assertion is false. Choose $x = 1$ and $y = 2$.

6.8 It is true. Proof: Let $x, y \in Q$ then $\frac{1}{2}x, \frac{1}{2}y \in Q$. If $x < y$ then $\frac{1}{2}x < \frac{1}{2}y$. Also $x = 1x = (\frac{1}{2} + \frac{1}{2})x = \frac{1}{2}x + \frac{1}{2}x$. So, by "Assumed Properties,"

$$x = \tfrac{1}{2}x + \tfrac{1}{2}x < \tfrac{1}{2}x + \tfrac{1}{2}y < \tfrac{1}{2}y + \tfrac{1}{2}y = y.$$

Let $z = \frac{1}{2}x + \frac{1}{2}y$ then $z \in Q$ and $x < z < y$. Since the x and y with $x < y$ were arbitrarily selected in Q and we showed $\exists z \in Q(x < z < y)$ then $(\forall x \in Q)[(\forall y \in Q)[x < y \implies (\exists z \in Q)[x < z < y]]]$.

6.9 Hint: Try $f(x) = x^{-\frac{2}{3}}$.

Chapter 7 p. 87:

7.1 $t_1 = 1$, $t_2 = 1$, $t_3 = t_2 + t_1 = 1 + 1 = 2$, $t_4 = t_3 + t_2 = 2 + 1 = 3$, $t_5 = t_4 + t_3 = 3 + 2 = 5$, $t_6 = t_5 + t_4 = 5 + 3 = 8$.

7.7 The conjecture should be: $B_n = \sum_{i=1}^{n} \frac{1}{i(i+1)} = \frac{n}{n+1}$ $\forall n \in N$.

7.9 Prove: $1 \cdot 2 + 2 \cdot 3 + 3 \cdot 4 + \cdots + n(n+1) = \frac{2}{3}(n+1)(n+2)$ $\forall n \in N$.
Proof: Let $P(n)$ be $1 \cdot 2 + 2 \cdot 3 + 3 \cdot 4 + \cdots + n(n+1) = \frac{n}{3}(n+1)(n+2)$.
Case 1: $(n = 1)$. The left side of $P(1)$ is $1 \cdot 2$ and the right side of $P(1)$ is $\frac{1}{3}(1+1)(1+2) = 2$. These are equal, so $P(1)$ is true.
Case 2: If $\exists k, P(k)$ is true then $1 \cdot 2 + 2 \cdot 3 + \cdots + k(k+1) = \frac{k}{3}(k+1)(k+2)$. The objective is to show $P(k+1)$ is true.

$$1 \cdot 2 + 2 \cdot 3 + 3 \cdot 4 + \cdots + k(k+1) + (k+1)(k+2) = \tfrac{k}{3}(k+1)(k+2) + (k+1)(k+2)$$
$$= (k+1)(k+2)\left[\tfrac{k}{3} + 1\right]$$
$$= (k+1)(k+2)\left[\tfrac{k+3}{3}\right]$$
$$= \tfrac{(k+1)}{3}\left[(k+1)+1\right]\left[(k+1)+2\right].$$

The first and last expressions in the equations above say exactly what $P(k+1)$ requires, so $P(n)$ is true for all $n \in N$.

7.11 The conjecture should be $B_n = (n+1)! - 1$.

7.15 Hint: Use $(1+x)^k(1+x) \ge (1+kx)(1+x) = 1 + (k+1)x + kx^2$.

7.17 Hint: $\sum_{i=1}^{k+1} i^3 = \sum_{i=1}^{k} i^3 + (k+1)^3 = \left[\frac{k(k+1)}{2}\right]^2 + (k+1)^3$

$$= (k+1)^2 \left[\frac{k^2}{2^2} + (k+1)\right].$$

7.21 Hint: Try S_k when $k = 2$.

Chapter 8, p. 104:

8.1 They are equivalent by the generalized commutative law for $\vee$.

8.4 c. $C = \{\frac{1}{2}, \frac{2}{3}, \frac{3}{4}, \ldots\} = \{\frac{n}{n+1} \mid n \in N\}$.

 e. $E = \{x \mid x \text{ is real and } 0 < x^2 < 1\}$.

 f. $F = \{0, 3, 8, 15, 24, \ldots\} = \{n^2 - 1 \mid n \in N\}$.

8.5 b. $B = \{y \mid y \in E_1 \wedge (y-6)(y-1) = 0\} = \{1, 6\}$.

 d. $\{z \mid z \in Z \wedge z^2 - 2 = 0\} = \emptyset$.

 h. $\{x \mid x \in Z \wedge x^2 - 3x \text{ is divisible by } 2\} = Z$. Do by cases.

8.6 $\emptyset \neq \{\emptyset\}$ because $\emptyset$ has no elements, but $\{\emptyset\}$ has the <u>set</u> $\emptyset$ as an element in it. Then also $\{\emptyset\} \neq \{\{\emptyset\}\}$.

8.7 a. $\sim(A \subseteq B) \equiv \sim \forall x[x \in A \Longrightarrow x \in B] \equiv \cdots \equiv \exists x[x \in A \wedge x \notin B]$.

8.8 a. Yes; b. No; c. No; d. No; e. No; h. No; i. Yes; j. No; k. Yes.

8.9 0, 1, 2, and 1.

8.11 Yes.

8.13 Use Tautology 25 in *Table 2.5*.

8.16 $\sim[A \subset B]$ iff $\sim[(A \subseteq B) \wedge (A \neq B)]$ iff $[(A \not\subseteq B) \vee (A = B)]$.

8.17 No. $A \subset \emptyset \Longrightarrow \exists x[x \in \emptyset \wedge x \notin A]$ which says $\emptyset$ is not empty.

8.18 $\sim(\forall x)[x \in A \Longrightarrow x \in B] \equiv \exists x[\sim(x \in A \Longrightarrow x \in B)]$

$$\equiv \exists[\sim(x \notin A \vee x \in B)]$$

$$\equiv \exists x[x \in A \wedge x \notin B].$$

8.19 $A \neq B$ iff $(\exists x \ni x \in A \wedge x \notin B) \vee (\exists x \ni x \in B \wedge x \notin A)$.

8.20 c. $C = \{x \in U \mid x \in T \wedge x \notin S\} = \{d, e\}$.

 d. $D = \{x \in U \mid x \in S \Longrightarrow x \in T\} = \{b, c, d, e, \ldots, z\} = U - \{a\}$.

8.21 b. $[a \in \{b, c\}] \equiv [(a = b) \vee (a = c)]$.

 d. $[\{a\} \in \{b, c\}] \equiv [b = \{a\} \vee c = \{a\}]$.

 g. $[\{b, c\} \in \{a\}] \equiv [a = \{b, c\}]$.

8.25 $\mathcal{P}(S)$ has 2^n elements.

Chapter 9, p. 115, 120:

9.1 b. True; c. False; f. False; g. False; h. False; j. False; m. False; n. False; o. True, but equality does not hold.

9.7 Hint: Use Tautology 20.

9.9 Hint: Use Exercise 2.8.

9.11   b. Hint: Use law of addition. Then $y \in A \Longrightarrow y \in A \lor y \in B$.

9.13   Hint: Use Tautology 18.

9.17   a. Hint: Use Tautology 20.

9.19   Hint: Use Tautology 15.

9.23   DeMorgan's law.

9.27   a. Syntax problems. $\forall$ and $\exists$ apply to predicates, not sets, e.g., if S is a set then $\forall x, S$ is nonsense. Disjoint means $C \cap D = \emptyset$.

      c. Ambiguous at best. Other major problems.

Chapter 10, p. 129:

10.1   a. $\bigcup S = \{x \mid \exists \lambda [\lambda \in \Lambda \land x \in A_\lambda]\} = \{a, b, c, d, e, f, g\}$.

      b. $\bigcap S = \{x \mid \forall \lambda [\lambda \in \Lambda \Longrightarrow x \in A_\lambda]\} = \{c\}$.

10.3   a. $\bigcup_{\lambda \in \Lambda} A_\lambda = \cdots = (-3, 3)$ and b. $\bigcap_{\lambda \in \Lambda} A_\lambda = \cdots = (-1, 1)$.

      c. $\bigcup_{\lambda \in \Lambda} A_\lambda = \cdots = E_1$ and d. $\bigcap_{\lambda \in \Lambda} A_\lambda = \cdots = (-1, 1)$.

      e. $\bigcup_{\lambda \in \Lambda} A_\lambda = \cdots = E_1$ and f. $\bigcap_{\lambda \in \Lambda} A_\lambda = \cdots = \{0\}$.

      g. $\bigcup_{\lambda \in \Lambda} A_\lambda = \cdots = (-c, c)$ and h. $\bigcap_{\lambda \in \Lambda} A_\lambda = \cdots = (-a, a)$.

10.5   Proof: Let A be a set, C a set of sets. If $\forall C[C \in C \Longrightarrow A \subseteq C]$. <u>Show</u>: $A \subseteq \bigcap_{C \in C} C$, i.e., show $A \subseteq \{x \mid \forall C[C \in C \Longrightarrow x \in C]\}$. If $y \in A$ then we want to show: $\forall C[C \in C \Longrightarrow y \in C]$. To do the latter, let $C_1 \in C$. From a premise, $C_1 \in C \Longrightarrow A \subseteq C_1$. So, $A \subseteq C_1$. Since $y \in A$ and $y \in A \Longrightarrow y \in C_1$, then $y \in C_1$ (why?). Hence $C_1 \in C \Longrightarrow y \in C_1$. This is exactly what we need to establish $\forall C[C \in C \Longrightarrow y \in C]$. Thus $y \in \bigcup_{C \in C} C$. Consequently, $A \subseteq \bigcap_{C \in C} C$. This generalizes Theorem 9.14.

10.7 U.

10.9   c. Distributive laws of $\land$ over $\lor$ and $\lor$ over $\land$.

Chapter 11, p. 138:

11.1 $\{(a,a),(a,b),(a,c),(a,d),(b,a),(b,b),(b,c),(b,d),(c,a),(c,b),(c,c),(c,d)\}$ is $A \times B$. $B \times A$ is similar.

11.2   NO.

11.3   a. The graph is the set of points in $E_1 \times E_1$ that are above the given parabola and below the given line.

11.5   a. No.   b. No, but explain why.

11.9   Hint: $A \neq \emptyset \land B \neq \emptyset \Longrightarrow \exists a \in A \land \exists b \in B$ so $(a, b) \in A \times B$.

Chapter 12, p. 147:

12.1 r is a p.o.r. on A iff $[a \in A \Longrightarrow (a, a) \in r] \land [(a, b) \in r \Longrightarrow (b, a) \in r] \land [(a, b) \in r \land (b, c) \in r \Longrightarrow (a, c) \in r]$.

12.3 f. $\text{dom}(r) = \{x \mid x, y \in E_1 \wedge y = \ln x\} = E_1^+$. And $\text{ran}(r) = \cdots = E_1$. $r^{-1} = \{(y, x) \mid (x, y) \in r\} = \cdots = \{(y, x) \mid x, y \in E_1 \wedge x = e^y\}$.

g. $\text{dom}(r) = \{x \mid (x, y) \in r\} = \{x \mid y = |x| \wedge [x, y \in Z]\} = Z$
$\text{ran}(r) = \{y \mid (x, y) \in r\} = \{y \mid x, y \in Z \wedge y = |x|\} = \{x \mid x \in Z \wedge x \geq 0\}$.

$$r^{-1} = \{(x, y) \mid (y, x) \in r\} = \{(x, y) \mid x, y \in Z \wedge x = |y|\}$$
$$= \{(x, y) \mid x \in Z^+ \wedge (y = x \vee y = -x)\}.$$

h. $A = Z$ and $r = \{(x, y) \mid x, y \in Z \wedge x^2 + y^2 = 25\}$. Then

$$r = \{(0, 5), (0, -5), (5, 0), (-5, 0), (3, 4), (4, 3), (3, -4),$$
$$(-4, 3), (4, -3), (-3, 4), (-3, -4), (-4, -3)\}.$$

Now $\text{dom}(r) = \{-5, -4, -3, 0, 3, 4, 5\} = \text{ran}(r)$, and
$r^{-1} = \{(x, y) \mid (y, x) \in r\} = r$.

j. $A = E_1$ and $r = \{(x, y) \mid x, y \in E_1 \wedge x^2 + y^2 = 25\}$.
$\text{dom}(r) = \{x \mid x^2 + y^2 = 25\} = [-5, 5] = \text{ran } r$.

$$r^{-1} = \{(x, y) \mid (y, x) \in r\} = \{(x, y) \mid y^2 + x^2 = 25\}$$
$$= \{(x, y) \mid x^2 + y^2 = 25\} = \{(x, y) \mid (x, y) \in r\} = r.$$

12.4 d. It is reflexive and symmetric, but not transitive.

f. The relation is reflexive, symmetric, and transitive.

h. The relation is an equivalence relation.

12.9 $\text{dom}(r) = E_1^+$ and $\text{ran}(r) = E_1$. r is from E_1^+.

12.15 a. $I_A \cup \{(a, b), (b, c)\}$ on $A = \{a, b, c\}$; b. $\neq$ on N; c. $<$ on N; d. "Is not disjoint from," for the set $\mathcal{P}(\{a, b, c\}) - \emptyset$; e. $\leq$ on N; f. $r = \emptyset$ on N; g. $=$ on E_1; h. $r = \{(a, b)\}$ on $\{a, b\}$.

12.27 a. Since R could be the empty set and still be symmetric, then there would be NO pair (x, y) that could belong to R.

b. R was given to be symmetric, not reflexive.

c. There are many things wrong. First, what is (lower case) r? There was no prior mention in the problem or in the solution. If it were a "typo" and supposed to be R, then "R is symmetric," does not say $(x, y) \wedge (y, x) \in R$. Instead, "$R$ is symmetric," says, $\forall x, y[(x, y) \in R \implies (y, x) \in R]$. It does NOT say $\wedge$. If the writer is trying to define a new r by $r = \{(x, y) \mid \forall x, y \in R \wedge (y, x) \in r\}$, that objective is not clear. Finally, if the writer had intended the statement to be $R = \{(x, y) \mid \forall(x, y) \in R \wedge (y, x) \in R\}$, then he or she would be trying to define R in terms of itself. This is circular.

d. A symmetric relation R does not have to have anything in it since it may be empty. Also, the connective is $\longrightarrow$ not $\wedge$.

Chapter 13, p. 157:

13.1   e. Reflexive: Let $x \in A$ then, since A is the power set of $\{a, b, c\}$, x is a subset of $\{a, b, c, \}$. So x and x have the same number of elements, hence x is r related to x. This says r is <u>reflexive.</u> Let $x, y \in A$. If x and y have the same number of elements then y and x have the same number of elements. This says r is <u>symmetric.</u> Let $x, y, z \in A$. If x and y have the same number of elements and y and z have the same number of elements then x and z have the same number of elements, so r is <u>transitive.</u>

By Theorem 13.1 the partition $\mathcal{P}$ induced by r is
$$\{[x] \mid x \in A\} = \{[\emptyset], [\{a\}], [\{a, b\}], [\{a, b, c\}]\},$$
where the cells are $[\emptyset] = \{\emptyset\}$; $[\{a\}] = \{\{a\}, \{b\}, \{c\}\}$;
$[\{a, b\}] = \{\{a, b\}, \{a, c\}, \{b, c\}\}$ and $[\{a, b, c\}] = \{\{a, b, c\}\}$.

13.2   a. After verifying $\mathcal{P}$ is a partition, etc. the r induced by $\mathcal{P}$ is found by using Equation (1). The result is
$$r = \{(a, a), (b, b), (c, c), (d, d), (a, b), (b, a)\}.$$

13.3   The equivalence classes are $[1] = \{1, 2\}$ and $[3] = \{3\}$. So $\mathcal{P} = \{[1], [3]\} = \{\{1, 2\}, \{3\}\}$. The equivalence relation induced by $\mathcal{P}$
is $r = \{(x, y) \mid \exists P (P \in \mathcal{P} \wedge x \in P \wedge y \in P)\}$
$= \{(1, 1), (2, 2), (3, 3), (1, 2), (2, 1)\}.$

13.5   a. Prove that *"congruence modulo n"* is an equivalence relation on the set Z of integers.
Proof: Let $n \in N$ be fixed and $x \in Z$ then $x - x = 0 = 0 \cdot n$ so $x \equiv x$ mod n. Thus it is reflexive. Next, suppose $x, y \in Z$ and $x \equiv y$ mod n then $\exists k \in Z \ni x - y = kn$, but then $y - x = -(x - y) = -(kn) = (-k)n$ so since $-k \in Z$, then $y \equiv x$ mod n. So it is symmetric. Finally, suppose $x, y, z \in Z$ and $x \equiv y$ mod n and $y \equiv z$ mod n then $\exists k \in Z \ni x - y = kn$ and $\exists h \in Z \ni y - z = hn$ so $x - z = x - y + y - z = kn + hn = (k + h)n$. Since $k + h \in Z$ then $x \equiv z$ mod n. Thus the relation is transitive. Consequently, it is an equivalence relation.

13.7   Proof: Suppose r is an equivalence relation on A and $[y_1]$ and $[y_2]$ are equivalence classes w.r.t. r. If $x \in [y_1] \cap [y_2]$ then $x \in [y_1]$ and $x \in [y_2]$. Thus $x \, r \, y_1 \wedge x \, r \, y_2$. So $y_1 \, r \, x \wedge x \, r \, y_2$. Therefore $y_1 \, r \, y_2$. Now suppose $y \in [y_1]$ then $y \, r \, y_1$. Since r is transitive and $y_1 \, r \, y_2$, then $y \, r \, y_2$. So $y \in [y_2]$. Thus $\forall x [x \in [y_1] \implies x \in [y_2]]$, i.e., $[y_1] \subseteq [y_2]$. By analogy $[y_2] \subseteq [y_1]$, so $[y_1] = [y_2]$.

13.9   Such a relation r is reflexive and symmetric, but not transitive.

13.11   d. $f_1'(x) = f_2'(x)$.

13.13   If $xry \land yrz$ then $(\exists B_1 \in \mathcal{P} \ni x \in B_1 \land y \in B_1) \land (\exists B_2 \in \mathcal{P} \ni y \in B_2 \land z \in B_2)$. There are two such sets B_1 and B_2 because we applied Rule ES twice, which requires that different variables be freed. The argument, as written, is essentially begging the question.

Chapter 14, p. 170:

14.1   a. f is a 1–1 function from A onto B.

   b. f is a 1–1 function from Z into Z.

   c. f is a 1–1 function from E_1 onto E_1.

   d. f is a 1–1 function from E_1^+ onto E_1.

   e. f is a not a function.

14.4   No. The function $f : E_1 \longrightarrow E_1$ given by $f(x) = e^x$ is 1 – 1 because $x_1 \neq x_2 \implies f(x_1) = e^{x_1} \neq e^{x_2} = f(x_2)$, but is not onto E_1 because $\sim \exists x \in E_1 \ni f(x) = -1$. The function

$$f(x) = \begin{cases} x, & \text{if } x \geq 0; \\ x+1, & \text{if } x < 0. \end{cases}$$

is onto E_1, but is not 1–1 since $f(-\frac{1}{2}) = f(\frac{1}{2})$.

14.5   Since $f(x) = x^3$ exists $\forall x \in E_1$, its domain is E_1. Since $\sqrt[3]{y}$ exists $\forall y \in E_1$, its range is E_1. Now f is 1–1 since $x_1 \neq x_2 \implies x_1^3 \neq x_2^3$.

14.11   Proof: Suppose all hypotheses and suppose $f = g$. Then $(x, y) \in f \iff (x, y) \in g \; \forall x, y$. The first thing we show is $A \subseteq C$. So if $a \in A$ then since $\text{dom}(f) = A$, $\exists b \in B \ni (a, b) \in f$. Since $f = g$ that pair $(a, b) \in g$, so $a \in \text{dom}(g)$. And since $C = \text{dom}(g)$, $a \in C$. Thus $A \subseteq C$. By analogy, $C \subseteq A$. So $A = C$.

   Now show $f(x) = g(x) \; \forall x \in A$. Since $A = C$, $\text{dom}(f) = \text{dom}(g) = A$. Let $x \in A$ then $\exists y \in B \ni (x, y) \in f$, i.e., $\ni y = f(x)$. Since $f = g$ then $(x, y) \in g$, i.e., $y = g(x)$. Thus $f(x) = y = g(x)$. Since $x \in A$ was arbitrary then $f(x) = g(x) \; \forall x \in A$.

   Conversely, show if $f(x) = g(x) \; \forall x \in A$ then $f = g$. To do this, let $(x, y) \in f$ then $y = f(x)$. Since $f(x) = g(x)$, $y = g(x)$. But then $(x, y) \in g$. Consequently, $f \subseteq g$. The converse part, i.e., $g \subseteq f$ is analogous, so $f = g$.

14.12   a. Yes. b. No. "f is not 1–1" implies "f^{-1} is not a function." For c, it says f^{-1} IS a relation from B onto A.

14.13   Define $f((a, b)) = (b, a) \; \forall (a, b) \in A \times B$. Show f is a function, 1–1, etc.

14.14 Define $g((a,(b,c))) = ((a,b),c)$. Show g is a $1-1$ function from $A \times (B \times C)$ onto $(A \times B) \times C$.

14.16 Let $A = \{a,b,c\}$ and S as indicated. Find f_S for each S.

    a. $S = A$ $f_s = \{(a,1),(b,1),(c,1)\}$

    b. $S = \{a,b\}$ $f_s = \{(a,1),(b,1),(c,0)\}$

    c. $S = \{b\}$ $f_s = \{(a,0),(b,1),(c,0)\}$

    d. $S = \emptyset$ $f_s = \{(a,0),(b,0),(c,0)\}$

14.19 f is a function, f is 1–1, and f is onto E_1.

14.20 $\emptyset \subseteq \emptyset \times A$ and $\emptyset$ is single valued, etc.

Chapter 15, p. 177:

15.1 f and g are each $1-1$ functions from $A = \{1,2,3,4,5\}$ onto A.

    e. $f^{-1} = \{(1,1),(3,2),(2,3),(5,4),(4,5)\}$.

    f. $g^{-1} = \{(2,1),(3,2),(4,3),(1,4),(5,5)\}$.

    g. $g \circ f = \{(1,2),(2,4),(3,3),(4,5),(5,1)\}$.

    h., i. $(g \circ f)^{-1} = \{(2,1),(4,2),(3,3),(5,4),(1,5)\} = f^{-1} \circ g^{-1}$.

15.2 $\{(a,x),(b,z),(c,y),(d,z)\}$.

15.3 $f|_S = \{(x,y) \mid (x,y) \in f \wedge x \in S\} = \{(x,x^2) \mid x \in \{1,3,5\}\}$
$= \{(1,1^2),(3,3^2),(5,5^2)\} = \{(1,1),(3,9),(5,25)\}$.

15.5   b. Proof: Suppose all hypotheses. First note $(x,x) \in I_A \iff I_A(x) = x$ $\forall x \in A$ and $f \circ I_A = f$ iff $(f \circ I_A)(x) = f(x)$ $\forall x \in A$, by Theorem 15.1 and Exercise 14.10. Since

$$(f \circ I_A)(x) = f(I_A(x)) = f(x), \quad \forall x \in A,$$

then $f \circ I_A = f$.

Chapter 16, p. 184:

16.1 $f[S] = \{f(x) \mid x \in S\} = \{f(x) \mid x \in \{1,2,3,4\}\}$
           $= \{f(1),f(2),f(3),f(4)\} = \{a,b,c\}$
    $f^{-1}[T] = \{x \mid f(x) \in T\} = \{x \mid f(x) \in \{b,c,d\}\} = \{2,3,4,5\}$
    $f^{-1}[\{b\}] = \{x \mid f(x) \in \{b\}\} = \{x \mid f(x) = b\} = \{2,4,5\}$
    $f^{-1}[\{d\}] = \emptyset$

16.3 Proof: Suppose $f : A \longrightarrow B$ and $S_1 \subseteq S_2 \subseteq A$ then the objective is to show $f[S_1] \subseteq f[S_2]$. If $y \in f[S_1]$ then $\exists x \in S_1 \ni f(x) = y$. But since $S_1 \subseteq S_2$, $x \in S_2$. But then $y = f(x) \in \{f(x) \mid x \in S_2\} = f[S_2]$. Since the y was arbitrarily selected, $f[S_1] \subseteq f[S_2]$.

16.6   b. Proof: Let $f : A \longrightarrow B$ and $S_1, S_2 \in \mathcal{P}(A)$ then the objective is to show $f[S_1 \cap S_2] \subseteq f[S_1] \cap f[S_2]$. To do that suppose

$y \in f[S_1 \cap S_2]$ then $\exists\, x \in S_1 \cap S_2 \ni f(x) = y$. Since $x \in S_1$, $y = f(x) \in f[S_1]$ and since $x \in S_2$, $y = f(x) \in f[S_2]$. But then $y \in f[S_1] \cap f[S_2]$. Hence $f[S_1 \cap S_2] \subseteq f[S_1] \cap f[S_2]$.

16.9 Proof: Suppose $f : A \longrightarrow B$ and $S_1, S_2 \in \mathcal{P}(A)$ and f is 1 – 1. Since containment goes one way whether f is 1 – 1 or not, we need only show containment the other way when f is 1 – 1. So let $y \in f[S_1] \cap f[S_2]$ then $y \in f[S_1]$ and $y \in f[S_2]$. Therefore $\exists\, s_1 \in S_1 \ni y = f(s_1)$ and $\exists\, s_2 \in S_2 \ni y = f(s_2)$. Since f is 1–1, $s_1 = s_2$, etc.

16.10 The equation $f[S_1 - S_2] = f[S_1] - f[S_2]$ is false.

Chapter 17, p. 193:

17.1 + is NOT 1–1 because + takes $(2,3)$ to 5 and $(4,1)$ to 5, but $(2,3) \neq (4,1)$.

17.3 a. Subtraction is a binary operation on Z, but it is neither commutative nor associative.

   b. Subtraction is not a binary operation on N because subtraction cannot be applied to $(2,5)$ with result in N.

17.5 b. Let $x, y \in A$. If $x > y$ then $x \circ y = x$ and $y \circ x = x$, as can be seen by interchanging x and y in the definition. Thus $x \circ y = y \circ x$. If $x < y$ then $x \circ y = y$ and $y \circ x = y$, hence $x \circ y = y \circ x$. If $x = y$ then $x \circ y = x$ and $y \circ x = y$, hence $x \circ y = x = y = y \circ x$. In any case $x \circ y = y \circ x$.

   c. An identity i would have to satisfy $x \circ i = i \circ x = x$. Since $x \in N$, $x \geq 1$. So $1 \circ x = x = x \circ 1$. Thus the answer is, yes, there is an identity w.r.t. $\circ$. It is 1.

17.7 Proof: Suppose $\exists\, j \in A \ni j \circ a = a \circ j = a$ $\forall a \in A$. Now i is an element of A, so $j \circ i = i \circ j = i$. Similarly, since i was a given identity, $i \circ a = a \circ i = a$ $\forall a \in A$. But since $j \in A$, $i \circ j = j \circ i = j$. Putting the two equations together, $i = i \circ j = j$, so any element j having the defining property of the identity i is that identity. So the identity is unique.

Chapter 18, p. 200:

18.1 b. Let M be a set such that $m = \#(M)$. Since $\emptyset \cap M = \emptyset$ then

$$0 + m = \#(\emptyset) + \#(M) = \#(\emptyset \cup M) = \#(M) = m.$$

   d. If $k < n$ then $\exists\, p \in N \ni k + p = n$. Since $p \in N$ then $p = 1$ or $p > 1$. If $p = 1$ then $k + p = k + 1 = n$. If $p \neq 1$ then $\exists\, r \ni p = 1 + r$. So $k + 1 + r = n$. Thus $k + 1 < n$. So in any case $k + 1 \leq n$.

18.2 a. By the definition of $<$ in W, we must show $\exists\, p \in N \ni 3+p = 5$.

$$3 + 2 = \#(\{0,1,2\}) + \#(\{0,1\}) \overset{\text{why?}}{=} \#(\{0,1,2\}) + \#(\{3,4\})$$
$$= \#(\{0,1,2,3,4\}) = 5. \quad \text{Since } 2 \in N,\ 3 < 5.$$
$$2 \cdot 3 = \#(\{0,1\}) \cdot \#(\{0,1,2\})$$
$$= \#\big(\{(0,0),(0,1),(0,2),(1,0),(1,1),(1,2)\}\big)$$
$$= \#(\{0,1,2,3,4,5\}) = 6.$$

18.5 Proof: Suppose hypotheses. Then $\exists\, p \in N \ni m + p = n$. So $mk + pk = (m+p)k = nk$. Since $pk \in N$ by Theorem 18.3a, $mk < nk$.

Chapter 19, p. 213:

19.1 If $(p,q),(r,s),(t,u) \in D$ and $(p,q) \sim (r,s)$ and $(r,s) \sim (t,u)$ then $p + s = q + r$ and $r + u = s + t$. Then

$$(p+u) + (s+t) = (p+s) + (t+u) = (q+r) + (t+u)$$
$$= (q+t) + (r+u) = (q+t) + (s+t).$$

Then $p + u = q + t$, but then $(p,q) = (t,u)$. Thus $\sim$ is transitive.

19.5 Proof: Suppose the hypotheses. (Since $z < 0$, we cannot multiply both sides of $x < y$ by z while reversing the inequality, since that begs the question.) By Theorems 19.3a and 19.2f, if $z < 0$, then $0 = z+(-z) < 0+(-z) = -z$, so $0 < -z$. That is $-z \in Z^+$. So by Theorem 19.3c, since $x < y$, we get $x(-z) < y(-z)$. By Theorem 19.6c, $-(xz) < -(yz)$. Now $-(xz)+(xz+yz) < -(yz)+(xz+yz)$ by using Theorem 19.3a. Applying commutative and associative laws, we get $yz < xz$, which was to be proven.

19.6 Proof: Suppose $x \in Z$. Now $0 \in Z$ by Theorem 19.2d. And $0x = x0$ by Theorem 19.2b. Since we have already shown that $x0 = 0$ then $0x = x0 = 0$. Thus the proof of 19.6 is complete.

19.8 If you did this: $(-x)(-y) = (-1)x(-1)y = (-1)(-1)xy = xy$; you'd be begging the question. Instead do this.
Proof: Suppose $x, y \in Z$.

$$(-x)(-y) = (-x)(-y) + 0 = (-x)(-y) + (0y)$$
$$= (-x)(-y) + [(-x+x)(y)]$$
$$= (-x)(-y) + [(-x)y + (xy)]$$
$$= [(-x)(-y) + (-x)y] + (xy)$$
$$= [(-x)(-y+y)] + xy$$
$$= [(-x)0] + xy = 0 + xy = xy.$$

19.12 a. Since the given set is a subset of Z^+ we know that if it is nonempty, it will have a least element by the W.O.P. Now $16 = 15 + 1$, so $16 > 15$. Then by Theorem 19.3c, $16 \cdot 16 > 15 \cdot 16$, but then $\exists\, n \in Z^+ \ni n^2 > 15n$, so the given set is nonempty. By W.O.P. it has a least element.

c. The given set contains: $-2, -3, -4, -5, \ldots$

19.17 Proof: Suppose $a, b \in Z^+$ and $a \mid b$. Since $a \mid b$ then $\exists\, x \in Z \ni ax = b$. Either $x \le 0$ or $x > 0$. If $x \le 0$ then, since $a > 0$, $b = ax \le a0 = 0$ which is a contradiction, so $x > 0$. But then $x \ge 1$, so $b = ax \ge a \cdot 1 = a$.

Chapter 20, p. 221:

20.1 a. $[2,3] + [-1,5] = [2 \cdot 5 + 3 \cdot (-1), 3 \cdot 5] = [7,15]$.

b. $[7,2] + [0,1] = [7 \cdot 1 + 2 \cdot 0, 2 \cdot 1] = [7,2]$.

e. $[7,6] - [9,8] = [7,6] + [-9,8] = [7 \cdot 8 - 9 \cdot 6, 6 \cdot 8] = [2,48]$ which is positive since $2 \cdot 48 > 0$. Thus $[7,6] > [9,8]$.

20.3 Proof: Let $x \in Q$, then $\exists\, (p,q) \in Z \times Z \ni q \ne 0 \wedge x = [p,q]$. By Theorem 19.2f, since $p \in Z$, $-p \in Z$ hence $[-p,q] \in Q$. Then $[p,q] + [-p,q] = [pq + (q(-p)), qq] = [pq - pq, q^2] = [0, q^2] = [0,1]$. But $[0,1]$ is the "zero" of Q by Theorem 20.2e. Choose $y = [-p,q]$ then $x + y = 0$.

20.7 Proof: Let $x \in Q$, then $-x \in Q$ and $(-x) + x = 0$, by Theorem 20.2. By Theorem 20.2e, $-(-x) \in Q$ and $-(-x) + (-x) = 0$. The objective is to show $(-(-x)) = x$.

$$-(-x) = -(-x) + 0 = -(-x) + [-x + x]$$
$$= [-(-x) + (-x)] + x = 0 + x = x.$$

20.10 Proof: This is analogous to the proof in Exercise 19.5.

20.11 Since $0 < 2 < 3$, then from Theorem 20.5f, $0 < 3^{-1} < 2^{-1}$.

20.13 b. Proof: If $x, y, z \in Q$, $x \ne 0$ and $xy = xz$ then x^{-1} exists and

$$y = 1 \cdot y = (x^{-1}x)y = x^{-1}(xy) = x^{-1}(xz) = (x^{-1}x)z = 1 \cdot z = z.$$

Alternately, if $x, y, z \in Q$, $x \ne 0$ and $xy = xz$ then $x(y + (-z)) = xy - xz = 0$. Since $x \ne 0$, x^{-1} exists and $x^{-1}x = 1$. Thus $y - z = 1 \cdot (y - z) = x^{-1}(x(y - z)) = x^{-1}0 = 0$. But then $y = z$.

Chapter 21, p. 226:

21.1 Suppose S has a least upper bound $m_1 \in E_1$ then $s \le m_1$ $\forall s \in S$. Also $\forall u$ (if u is an upper bound of S then $m_1 \le u$). If $\exists m_2 \in E_1$ such that m_2 is a least upper bound, then m_2 is an upper bound of S, so $m_1 \le m_2$. Reversing the roles, $m_2 \le m_1$. Then $m_1 = m_2$.

21.3 Since $s \leq 2 \ \forall s \in S$, 2 is an upper bound for S. If $\text{LUB}(S) \neq 2$ then $\exists \ s_1 \in S \ni 1 < s_1 < 2$ and s_1 is an upper bound of S. Using the ideas in the solution to Exercise 6.8, $s_1 < \frac{s_1+2}{2} < 2$. Since $\frac{s_1+2}{2} \in Q$ and $1 < s_1 < \frac{s_1+2}{2} < 2$, then $\frac{s_1+2}{2} \in S$. Then s_1 is not an upper bound of S. This contradiction says $2 = \text{LUB}(S)$.

21.5 0 is the greatest lower bound of the given set. Use ideas similar to Exercise 21.3 to justify this assertion.

Chapter 22, p. 232:

22.1 a. False; b. False; c. False; d. False; e. False; f. False; g. True; h. True; i. True.

22.5 Prove Theorem 22.3c, i.e., $\forall x \in E_1, x \cdot 0 = 0 = 0 \cdot x$.
Proof: From Axiom A5, $0 \in E_1$ and $a + 0 = a \ \forall a \in E_1$. Thus $0 + 0 = 0$. Let $x \in E_1$ then $0x = (0+0)x = 0x + 0x$. Then

$$0 = 0x + (-(0x)) = [0x + 0x] + (-(0x))$$
$$= 0x + [0x + (-(0x))] = 0x + 0 = 0x = x0.$$

22.10 One way to do this is to use Theorem 22.5h. Since $0 \leq x^2 \ \forall x \in E_1$, then $0 \leq 1^2$. Now $1^2 = 1 \cdot 1 = 1$, so $0 \leq 1$. We need only rule out $0 = 1$, but this is assured by A7.

22.18 Proof: We do only the ab part. Suppose $a, b \in E_1$ and $a \in Q$ and $b \in E_1 - Q$ and $a \neq 0$. Then $a^{-1} \in Q$ and $aa^{-1} = a^{-1}a = 1$. We want to show $ab \in E_1 - Q$, so suppose on the other hand that $ab \notin E_1 - Q$ then $ab \in Q$. Then

$$b = 1 \cdot b = (a^{-1}a) \cdot b = a^{-1}(ab).$$

Now $a^{-1} \in Q$ and $ab \in Q$ from the assumption previously mentioned, so $a^{-1}(ab) \in Q$ by Theorem 20.2a. But this is b, so $b \in Q$. This is a contradiction since $b \notin Q$. Therefore $ab \in E_1 - Q$.

Chapter 23, p. 242:

23.1 Proof: Suppose the hypotheses. We do the $N \cong D$ part. Now $D - \{2n - 1 \mid n \in N\}$. Clearly D is a subset of N. Define g by $g(n) = 2n - 1 \ \forall n \in N$ then g is a relation from N to N. Now $n_1 = n_2$ iff $2n_1 = 2n_2$ iff $2n_1 - 1 = 2n_2 - 1$ iff $g(n_1) = g(n_2)$. Thus g is a function from N to N and it is 1-1. Let $d \in D$ then $d = 2n_0 - 1$ for some n_0. For this n_0, $g(n_0) = 2n_0 - 1 = d$, so g is onto D. Thus $N \cong D$.

23.3 Proof: Suppose B is a set, $S \subseteq B$ and S is infinite. B is either finite or infinite. If B were finite then by Theorem 23.5, S would be finite which it is not. So B is not finite, hence B is infinite.

23.7   Proof: Suppose A were infinite and $A - \{a\}$ finite. Then $(A - \{a\}) \cup \{a\}$ would be finite by the lemma. But $(A - \{a\}) \cup \{a\} = A$, so A would be finite. This contradiction says $A - \{a\}$ is infinite.

Chapter 24, p. 252:

24.1   Show: f is onto $\bigcup_{n \in N} A_n$. To do this, let $y \in \bigcup_{n \in N} A_n$ then $\exists\, n \in N \ni y \in A_n$ by definition of $\cup$. Since $y \in A_n$ and $A_n = \{a_{n1}, a_{n2}, a_{n3}, \ldots, a_{nm}, \ldots\}$ then $y = a_{nm}$ for some $m \in N$. Now $(n, m) \in N \times N$ and $f((n, m)) = a_{nm} = y$, so f is onto $\bigcup_{n \in N} A_n$.

24.5   Proof: Suppose A is countable and $B \subseteq A$ and B is denumerable. Since A is countable then A is finite or denumerable. If A were finite then, by Theorem 23.5, B would be finite. But B is denumerable, so this is a contradiction. Therefore A is not finite. The only other possibility is that A is denumerable.

24.8 p and q are natural numbers, so $p \geq 1$ and $q \geq 1$. Thus $p + q - 1$ and $p + q - 2$ are consecutive non-negative integers, so one of them is even. Hence the indicated quotient is a non-negative integer. Adding p yields a natural number.

Chapter 25, p. 257:

25.1   Here is one of several ways to do this. Observe that $(0, 1) \subseteq [0, 1]$ and $(0, 1) \cong E_1$. Since E_1 is uncountable by the corollary to Theorem 25.1, E_1 is not finite. Hence $(0, 1)$ is not finite. By the Corollary to Theorem 23.5, $[0, 1]$ is not finite.

25.3   d. $A = (0, 1)$. Now $B = (0, \frac{1}{2}) \subset (0, 1)$. The function $f : A \longrightarrow B$ given by $f(x) = \frac{1}{2}x \; \forall x \in (0, 1)$ is a 1–1 function from A onto B. So $B = (0, \frac{1}{2})$ is an example of a proper subset of A that is in 1–1 correspondence with A.

     e. $A = E_1$. An argument can be made using Exercise 25.3d and the equivalence $E_1 \cong (0, 1)$.

     f. Let $A = \{f \mid f : E_1 \longrightarrow E_1 \wedge f \text{ is differentiable}\}$. Using Theorem 25.5, if $B = A - \{f_0\}$ where $f_0(x) = x$ then $B \cong A$, but $B \subset A$.

25.5 $0.134999 \cdots = 0.134 + .000999 \cdots$. And $0.000999 \cdots$ is a geometric series with first term $\frac{9}{10000}$ and common ratio $\frac{1}{10}$.

$$0.134999 \cdots = \frac{134}{1000} + \frac{\frac{9}{10000}}{1 - \frac{1}{10}} = \frac{134}{1000} + \frac{1}{1000} = \frac{135}{1000} = 0.135.$$

25.9   Proof: Since $n\sqrt{2} \neq m\sqrt{2}$ iff $m \neq n$ and since $k\sqrt{2}$ is irrational $\forall k \in N$, the set $\{\sqrt{2}, 2\sqrt{2}, 3\sqrt{2}, \ldots, k\sqrt{2}, \ldots\}$ is an infinite subset of $E_1 - Q$. Therefore $E_1 - Q$ is denumerable or uncountable. If

$E_1 - Q$ were not uncountable then it would be denumerable. Now Q is denumerable. Thus $(E_1 - Q) \cup Q$ would be denumerable. But $(E_1 - Q) \cup Q = E_1$ is not denumerable by the corollary to Theorem 25.1. Thus the assumption $E_1 - Q$ is not uncountable is false. Hence $E_1 - Q$ is uncountable.

25.11 Proof: If $\exists\, A \ni A$ is finite and $B \subset A$ and $A \cong B$ then $\#(A) = \#(B)$ and $B \cap (A - B) = \emptyset$ and $B \cup (A - B) = A$. Therefore

$$\#(A) + \#(A - B) = \#(B) + \#(A - B) = \#(A) = \#(A) + 0.$$

By Theorem 18.3f, $\#(A - B) = 0$, so $A - B = \emptyset$. Thus $A = B$. Thus the assertion is established.

Chapter 26, p. 271:

26.1 $0.111\cdots = .1 + .01 + .001 + \cdots = \frac{1}{2} + \frac{1}{4} + \frac{1}{8} + \cdots = 1.$

$0.10101\cdots = \frac{1}{2} + \frac{1}{8} + \frac{1}{32} + \cdots = \frac{\frac{1}{2}}{1 - \frac{1}{4}} = \frac{2}{3}.$

26.5 Proof: Let A be a set such that $\#(A) = a$ then
$$a \cdot 0 = \#(A) \cdot \#(\emptyset) = \#(A \times \emptyset) = \#(\emptyset) = 0.$$

26.6 See Exercise 14.15.
$$a \cdot 1 = \#(A) \cdot \#(\{0\}) = \#(A \times \{0\}) \overset{why}{=} \#(A) = a.$$

26.7 $\aleph_0 = \#(N)$ and $n = \#(N_n)$, but $N \cap N_n \neq \emptyset$. However, $N \cong \{n+1, n+2, n+3, \ldots, n+k, \ldots\}$ which is disjoint from N_n. So

$$\begin{aligned}
\aleph_0 + n &= \#(N) + \#(N_n) \\
&= \#(\{n+1, n+2, \ldots, n+k, \ldots\}) + \#(N_n) \\
&= \#(\{n+1, n+2, n+3, \ldots, n+k, \ldots\} \cup N_n) \\
&= \#(\{1, 2, \ldots, n\} \cup \{n+1, n+2, \ldots, n+k, \ldots\}) \\
&= \#(N) = \aleph_0
\end{aligned}$$

26.11 $\aleph_0^{\aleph_0} = (2\aleph_0)^{\aleph_0} = 2^{\aleph_0} \cdot \aleph_0^{\aleph_0} = c \cdot \aleph_0^{\aleph_0} = c^{\aleph_0} \cdot \aleph_0^{\aleph_0} = (c\aleph_0)^{\aleph_0} = c^{\aleph_0} = c.$

26.16 Proof: Suppose $\nu = \#(A)$. By Theorem 26.4, $\#(A) < \#(\mathcal{P}(A))$, so by Theorem 20.8, $\nu = \#(A) < 2^{\#(A)} = 2^\nu.$

26.17 By Theorem 26.8, $2^\nu = 2^{\#(A)} = \#(\mathcal{P}(A))$, so $2^{\aleph_0} = \#(\mathcal{P}(N)).$

26.21 $c^c.$

Chapter 27, p. 282:

27.1 $r = \{(a,a),(b,b),(c,c),(a,b),(a,c),(b,c)\}$ is reflexive, antisymmetric, and transitive, so r is a partial order relation on $A = \{a,b,c\}$. $r^{-1} = I_A \cup \{(b,a),(c,a),(c,b)\}$ is also a partial order relation.

27.3 Let $(N, |)$ be a poset and $B = \{x \in N \mid x \neq 1\}$. The minimal elements of B w.r.t. $|$ are the positive prime numbers, because if p is prime then $n \mid p \Longrightarrow n = p$. There are no minimum, no maximal, and no maximum elements of B w.r.t. $|$.

27.5 No. Let $A = \mathcal{P}(N)$ be partially ordered by $\subseteq$. Let
$$B = \{\{0\}, \{0,1\}, \{1\}, \{1,2\}, \{1,2,3\}, \ldots\}.$$
Then $\{0,1\}$ is maximal because any set in B containing $\{0,1\}$ is equal to $\{0,1\}$ and no other set in B is maximal. But $\{0,1\}$ is not a maximum element in B since $\sim \forall S \in B(S \subseteq \{0,1\})$.

27.9 E_1 is a partially ordered set w.r.t. $\leq$. 2 is comparable to 5 and 5 is comparable to 2, but $2 \neq 5$ so the relation, "is comparable to," is not antisymmetric.

Chapter 28, p. 287:

28.1 a. 12 and 15 are both maximal in B w.r.t. $|$. B has no maximum elements.

  b. 12 and 15 are both minimal in B w.r.t. $|$. B has no minimum elements.

  c. 60, 120, 180 are upper bounds of B in A w.r.t. $|$. Also 1 and 3 are lower bounds of B in A w.r.t. $|$.

  d. 60 is the LUB(B) in A w.r.t. $|$ and 3 is the GLB(B) in A.

  e. B is not a chain w.r.t $|$. Any set S containing B will contain 12 and 15 which are not comparable.

  f. GCD(B).

  g. LCM(B).

  h. No[2].

28.7 Yes. $\{\emptyset, \{a\}, \{a,c\}, \{a,b,c\}\}$. Yes.

Chapter 29, p. 297:

29.1 a. Since $S_0 \in S$ and $s_0 \in S_0$ then $\{(S_0, s_0)\}$ is a set f_0 of ordered pairs with domain $\{S_0\} \subseteq S$ and $s_0 = f(S_0) \in S_0 = \bigcup_{S \in \{S_0\}} S \; \forall S_0 \in \{S_0\}$. And since $(S_0, s_0), (S_1, s_1) \in f_0$ and $S_0 = S_1 \Longrightarrow s_0 = s_1$, we know f_0 is a function in $\mathcal{F}$.

  b. $\mathcal{F}$ is a set of sets and $\subseteq$ is always a reflexive, antisymmetric, and transitive on a nonempty set of sets, so $(\mathcal{F}, \subseteq)$ is a poset.

  c. Suppose $(S_0, s_0), (S_1, s_1) \in F$ and $S_0 = S_1$ then $\exists f_0, f_1 \in C \ni (S_0, s_0) \in f_0$ and $(S_1, s_1) \in f_1$. But C is a chain so either $f_0 \subseteq f_1$ or $f_1 \subseteq f_0$. Suppose WLOG $f_0 \subseteq f_1$ then $(S_0, s_0), (S_1, s_1) \in f_1$. Since f_1 is a function and $S_0 = S_1$ then $s_0 = s_1$. So F is a function.

   e. Clearly $\mathcal{C} \subseteq \mathcal{C} \cup \{F'\}$. Now $S_1 \notin \text{dom}(F)$ as a result of an assumption in Part 2, so the containment is proper. Next we must show $\mathcal{C} \cup \{F'\}$ is a chain in $\mathcal{F}$. It is in $\mathcal{F}$ as observed in the proof, so let $f_1, f_2 \in \mathcal{C} \cup \{F'\})\}$. If both f_1 and $f_2 \in \mathcal{C}$ then f_1 and f_2 are comparable. Otherwise, one or both of f_1 and f_2 contains the pair (S_1, s_1). In either of the latter two cases, f_1 is comparable to f_2, so $\mathcal{C} \cup \{F'\}$ is a chain.

29.3 a. If $A = \emptyset$ then $\exists\, B_0 \subseteq B \ni A \cong B_0$. That $B_0 = \emptyset$. The case where $B = \emptyset$ is analogous.

   c. From the proof, $f_0 : A_0 \longrightarrow B_0$, $a \notin \text{dom}(f_0)$ and $b \notin \text{ran}(f_0)$. Then $f_0 \cup \{(a, b)\} \subseteq (A_0 \cup \{a\}) \times (B_0 \cup \{b\})$. $f_0 \cup \{(a, b)\}$ is a function from $(A_0 \cup \{a\})$ to $(B_0 \cup \{b\})$ by Exercise 14.17.

   d. Now $f_0 \in \mathcal{F}$, so f_0 is a 1–1 function from A_0 onto B_0. If $A_0 \neq A$ then $B_0 = B$. Then $f_0 : A_0 \longrightarrow B$ is 1–1 and onto B, so $A_0 \cong B$. The conclusion of the lemma is established.

Chapter 30, p. 303:

30.1 Since $(N, \,|\,)$ is a poset, then the initial segment of N determined by 24 is $S_{24} = \{n \in N \mid n \,|\, 24 \land n \neq 24\} = \{1, 2, 3, 4, 6, 8, 12\}$.

30.3 a. $f(x) = x^3$ is order preserving because $x \leq y \Longrightarrow x^3 \leq y^3$.

   b. $f(x) = \sin x$ is not order preserving because $\frac{\pi}{3} \leq \frac{3\pi}{4}$ but $\sin \frac{\pi}{3} \not\leq \sin \frac{3\pi}{4}$.

   c. $f(x) = e^x$ is order preserving because $x \leq y \Longrightarrow e^x \leq e^y$.

   d. By Part c, since this $f : E_1 \longrightarrow E_1^+$ is 1–1 and onto E_1^+ then $E_1 \approx E_1^+$.

30.5 $B_{\{a,b,c,d\}} = \{\{a\}, \{a, b\}, \{a, b, c\}\}$ is order isomorphic to A' so $A' \prec B$.

30.7 $f = \{(n, 2^{n-1}) \mid n \in N\}$ is an order isomorphism.

INDEX

A

$a, b \in S$, 95

A.C., means
  Axiom of Choice, 289
abbreviated truth table, 11
absolute value, 56
addition of
  cardinal numbers, 196, 259
  integers, 203
  natural numbers, 197
  rational numbers, 216
  real numbers, 227
additive identity, 197, 218
aleph null, $\aleph_0$, 260
algebra of statements, 23, 117
algebraic number, 252
antecedent, 12, 20
antisymmetric, 147, 275
Archimedean order, 230
arguments, 20
associative operation, 190
axiom of
  choice, 289
  extent, 96
  pairing, 100
  power sets, 102
  specification, 99
  unions, 129
atomic statement, 8

B

begging the question, 21, 57
biconditional, 7
binary operation, 189
bound variable, 38

C

C.P., conditional proof, 42
cancellation law of
  $+$ in Z and Q, 208, 222
Cantor, Georg, 93, 245, 263
cardinal number, 196
  of the continuum, 260
cardinally equivalent, 196, 238
Cartesian product, 136
cell (of a partition), 151
chain, 287
characteristic function, 169, 268
closed operation, 189
closure, 204, 208, 217
codomain of a function, 161
Cohen, Paul, 265, 290
commutative operation, 190
comparable, 277
complement of a set, 119
complete ordered field, 226
composition of functions, 173
compound predicate, 32
compound statement, 4
conclusion of argument, 12, 20
conditional, 7
conditional proof, 42
congruence mod n, notation
  for, $\_\_ \equiv \_\_$ mod n, 158
conjecture, 55
conjunction of statements 5
connective, logical, 4, 114
  set theoretic, 109, 115
consequent, 12, 20
constant, 30
continuum hypothesis, 264

327